Sustainable Development Goals Series

The **Sustainable Development Goals Series** is Springer Nature's inaugural cross-imprint book series that addresses and supports the United Nations' seventeen Sustainable Development Goals. The series fosters comprehensive research focused on these global targets and endeavours to address some of society's greatest grand challenges. The SDGs are inherently multidisciplinary, and they bring people working across different fields together and working towards a common goal. In this spirit, the Sustainable Development Goals series is the first at Springer Nature to publish books under both the Springer and Palgrave Macmillan imprints, bringing the strengths of our imprints together.

The Sustainable Development Goals Series is organized into eighteen subseries: one subseries based around each of the seventeen respective Sustainable Development Goals, and an eighteenth subseries, "Connecting the Goals", which serves as a home for volumes addressing multiple goals or studying the SDGs as a whole. Each subseries is guided by an expert Subseries Advisor with years or decades of experience studying and addressing core components of their respective Goal.

The SDG Series has a remit as broad as the SDGs themselves, and contributions are welcome from scientists, academics, policymakers, and researchers working in fields related to any of the seventeen goals. If you are interested in contributing a monograph or curated volume to the series, please contact the Publishers: Zachary Romano [Springer; zachary.romano@springer.com] and Rachael Ballard [Palgrave Macmillan; rachael.ballard@palgrave.com].

Loreen Maseno
David Andrew Omona
Ezra Chitando • Sophia Chirongoma
Editors

Religion, Climate Change, and Food Security in Africa

palgrave
macmillan

Editors
Loreen Maseno
Department of Religion, Theology, Philosophy
Maseno University
Maseno, Kenya

Ezra Chitando
University of Zimbabwe
Mount Pleasant, Zimbabwe

David Andrew Omona
Uganda Christian University
Mukono, Uganda

Sophia Chirongoma
Midlands State University
Zvishavane, Zimbabwe

ISSN 2523-3084 ISSN 2523-3092 (electronic)
Sustainable Development Goals Series
ISBN 978-3-031-50391-7 ISBN 978-3-031-50392-4 (eBook)
https://doi.org/10.1007/978-3-031-50392-4

This Palgrave Macmillan imprint is published by the registered company Springer Nature Switzerland AG.
The registered company address is: Gewerbestrasse 11, 6330 Cham, Switzerland

Paper in this product is recyclable.

Contents

Notes on Contributors

Seyram B. Amenyedzi is a Reverend Minister, Theologian, Missiologist, Research Methodologist and Qualitative Data Analyst. She has a strong passion to help marginalized people in society especially persons with disabilities, women, youth and children. She obtained her PhD from Stellenbosch University. Her research interests include Missiology, Research Methodology, Disability, Womanism and Afrocentrism.

Kudzai Biri is an Associate Professor in the Religion Department at Hope College, Michigan in the United States of America. She was appointed as Extraordinary Professor by the Department of Religion and Theology, University of Western Cape, South Africa. Her research and teaching interests include Pentecostalism, religion and inequality, religion and politics, religion and violence, gender and religion and migration.

Mwawi Chilongozi is the Secretary General of the Church of Central Africa Presbyterian (CCAP) General Assembly and an ordained minister of CCAP, Synod of Livingstonia. She is also a Parish Minister of Thanthwe CCAP in Lilongwe, Malawi. She is a research fellow at the Faculty of Theology at Stellenbosch University in South Africa. Her research interests include theology, health, and gender.

Sophia Chirongoma is a Senior Lecturer in the Religious Studies Department at Midlands State University, Zimbabwe. She is also an Academic Associate/Research Fellow at the Research Institute for Theology and Religion (RITR) in the College of Human Sciences, University of South Africa (UNISA).

Ezra Chitando is a Professor of Religious Studies at the University of Zimbabwe and Theology Consultant on HIV for the World Council of Churches. He is also an extraordinary professor at the Desmond Tutu Centre for Religion and Social Justice. He has diverse research and publication interests.

Nicholas Miyoba Haambokoma is Lecturer, Department of Religious and Cultural Studies at the University of Zambia. His research and teaching interests include Religious Education, Teacher Education, Religion and Environment, and Sex Education.

Sokfa F. John is a postdoctoral research fellow at the SARChI in Sustainable Local Livelihoods, University of KwaZulu-Natal. He is also an associate of the Center for Mediation in Africa, University of Pretoria. His research interests are in ethnoreligious conflicts, peace-building and mediation, digital conflicts, and digital cultures.

B. R. Gaie Joseph is Associate Professor at the Department of Theology and Religious Studies, University of Botswana where he teaches Ethics, Epistemology and Metaphysics based courses. He completed his postdoctoral studies in research ethics at the John's Hopkins University. His research interests include Theory and Applied Ethics (medical, political, professional, Religious) and Botho Ethics.

Gertrude A. Kapuma is Senior Lecturer in Pastoral Care and Trauma Counselling at Zomba Theological University. She is the Head of Practical Theology Department, Ordained Minister of the Church of Central Africa Presbyterian, Deputy Secretary General of the CCAP General Assembly, and also the Country Coordinator of Malawi Circle Chapter.

Appiah Kubi is a practising Medical Doctor at the Korle-Bu Teaching Hospital in Korle-Bu—Accra, Ghana, where he has worked for several years. Currently he is an M. Phil candidate pursuing religious studies at the Central University, Ghana.

Tshenolo J. Madigele is a lecturer, Department of Theology and Religious Studies, University of Botswana. Her teaching areas include Systematic Theology and Christian Moral Ethics. Some of her research focuses on community participation in environmental policy formulation and youth behaviour in relation to environmental responsibility. Additionally, Madigele has contributed to the understanding of youth behaviour and environmental responsibility through her article titled "Youth and environmentally responsible consumer behaviour: The case of Plastic Bag Usage in Botswana" Which was published in PULA: Botswana Journal of African Studies.

Newton Kahumbi Maina is Senior Lecturer, Department of Philosophy and Religious Studies, Kenyatta University, Nairobi. His research and teachings interests include Islamic Studies, Gender, Religion and Culture, and Christian–Muslim Relations His latest publications are on Digital Health and Maternal Care for Muslim refugee women, Shirazi civilization, Freedom of Religion and History of Christian–Muslim relations.

King'asia Mamati is a Doctoral researcher in Social and Cultural anthropology at the University of Cologne. His research interests include human-environment relationships, indigenous peoples, local knowledge systems, environmental change and religion.

Loreen Maseno is Senior Lecturer, Department of Religion, Theology and Philosophy, Maseno University, Kenya and Research Fellow, Faculty of Theology, University of Pretoria, South Africa. She is an Ambassador Scientist for the Alexander Von Humboldt Foundation. Her research and teaching interests include Theology and climate change, feminist Ecology, gender, Pentecostalism, Religion and Inequality, food and Spirituality, Virtual Religion and Kinship.

Elizabeth P. Motswapong is Senior Lecturer in World Religions/ Comparative Religions, (specialization in Hindu Studies) in the Dept of Theology and Religious Studies at the University of Botswana. She has published widely on Gender Studies, (Women to Queer Studies), and Comparative religions. Her research interests lie predominantly in comparative studies especially in religion and spirituality between Asian religions and Botswana Indigenous religions.

Tenson Muyambo is Lecturer, Department of Teacher Development, Great Zimbabwe University, Zimbabwe and Research Fellow, Research Institute for Theology and Religion, College of Human Sciences, University of South Africa, South Africa. He is a member of African Consortium for Law and Religion Studies. His research and teaching interests include Religion and gender, indigenous knowledge, pandemics, environment and education.

Nelly Mwale is Senior Lecturer, Department of Religious and Cultural Studies at the University of Zambia. Her research and teaching interests include Religion in the public sphere, religion and education, Religion and Development, Religion and Environment, Church History and African Indigenous Religions.

Linda Naicker is a Researcher based at the University of South Africa. Prior to this, she was a freelance Research Consultant at KwaZulu-Natal Christian Council. She is Scholar and Activist. Her research interests and publications focus on the intersections of religion, race, class and gender.

George C. Nche is a Global Excellence and Stature (GES) Senior Postdoctoral Research Fellow in the Department of Religion Studies, University of Johannesburg (UJ), South Africa. His major research focus is on Religion and Ecology where he explores the intersection between religious institutions/beliefs/values/actors and climate change/environmental sustainability. Other research foci are environmental ethics, religion and social issues.

Simon Gisege Omare is Senior Lecturer in the Department of Philosophy, Religion and Theology at Moi University, Kenya. He holds a PhD in Religious Studies. His area of interest is Sociology of Religion. He is a Senior Lecturer at Moi University, Kenya. Simon is the author of two books, over 12 articles in journals and 10 chapters in books. (See https://scholar.google.com/citations?user=JP78Z54AAAAJ). Currently, he is writing a book on African worldview and Intelligent Design.

Rose Omari is a Principal Research Scientist and Deputy Director at the Science and Technology Policy Research Institute, Council for Scientific and Industrial Research (CSIR-STEPRI), Ghana. Her research interests include food, nutrition, science, technology and innovation policy analysis, consumer behaviour studies with emphasis on food safety, nutrition and public health, and gender in Science, Technology, Engineering and Mathematics (STEM) studies.

David Andrew Omona holds a PhD in Political Studies with a focus on International Relations. Besides, he holds other qualifications in various fields of academia. Currently he is an Associate Professor of Ethics and International Relations, and Dean, School of Social Sciences at the Uganda Christian University. He has carried out research and published several journal articles and book chapters in his area of expertise.

Sylvia Owusu-Ansah is Dean of Theology (Perez University College); Head Pastor (Revival Temple-Perez Chapel International). Her research interests include gender studies, interreligious conflict mediation, missions-related studies, cross-cultural communication, and dialogue. Among her publications are the book chapter 'The Role of Interreligious Collaboration in Conflict Prevention and Peaceful Multi-Religious Co-Existence: A Case Study of Northern Ghana' (2018).

Beatrice-Joy Owusu-Ansah is a PhD candidate in Plant breeding at the West Africa Centre for Crop Improvement (WACCI). With a background in Genetics and Plant biotechnology, she is passionate about promoting global food security, especially in Africa. She is also keen on advocating the role of gender and religion in achieving global food security and sustainable development.

Mercy Shumbamhini is a Roman Catholic Nun of the Congregation of Jesus. She is a clinical social worker, theologian, spiritual director, narrative therapist, safeguarding consultant, lecturer, researcher and writer. Currently, she works at Arrupe Jesuit University, Zimbabwe as a development officer and lecturer. Shumbamhini's greatest desire is to give those at the margins especially women and children hope and dignity.

D. Saudu Siamisang is a teacher of Religious Education in the Department of Religious Education at Maun Secondary School. He was awarded his Master of Arts in Theology and Religious Studies in April 2023 by the University of Botswana.Saudu's interest lies in the potential of belief systems to improve insights into the political spheres and societal development. He is currently working on a study on intercultural, social equity, religion, community development, and advancement.

Fortune Sibanda is a Professor of Religious Studies, Department of Theology and Religious Studies, University of Eswatini and Department of Philosophy and Religious Studies, Great Zimbabwe University. Sibanda is also a Research Fellow in the Research Institute for Theology and Religion, UNISA, South Africa. His research interests include religion and peace-building, law, health, climate change, gender, human rights, marginality, among others. Sibanda is an ACLARS Board member and also sits on the ACLARS Publication Committee and is a Southern African Region Representative.

Obvious Vengeyi is Professor of biblical studies in the Department of Philosophy Religion and Ethics at the University of Zimbabwe. His seminal book, Aluta Continua Biblical Hermeneutics for Liberation (2013) addresses central topics ranging from land and poverty, to religion and social justice. Although he acknowledges the various interpretations as socially relevant, his writings promote social, political, cultural and economic location of the biblical text before appropriating it for whatever cause, however noble.

List of Figures

List of Tables

PART I

Religion and Climate Change: SDGs and Dimensions of Food Security in Africa

CHAPTER 1

Religion, Climate Change, and Food Security in Africa

Loreen Maseno and Ezra Chitando

Introduction

Religion remains a significant factor in Africa, with some statistics indicating that 95% of the continent's citizens identify with a religion (Afrobarometer 2022). The religious demography of Africa suggests that by 2022 Christians constituted 45%, about 40% Muslims, and 15% other religions which include Traditional African religions, Judaism, Bahá'í Faith, Hinduism, Buddhism, and Rastafarianism (Pew Research; World Christian Database). Religion serves as a guide to belief and action, permeating all aspects of life. As Magesa (2013) confirms, following Mbiti's (1969: 1) oft-cited aphorism that "Africans are notoriously religious," religion matters in Africa (see also Sanni 2016). Consequently, religion has

L. Maseno (✉)
Department of Religion, Theology and Philosophy, Maseno University, Maseno, Kenya

E. Chitando
Department of Philosophy, Religion and Ethics, University of Zimbabwe, Harare, Zimbabwe

L. Maseno et al. (eds.), *Religion, Climate Change, and Food Security in Africa*, Sustainable Development Goals Series,
https://doi.org/10.1007/978-3-031-50392-4_1

something (or even much!) to say about what and how Africans eat, how they live, die, and the anticipated life beyond the demise of the physical body. Given this centrality of religion to life in Africa, it is critical to reflect on the role of religion in equipping Africa to respond to the climate emergency and the accompanying food insecurity threatening Africa. Images of Africa as an emaciated and hungry continent, perpetually seeking handouts from the rest of humanity, are not consistent with the vision of healthy and well-nourished citizens articulated in the African Union's Agenda 2063 (African Union 2015: 15) and the United Nations' Sustainable Development Goal (SDG) 2, "Zero Hunger."[1] The global community also has the responsibility of meeting these objectives. The following observation is instructive:

> Success towards these objectives requires: (i) Full engagement of religious leaders at senior levels in highlighting the ethical issues at stake in the famine emergency, building on the teachings of religious traditions, individually and collectively; (ii) specific engagement of religious leaders in efforts to negotiate access to areas acutely affected by famine; and (iii) cooperative (track two) peacemaking efforts with religious communities in famine affected areas. (Marshall et al. 2018: 5)

Achieving food security in the face of climate change is crucial for Africa's survival and sovereignty. It is scandalous for African leaders to beg for the Russia-Ukraine war to end because it threatens the continent's food security. African scholars of religion and theology have, alongside scholars and intellectuals from other disciplines, the responsibility of reflecting on how the continent can ensure that it survives the twin challenges of climate change and food insecurity, while achieving food sovereignty. By identifying the challenges that faith communities face in responding to climate change and food insecurity in Africa (see, e.g., Umuziranenge 2016), they can contribute toward the transformation, empowerment, and development of the continent. This will ensure that there will be a multidisciplinary approach toward responding to climate change and food security (see also Sirba and Chimdessa 2021).

Religious actors from the diverse religious communities in Africa are required to invest in responding to climate change and the accompanying threat of food insecurity. While reflections from within Christianity have

[1] https://sdgs.un.org/goals/goal2.

tended to dominate the field of religion and eco-justice (see, e.g., Beros et al. 2022), there is a lot of potential for other religions (see, e.g., Soudini 2022) to make effective contributions toward responding to the challenges that the continent faces in terms of the climate emergency and food insecurity. As contributors to this volume highlight, religion is well-placed to transform Africa's situation and to confound critics.

The Context

There is a growing realization that the earth is clearly warming at a worrying pace. Goal 13 of the SDGs indicates that climate change is an existential crisis, while Goal 2 seeks to achieve "Zero Hunger," achieve food security and improved nutrition, and promote sustainable agriculture. Climate change has far-reaching effects for agriculture and other livelihood activities which ensure the availability, suitability, distribution, and accessibility of food. The Intergovernmental Panel on Climate Change notes that human emissions of gases such as carbon dioxide (CO_2) which trap heat onto earth have been recorded as excess over the years (IPCC 2001). Many scholars agree that this is human-induced climate change owing to the emissions which are said to enable those in developed countries to live in better economic standards with relative security and comfort.

Although the phenomenon of climate change is global, its impacts are varied. The tropical arid and semi-arid areas of Africa are most vulnerable owing to the fact that over 70% of the many people derive their livelihoods from agriculture and natural-based resource activities (Antwi 2013). Many in the rural areas will not even have sufficient food in the face of drought and poor rainfall. Agriculture in Africa is dependent on good weather patterns and farm inputs. However, the climate changes experienced in Africa have affected the planting seasons, the harvests, the preservation and storage of yield, and food security in general. It is, therefore, clear that human-induced climate change affects people in Africa and other developing countries disproportionately, leading to extreme positions of poverty, hunger, and famine. In general, livelihood options are dependent on climate-sensitive factors that make the people of Africa vulnerable. Indeed, it is the countries in Africa that have less capacities to adapt which will be hit particularly hard by the negative effects of climate change such as decreased yields (Lobell et al. 2011).

Unfolding at the time of writing was the East Africa drought and food crisis affecting the Horn of Africa. This crisis is reported to have evolved into a large-scale humanitarian crisis that is directly affecting Ethiopia, Kenya, and Somaliland. In a region already reeling from numerous shocks—climate change, internal conflicts, the war in Ukraine causing detrimental impact on food and fuel accessibility, the socio-economic effects of the COVID-19 pandemic, and destructive desert locusts—urgent measures were needed to avoid a catastrophe.

This drought and food crisis has already impacted several countries; however, the situation has since deteriorated further, especially due to the impacts of *La Nina* coupled with climate change effects. The fourth consecutive rainy season has failed to provide relief with low and erratic rainfall leading to a climatic event not seen in the last 40 years. According to the UNICEF press release of 22 April 2022, at least 20 million people, including 10 million children, are facing severe drought conditions—a 38% increase in people and children in need since February 2022. The World Food Programme is on record that severe water shortages and significant vegetation deficits have led to 3 million livestock deaths in Kenya and Ethiopia.

Scholars point to the main issues in climate change discourse and its implication to the food security equation. According to the Food and Agriculture Organisation (FAO), agriculture-based livelihood systems that are already vulnerable to food insecurity face immediate risk of increased crop failure, new patterns of pests and diseases, lack of appropriate seeds and planting material, and loss of livestock (FAO 2008: ii). This means that climate change as it impacts Africa affects all four dimensions of food security: food availability, food accessibility, food utilization, and food systems stability. At the same time, the impacts of droughts on livestock numbers in African countries (FAO 2008: 22) clearly depict the losses of national herds. This links directly to the effects of climate change such as poverty and inequality (World Bank 2016). The following observation remains valid:

> With a change in the climate, agricultural production, prices and infrastructure will change, limiting the amount and quality of food produced. Again linked to the situation around nutrition in a family, a country or the world is the state of health. With inappropriate or too little food health conditions of people are getting worse. Supplementary climate changes like increasing temperature, floods or droughts are threatening human physical and psychological circumstances. (Wlokas 2008: 12)

It is becoming increasingly clear that the world will not be able to meet the target of SDG 2, namely, "Zero Hunger." The devastating impact of climate change, conflict, growing inequality, and, more recently, the COVID-19 pandemic and the war in Ukraine constitute some of the major factors. The exploitative system of global capitalism, poor policies and practices, as well as the exclusion of women also constitute some of the key factors. The struggles that the world is facing are clearly articulated in the *State of the Food Security and Nutrition Report 2023*. Among other strategic observations, the report makes the following sober submissions:

- A much larger proportion of the population in Africa faces hunger compared to the other regions of the world—nearly 20 percent compared with 8.5 percent in Asia, 6.5 percent in Latin America and the Caribbean, and 7.0 percent in Oceania.
- It is projected that almost 600 million people will be chronically undernourished in 2030, pointing to the immense challenge of achieving the SDG target to eradicate hunger. This is about 119 million more than in a scenario in which neither the pandemic nor the war in Ukraine had occurred nor around 23 million more than if the war in Ukraine had not happened. Most progress is expected to occur in Asia, whereas no progress is foreseen in Latin America and the Caribbean, and hunger is projected to increase significantly in Africa by 2030. (FAO, IFAD, UNICEF, WFP and WHO 2023: 6)

Conceptualizing Definitions and Relationships

This section seeks to contribute toward clarifying some of the key terms in this volume. These definitions are in the context of unpacking the impact of climate change on food systems and food security. They also consider appropriate strategies for mitigating and adapting to climate change in several key domains of religion and food security.

It has been argued elsewhere that there is no internationally agreed definition of the term "climate change" since climate change could cut across the reference to, first, long-term changes in average weather conditions (WMO); second, all changes in the climate system, including the drivers of change, the changes themselves, and their effects (GCOS); or, third, only human-induced changes in the climate system (UNFCCC).

Climate Change

Nevertheless, in order to proceed, we consider that, according to the United Nations Framework Convention on Climate Change (UNFCCC), the term "climate change" refers to any change in climate over time due to either natural variability or human activity, which in the end alters the composition of the global atmosphere and is in addition to natural climate variability observed over comparable time periods. According to the Intergovernmental Panel on Climate Change IPCC (2001), climate change is any change in climate over a time period that alters the composition of the global atmosphere and this change might be due to natural climate variability or a result of human activity. Scientists largely agree that the culpable human activities include the burning of fossil fuels, natural causes, industrialization, and changes in land use which have continued to modify the concentrations of atmospheric properties of the surface that absorb or scatter radiant energy (Yadav et al. 2019: 2). But changes in climate, particularly increases in temperature, have already affected a wide range of physical and biological systems in many aquatic, terrestrial, and marine environments worldwide. It has been shown repeatedly that climate change will increase the risks of extinction of more vulnerable species and loss of biodiversity. The extent of damage or loss and the number of systems affected would increase with the magnitude and rate of climate change (Yadav et al. 2019). It is also to be noted that according to the Intergovernmental Panel on Climate Change's (IPCC) Fourth Assessment Report (AR4), Africa is the continent most vulnerable to climate change and climate variability (Chitando et al. 2022).

Food Security and Food Sovereignty

Whereas food systems include the production, transport, processing, packaging, storage, retail, consumption, loss and waste of food, food security is the outcome of food system processes all along the food chain. According to the Food and Agriculture Organization (FAO) food security happens when all people, at all times, have physical, social, and economic access to sufficient, safe, and nutritious food that meets their dietary needs and food preferences for an active and healthy life (FAO, 1996). Many activists argue that food security does not go as far as food sovereignty. For them, food security describes a condition regarding access to adequate food, while food sovereignty is more explicitly a political agenda for how to address inadequate access to food and land rights. Cognizant of the fact that food security is concerned with the adequate protection and unequal distribution of existing food systems, food sovereignty being rooted in local and grassroots movements advocates for a democratic food system, localized food systems and in so doing places value on food providers. Such a system is deemed to involve collaborative inputs from citizens and producers (Clapp 2014). According to Polzin, food sovereignty is therefore a broader endeavor which adds a human dimension to food security. It also embodies the right to food because it acknowledges that one should have a say in their relationship with food (Polzin 2018). According to Jarosz, the discourses between food security and food sovereignty show the significant contributions of indigenous scholars to food sovereignty discourses who continue to make clear the linkages to racialized oppression, which dismantled indigenous food-ways through agricultural modernization and urbanization (Jarosz 2014). However, it is also argued that the dualism between food security and food sovereignty needs not lead to polarizing positions. Rather, as Clapp argues, the juxtaposition of food security and food sovereignty as competing terms is in many ways more unclear than helpful to policy dialogue that attends to issues of hunger and the global food system (Clapp 2014). Thus, although most contributors to this volume make references to food security, it is instructive to note that they include food sovereignty in their purview.

Religion, Nutrition, and Gender

Closely linked to this is the discussion on "Religion, Nutrition, and Gender." Scholars continue to show that on matters of food, there are components of cultural life and the rights of women (who are disproportionately affected by the subordination of female farm workers). This means that many women may not access adequate nutritional elements they need for their changing bodies as mothers, menopausal, teenage girls, and so on by reason of culture, their domestic work, and indigenous practices. According to World Vision, the projections for women and girls when it comes to nutrition are alarming. They predict that an additional 4.8 million women will suffer from anemia, over and above the 1 billion women and girls who are currently affected. Further to that, women and girls remain more likely to be malnourished, making up 60% of the world's people living with chronic hunger. These women shoulder the burden of domestic care responsibilities, suffer serious mental health challenges, and face alarming increases in sexual- and gender-based violence. In general, it has been shown that gender inequality is both a cause and a consequence of malnutrition, with the potential of confining women and girls in a vicious multigenerational cycle of poverty and unmet potential (World Vision Canada 2020).

Climate-Related Conflicts and Food Insecurity

We take note that while climate-related conflicts contribute to food insecurity, the impact is also vice-versa—food insecurity also impacts climate-related conflicts. The vibrant relationship between climate-related conflicts and food production and distribution is profound in many areas of Africa (Tarusarira 2021). Although much of the intersection of religion and climate change focuses on how religion influences pro- and anti-environment beliefs, attitudes, and behaviors toward mitigation and adaptation efforts, we also see that conflicts emerge when climate change exceeds communities' adaptive capacities or adaptation strategies go wrong. In all, climate-related conflicts impact food production and distribution.

The Pillars of Food Security

Part I of the book wrestles with the four pillars of food security such as food availability, access, utilization, and stability. Climate change immediately affects food availability, access, and stability since it creates extreme weather, natural disasters, droughts, increase in pests and disease which thrive in hotter conditions, soil degradation, desertification, loss of biodiversity, shortage of water, floods, and rising sea levels, among other things (Levasseur 2021; Matsa 2021; Serventi 2022). It follows that climate change will affect food security through its impacts on all elements of global, national, and local food systems.

Climate change will first affect the people and food systems that are already vulnerable, but over time the geographic distribution of risk and vulnerability may shift. Climate change has an impact on food access. Access to food refers to the ability of individuals, communities, and countries to purchase sufficient quantities of quality food as per their demand. The strongest impact of climate change on the economic output and access is expected for sub-Saharan Africa, which is also feared to suffer the largest contraction of agricultural incomes due to climate change (Yadav et al. 2019). Agriculture is the main source of food as well as income in many developing regions of the world. Climate change poses a serious threat to food access for both rural and urban populations by reducing agricultural production and incomes, increasing risks, and disrupting markets (Olsson et al. 2014). In all, food security and climate change integrate in various ways as this book considers various challenges posed by changing climate, crop productivity, population increase, increasing temperature, disruptions in rainfall patterns, drought, and their combined impact on planting environments, crop adaptation, production, and management and the continued demand for food security.

Africa's Greater Vulnerability to Climate Change

Although the phenomenon of climate change is global, its impacts are varied and mainly detrimental to the largely tropical arid and semi-arid areas as in sub-Sahara Africa that is most vulnerable. Studies show very clearly that about 70% of the people of Africa live in rural areas and mostly depend largely on agriculture and natural resource-based activities for their livelihoods. Most of these livelihood options are dependent on climate-sensitive factors, making the people vulnerable. Therefore, for Africa that is more dependent on climate-sensitive resources, any inordinate change in climate will surely and negatively impact on its natural ecological systems to retard capacity for food security. It is still on this continent that the adage, Africans are notoriously religious, has been expressed.

Tentative Definition of Religion

The definition of religion has undergone broad changes among scholars over the last generation. Previously, religion was defined to be systems of ideas to which believers assented: the belief and worship of supernatural powers that affect human being's daily lives. It was also said to be a collection of cultural systems, beliefs, and worldview that it relates humanity to an order of its occurrence. Ninian Smart (1969) went further and introduced the seven dimensions that constituted his dimensional model of religion: ritual, mythological, doctrinal, ethical, social, material, and experiential. This model was considered a basic framework for exploring and evaluating religion. In recent times, religion has come to be understood as embodied practices that cultivate relations among people, non-human forces—nature, spirits, ancestors, saints, gods—resulting in communities, places, and feelings that shape those who take part (Hoover 2006). Commonly in this volume, the latter understanding is utilized as our operationalized definition, save for one chapter where we provide for a creative interlude in which religion is utilized as a transforming structure and process. This interlude serves to jog our conceptualization of definitions and stretch our frames of reference with regards to religion. We further ask, how then does the religiosity expressed on the continent in any way interface with these challenges of climate change and food security? To be sure, we operate with an inclusive, open, and pluralistic understanding of religion. We consider African Traditional/Indigenous Religions (see, e.g., Chitando 1997), Christianity (in its diverse forms and expressions), Hinduism, Islam, Rastafari, and other spiritualties to be integral to the definition of religion in Africa. We are convinced that our inclusive and expansive approach facilitates a broader understanding of the interface between religion, climate change, and food security/sovereignty in Africa.

Scholarship on Religion, Climate, and Food Security

According to Raj et al. (2022), climate change impacts agricultural production, supply chains, and pricing. Further, food production is projected to decline in tropical regions (Hertel and Lobell 2014) and food yields are set to decline owing to global warming (Wolfe 2013). The impact of this is that more than 120 million people will become undernourished (Raj et al. 2022). Studies on food security and food needs are shifting a gaze to not only changes in climate or demographics but also to food production systems, access, and adaptation (Porter et al. 2017).

Food security and climate change have been well extrapolated and their nexus clearly spelled out (Lobell and Burke 2010). Many scholars agree that climate change is already affecting food security through increasing temperatures, changing precipitation patterns, and greater frequency of some extreme events (Schmidtko et al. 2017). In many countries, for example the Horn of Africa, when the temperatures have increased, there is a drastic decrease in crop yields such as maize and wheat. But at the same time, when rainfall patterns change drastically, crops have been seen to fail and farmers suffer huge losses (Huntington et al. 2017). Moreover, climate change has continued to affect food security in the arid lands of countries. On the other hand while climate change is affecting food security, food security is becoming increasingly affected by climate change. This is because, as scholars have shown that cereal prices are set to increase in coming years due to climate change, this would impact consumers globally through higher food prices. When food prices are high there is high likelihood that malnutrition will be experienced in various pockets of the world, together with risks of food system disruptions. In general, climate change impacts food security terms of food availability and production (Iizumi et al. 2018). For example, in Kenya Ketiem et al. (2017) note that in recent years, yields of important food crops such as maize, wheat, sorghum, and mangoes have declined and are not readily available as before, thereby increasing food insecurity.

Studies show that mitigation, adaptation, and resilience are key for food security in the face of climate change vulnerabilities (Hodgkin and Bordoni 2012). The increased use of agricultural biodiversity and genetic resources will play an important role in improving both adaptability and resilience of agricultural systems. At the same time, deliberate crop diversification can increase adaptation and resilience in a range of ways (Lin 2011).

Crop failure, pests and crop diseases have had a lasting impact on food security. In Africa, these have been shown to be leading factors that have sustained food insecurity across the continent. Therefore, there is a need for enhanced capacity to suppress pest and disease outbreaks so as to maintain sufficient crop production. Cutting across these challenges facing the Anthropocene, recent scholarship stresses the place of indigenous knowledge systems (Maseno and Mamati 2022) and religious knowledge, which play a role in society.

This Volume

Cognizant of the importance of the issues outlined above, this book brings together scholars and practitioners to reflect on the questions, "How does religion interact with climate change and food security in Africa? What precisely does human induced climate change mean for practitioners of religion in Africa and how can its effects be mitigated? How can religion in Africa be a resource to counter climate change and its effects (SDG 13), achieve zero hunger and promote food security in order to meet the goal of SDG 2?" Although there are no easy answers to these questions, there are very few scholarly reflections and voices from Africa on the interface between religion, climate change, and food security. A few studies combine these three factors and even more acutely, fewer in African settings.

The Global Hunger Index (GHI) tracks and measures hunger at a global level. The 2019 climate change focus on Global Hunger Index report notes that climate change is an increasing threat to the world's hungry and vulnerable people who require immediate action. This report is clear that Africa, south of the Sahara, has one of the highest (GHI) scores in the world at 28.4 driven up my undernourishment and child stunting and child mortality rates (Grebmer and Bernstein 2020; Wiesmann et al. 2015). Although there are no obvious reasons why the pace of climate change caused by human activity should or should not match the pace with which we are able to adapt food production systems, the implications of looming food shortages and hunger in Africa remain an acute priority. Searchinger et al. argue that although there is no silver bullet, there are plausible solutions, what he calls "a five course menu," to ensure that we can feed the nearly 19 billion people on the planet by 2050 without destroying the planet itself (Searchinger et al. 2019). The Intergovernmental Panel on Climate Change (IPCC) report notes that climate change continues to affect land use, desertification, and cumulatively food and commodity prices (IPCC 2019). It is, however, clear that both rainfall and temperature variability impact negatively on food production, water resources, biodiversity, human and livestock populations (Antwi 2013).

This volume seeks to fill the *lacuna* in scholarship by focusing on the extent to which religion can contribute toward addressing climate change and food security in Africa. While the interface between climate change and food security has, and continues to, receive(d) attention, the impact of religion on climate change and food security in Africa remains under-researched and theorized. Thus, for example, there has been an analysis of the relationship between climate change and food security, such as by Islam (2017), who notes that climate change causes adverse impacts on both ecosystems and human societies as it increases the incidence of floods, droughts, and other climate-related hazards and that this calls for serious interrogation in most African contexts.

Behnassi et al. (2022) argue that the resilience of food security is threatened in the face of environmental and climate risks and global pandemics such as the COVID-19 crisis. Ebert et al. (2018) note that climate change has already started affecting agricultural growth and represents one of the greatest environmental and societal threats facing the planet. Many books in this field do not dedicate sustained attention to religion and food security, thus there is a need to plug this gap by addressing "religion, climate change and food security in Africa" in a more direct way.

Our Main Contribution

We consider, in detail, the extent to which religion and climate change both *create* and *affect food security* in specific African settings. The themes in this book are approached from diverse disciplinary and methodological angles to cover four main aspects: first; to probe the potential role of religion in Africa in accelerating the achievement of SDGs 2 and 13. Second, to problematize the influence of religion and the challenges it poses toward responding to the climate emergency and the elimination of hunger in Africa. Third, to approach the religions of Africa in their plurality, gaging their strategic significance in light of the two emergencies. Fourth, to probe religious teachings, practices, personalities, and institutions in Africa in the wake of SDGs 13 and 2 as they wrestle with the interplay between religion, climate change, and the dimensions of food security in Africa.

Chapters in this Volume

The parts of the book are arranged in the order of dimensions of food security, including food availability, food accessibility, food utilization, and food systems stability in the context of countries of Africa. In this volume, we take note of the strategic role of the Circle of Concerned African Women Theologians (the Circle) in addressing the theme of religion and women's rights, dignity, and well-being. The Circle was founded in 1989 with a clear agenda for women to empower themselves, other women, and men through research, reading, writing, and publishing in the areas of religion, culture, and theology.

Circle members continue to interrogate how religions/cultures construct and impact women, as well as finding ways of re-interpreting religions/cultures for the empowerment of women. In full view of this, eleven Circle members were invited to contribute to this volume, namely Sophia Chirongoma, Kudzai Biri, and Mercy Shumbamhini from Zimbabwe; Seyram Amenyedzi and Sylvia Owusu-Ansah from Ghana; Loreen Maseno from Kenya; Nelly Mwale from Zambia; Tshenolo Madigele and Elizabeth Motswapong from Botswana; Linda Naicker from South Africa; and Getrude Kapuma from Malawi. This volume, therefore, demonstrates the Circle's ongoing commitment to strategic scholarly output on themes that have a bearing on the health and well-being of African women.

In Part I, the contributors respond to the theme of "Religion and Climate Change, SDGs, and Dimensions of Food Security in Africa." In Chap. 2, David Andrew Omona, "Prophetic Action, Climate Change, Food Security and SDG 2 in Africa," weaves a tapestry of the church's religious and prophetic role informed by SDG 2 to conscientize the masses by "talking forth." "Talking forth" should include addressing the ills of climate change and food insecurity and other issues that are detrimental to the people of God in a given time.

Chapter 3, by Sophia Chirongoma and Mercy Shumbamhini, "Interfacing Gender, Religion, Food Security, and Climate Change in Africa," underlines that Africa is most vulnerable because over 70% of the people derive their livelihoods from agriculture. Women play a vital role in food security and climate change mitigation; hence, the agency of rural female farmers is essential for enhancing agricultural productivity and realizing the Sustainable Development Goals (SDGs), including ensuring food security (SDG 2) and addressing the perils of climate change (SDG 13). Positioned within the ecofeminist approach to climate change, this chapter argues that climate change is dramatically altering the planet and affecting human livelihoods and food security in ways that elicit religious responses. It shows how religion in Africa can be a resource to counter climate change and its effects (SDG 13), achieve zero hunger, no poverty, and promote food security in order to meet SDGs 1 and 2.

Chapter 4, by Sokfa John, "Food Availability, Accessibility, and Climate Change in Africa," adds to the debate by examining the potential of religion in Africa to improve food availability and accessibility in rethinking religion as a transforming structure and process. From a perspective of the sustainable livelihoods framework, this places religion on the same level with other institutions, processes, and entities—private, public, legal, cultural, and others—which have the power to positively or negatively transform the vulnerability context of food availability and access in Africa, as well as the choices, assets, and conditions that shape food availability and access, including responses to climate change and climate justice. The chapter sums up by showing how religion, as a transforming structure and process, can impactfully undermine or accelerate availability and access to food in Africa.

Chapter 5, by Seyram Amenyedzi and Rose Omari, "Religious Perspectives on Climate Change and Food Security in Ghana," brings to our sharp attention the religious (particularly Christian) dimensions to climate change and food security in Ghana. This chapter anchors on an ecotheological framework in which theological discourses on nature and which raise concern about the state of the environment are interrogated. Accounts from the Bible are therefore unpacked to aid reflection on efforts toward achieving food security.

The collection of chapters in Part II of the book seeks to address aspects of climate change and food insecurity such as crop failure, pests and crop diseases. Chapter 6, by Fortune Sibanda, "Rastafari, Climate Change and Crop Failure in Africa," posits that in spite of being marginalized and misunderstood, Rastafari environmental ethic guided by a "green philosophy" and dependence on *Ital* (natural) food is consistent with organic farming, food security, and living in harmony with Mother Earth in contrast to the toxic and greed-driven consumerist "Babylon *shitstem*" (Western corrupt and oppressive system). Using a de-colonial theoretical framework, the author established that Rastafarians serve as a strong resource for supporting organic farming, food security, and climate restabilization within a framework of their environmental I-consciousness.

Chapter 7, "Catholicism, Climate Change and Pests in Africa," by George Nche, brings into sharp focus the threat posed by insect pests to food production in Africa. Given that temperature is the main environmental factor affecting insect population dynamics, the nexus with climate change is clearly shown in the chapter, pointing to the stark reality that climate change will lead to the proliferation and survival capacity of pests. This calls for more innovative and affordable context-specific pest management strategies for the continent. But in the chapter the contributions of the Catholic Church to the global fight against hunger and food scarcity, especially in Africa, are well documented. Drawing from ecclesiastical documents and scholarly material, the chapter considers the role of the Catholic Church in addressing the threat posed by climate change-induced pests in Africa to highlight implications for the Catholic Church and policymakers in Africa.

Chapter 8, by Obvious Vengeyi, "Farming God's Way and Climate Change in Zimbabwe," demonstrates how Foundations for Farming, a Christian organization in Zimbabwe whose major concern is poverty alleviation, deliberately deploys biblical verses in Zimbabwe to craft a method of farming called Farming God's Way. At the center of this chapter are the Bible and its potency in the establishment of the principles of farming for poverty alleviation. The Bible is replete with references of stories and parables about subjects such as seeds, crops, trees, farmers, seasons, weather, hunger, pests, and diseases, among others. Ultimately, the chapter considers the role of the Bible in the diagnosis by Foundations for Farming of the causes of peasant poverty in Zimbabwe and Africa at large.

Chapter 9, "Crop Diseases and Food Insecurity in Africa: A Hindu Perspective," by Elizabeth Motswapong, Joseph Gaie, and Siamisang Saudu, foregrounds Hinduism in the efforts toward food security and climate change, which resonate with the Sustainable Development Goals' commitment to reach Zero Hunger by 2030. It argues that Hinduism has continued to be an active player in the fight against food insecurity and crop disease in Africa owing to its environmentally sensitive philosophy. This is partly drawn from sacred texts, such as the Mahabharata, the Ramayana, and the Vedas which contain the earliest messages for the preservation of the environment and ecological balance. In addition, the teaching of non-violence or *ahimsa* has also catapulted the ethics of symbiotic relation between human beings and environment and serves as a resource in addressing crop disease as one of the causes of food insecurity in Africa.

Chapter 10, "Crop Diseases, Pests and Food Insecurity in Africa: An Islamic Perspective," by Newton Kahumbi Maina, makes clear that rising temperatures have also contributed to the emergence and spread of crop diseases, pests and pathogens which have exacerbated global food insecurity. Owing to the vulnerabilities of Muslim regions to climate change, the 57-member Organization of Islamic Cooperation (OIC) drafted the "Islamic Declaration on Climate Change" which called on oil-producing countries to phase out greenhouse gas emissions and to commit to a zero emissions strategy. The chapter probes the role of Islam on food security and climate change in Africa, specifically in the regions where Islam holds sway, such as in North Africa, the Horn of Africa, and the Sahel regions. Kahumbi unearths the following: first, how can Islam be used to combat climate change and its impacts in line with Sustainable Development Goal (SDG) 13; second, the ways in which Islam could be an avenue to end hunger, achieve food security and improved nutrition, and promote sustainable agriculture in Africa, as per SDG 2. Drawing heavily from the two main scriptural sources of Islam—the Qur'an and Sunna (teachings of Prophet Muhammad)—and interpretations of scholars on the two main sources, the chapter explores the interplay between Islam, food security, and climate change in Africa. The chapter demonstrates how the two sources provide principles that govern climate change and food security and how, if well operationalized, Africa can make strides on both food security and climate change.

Part III of the book covers aspects of human flourishing that resonate with climate change, religion, and food security. Chapter 11, "Indigenous Knowledge, Food (In)Security and Climate Change: A Case Study of *Bota Reshupa* (Herbal Porridge) Among the Ndau of South-Eastern Zimbabwe," by Tenson Muyambo, considers how indigenous knowledge systems from the Ndau people in South-East Zimbabwe could be used to ensure food security in the face of climate change. Herbal shrubs used in the preparation of *bota reshupa* are almost extinct, yet the porridge is laden with food elements vital for immunity and disease resistance. This has threatened people's health and food security, thereby putting the attainment of Sustainable Development Goals 2, 3, and 13 into jeopardy. The chapter recommends that although forests that harbor herbal shrubs essential for the preparation of *bota reshupa* are under threat from climate change, the Ndau indigenous knowledge systems' practitioners can still employ taboos to ensure the protection and maintenance of forests and the establishment of herbal gardens as a means to mitigate climate change and food insecurity in African communities.

Chapter 12, "Indigenous Knowledge Systems, Climate Change and Food Security: Perspectives from Bungoma County, Kenya," by Mamati King'asia and Simon Gisege Omare, argues that the revitalization of indigenous knowledge is vital in promoting food security. The chapter is in line with the Sustainable Development Goal (SDGs) 2, which seeks to achieve Zero Hunger, achieve food security and improved nutrition, and promote sustainable agriculture. It maintains that during Kenya's precolonial period, food production and nutrition heavily relied on indigenous knowledge to safeguard the well-being of the individual and the community. These indigenous knowledge systems have declined due to colonialism, modernity, and the commercialization of food. The chapter recommends that indigenous knowledge systems be maintained, in order to promote food security since they are readily accepted by the local communities. Chapter 13, "African Women, Religion and Food Security in the Context of Pandemics," by Ezra Chitando, engages with the themes of women, climate change, food security/sovereignty, and pandemics in Africa. It posits that all these challenges and themes are gendered. Consequently, climate change, food insecurity, and pandemics have a more devastating impact on women and girls than men and boys. Yet, African women have not surrendered to their fate. They continue in strategic ways to provide effective leadership, particularly from within their faith contexts, to meet the SDGs and Agenda 2063.

Part IV of the book focusses on shifting vulnerabilities in relation to religion, climate change, and food security. Chapter 14, by Gertrude Kapuma and Mwawi Chilongozi, "Farming God's Way to Avert Crop Failure and Pests in Malawi," describes how most farmers in Malawi do not follow best agriculture practices, leading to soil erosion and weak soils for crops. In order to add fertility to soil, farmers in Malawi have resorted to using chemical fertilizers and hybrid seeds to get better yield. Yet the majority of the rural farmers cannot afford to buy such farm inputs as the prices go up every year. The authors show how "farming God's way" is a way to help bring back lost food nutrients to the soil and is a solution to overcome food insecurity in the rural areas of Malawi.

Chapter 15, "Religious and Cultural Drivers of the Feminization of Food Insecurity in South Africa: Contextualizing Ruth," by Linda Naicker, takes us to South Africa and demonstrates that urban food insecurity in South Africa is rising at a rapid rate. The locus of poverty and food insecurity is shifting as a result of rapid urbanization and poor governmental policy interventions. Yet, research shows the way women and children are disproportionately affected by climate-induced food insecurity. This chapter holds that the perpetuation of the food insecurity crisis in South Africa is a result of a myriad of factors, inclusive of structural and systemic oppression, climate change, and the cultural and religious arrangements of society. Re-reading the biblical book of Ruth through the lens of African Women's Theology, the chapter examines the roles of Naomi, Ruth, and Boaz in the biblical narrative in order to probe how religious underpinnings of women serve to perpetuate the oppression and food insecurity status of under-resourced urban women.

Chapter 16, "Faith-Based Organisations and Food Security and Sovereignty in Africa: A Critical Review," by Tshenolo Madigele and Ezra Chitando, contributes to the reflections on faith-based organizations (FBOs) and development by addressing a somewhat neglected theme, namely, FBOs and food security and sovereignty in Africa. It deploys a documentary review to appreciate the interface between selected FBOs and food security in Africa. The chapter highlights the strengths of FBOs in responding to the food crisis/insecurity in Africa. The chapter also summarizes the activities of these FBOs in promoting food security in Africa. It critiques the interventions by FBOs and concludes by underscoring the centrality of food security and sovereignty to sustainable development in Africa.

Part V marks the last section of the book and covers aspects of resilience in the face of climate change and food security from a religious perspective. Chapter 17, by Kudzai Biri, “Pentecostalism, Theology of Survival and Food Security in Zimbabwe,” examines controversial Zimbabwean Pentecostal beliefs and practices enshrined in their theology of survival. The deployment of the theology of survival that revolves around food security is located within the matrix of the gospel of prosperity, dominion, and culminates in excessive accumulation. The argument in this chapter is that Pentecostal theology of survival adopts a capitalist orientation that has covert structures of oppression and exploitation, thus blurring the division between the blessings of God and greed. The chapter analyzes sermons and practices in selected leading international Pentecostal denominational churches, both older and newer. It clearly demonstrates that the theology of survival and its tenets have both political overtones and strong economic bearings on livelihoods and sustainable development. This emboldens the accumulation of property in the name of holiness, thrives in circumstances which promote varied modes of capitalism, and consequently widens the gap between the rich and the poor, leaving the masses in abject poverty.

Chapter 18, by Sylvia Owusu-Ansah, Beatrice-Joy Owusu-Ansah and Appiah Kubi, “Religion, Climate Change, and Food Security in Africa: Religion, Food Security, and Resilience of Rural People in Ghana,” posits that for most rural people whose main source of income is farming, climate change is one of the greatest threats to livelihoods and food security. Higher temperatures, frequent flooding, drought, and erratic rainfall patterns among others are major challenges to small-scale farming. There is a need for resilience of rural people to help them adapt to external and internal factors while transforming existing conditions to become more food secure. Resilience in the context of the rural Ghana encompasses three main aspects: adaptability, transformation, and persistence. Using the biblical pattern in the book of *Genesis* chapters 12 and 26 regarding the farming activities of Abraham and Isaac in times of drought as a basis, the authors draw semblance in modern-day difficulties facing agriculture in terms of climate change and its associated problems. Chapter 19, “Religion, Food security, and Climate Change Mitigation: A Case of Luangwa Valley Women of Eastern Zambia,” by Nelly Mwale and Nicholas Haambokoma, brings the volume to a close. It explores how Zambian women from the Luangwa Valley utilize indigenous knowledge systems to mitigate climate change and promote food security. It shows that in a context characterized by recurrent impacts of different ongoing climatic changes and food insecurity, women use indigenous knowledge to address climate change and food insecurity through conservation farming which restores soil health, boosts production, and enhances food security. The chapter contributes to discourses of religion, food security, and climate change by showcasing that rural women remain active agents of change through their utilization of indigenous beliefs and practices to address the environmental crisis and in turn foster food security and contribute to promoting health organic foods that are of nutritional value to people.

Conclusion

This volume breaks new ground by exploring the nexus between religions, climate change, and food security in Africa. It brings together contributors from different African contexts and specializations, mostly within African theology, religious, and biblical studies, to reflect on how religion (in its broadest sense) can contribute to effective responses to climate change and food security in Africa. The volume confirms the changing role of African scholars in relation to engaging with the continent's most pressing existential challenges. Where previously many African scholars could spend time in speculative pursuits (which still have their place in academia), this volume sees African scholars focusing on the climate emergency and the threat of food insecurity. It is envisaged that religious leaders, governments, non-governmental organizations (NGOs), policymakers, development practitioners, and others will interact with the reflections and recommendations put forward in this volume. This will be a significant step in Africa's struggle against climate change, hunger, as well as place it more strategically to meet the targets of SDG 2, "Zero Hunger," and SDG 13, "Climate action."

Indeed, the development agendas outlined in the economic blueprints of the different African countries, the continental Agenda 2063, and the global SDG Agenda will not succeed without Africa achieving food security and food sovereignty. Further, Africans must be strategic and organize effectively to overcome the twin threats of climate change and food insecurity/food dependence. It is scandalous that one of the world's most fertile continents, Africa, must rely on food handouts from others, including its former (and current) oppressors. Religion, informed by the quest to promote holistic liberation, must contribute toward Africa's food security and food sovereignty. We hope that diverse other actors, including political leaders, faith leaders, civil society activists, development specialists, researchers from diverse disciplines, and others, shall also invest in the quest to ensure that African Union's vision of healthy and well-nourished citizens will be realized yesterday, as tomorrow is already too late.

References

Afrobarometer. 2022. For religious leaders in Africa, popular trust may present opportunity, challenge in time of crisis. Dispatch No. 536. Accra: Afrobarometer.

African Union. 2015. Agenda 2063: The Africa We Want: Framework Document. Addis Ababa: African Union. Available at https://au.int/sites/default/files/documents/33126-doc-framework_document_book.pdf.

Antwi, A. 2013. "Climate Change and Food Security: An overview about the issue". Published report. Friedrich Ebert Stiftung.

Beros, D. et al. Eds. 2022. *International Handbook on Creation Care & Eco-Diakonia: Concepts and Theological Perspectives of Churches from the Global South*. Oxford: Regnum Books International.

Behnassi, M., Baig, M.B., Sraïri, M.T., Alsheikh, A.A. and Risheh, A.W.A.A., 2022. *Food Security and Climate-Smart Food Systems*. Springer International Publishing.

Chitando, E. 1997. *A curse of the western heritage*? Imagining religion in an African context. *Journal for the Study of Religion* 10 (2): 75–98.

Chitando, E, E. Conradie and S. Kilonzo. Eds. 2022. *African Perspectives on Religion and Climate Change*, New York: Routledge.

Clapp, J. 2014. Food security and food sovereignty: time to get past the binary. *Dialogues in Human Geography* 4 (2): 206–211.

Ebert, A., R. Redden, D. Hunter, J. Hatfield and Shyam Singh Yadav (eds.) 2018. *Food Security and climate change*. United Kingdom: John Wiley and sons.

FAO, IFAD, UNICEF, WFP and WHO. 2023. *The State of Food Security and Nutrition in the World 2023. Urbanization, agrifood systems transformation and healthy diets across the rural–urban continuum*. Rome: FAO. https://doi.org/10.4060/cc3017en. Accessed 20.7.2023.

FAO. 1996. *Rome Declaration on World Food Security and World Food Summit Plan of Action*. Available online at: http://www.fao.org/DOCREP/003/W3613E/W3613E00.HTM.

FAO. 2008. The State of Food and Agriculture. Online Report. https://www.fao.org/documents/card/en/c/ec403d63-9ced-5a72-bcbc-b0ee69373437.

Grebmer, K. & Bernstein, J. 2020. Global Hunger Index 2020 One Decade to Zero Hunger—Linking Health and Sustainable Food Systems. Dublin/Bonn: Chatham House.

Hertel W.T. & D.B. Lobell. 2014. Agricultural adaptation to climate change in rich and poor countries: Current modeling practice and potential for empirical contributions. *Energy Economics* 46: 562–575. https://doi.org/10.1016/j.eneco.2014.04.014.

Hodgkin, T. and Bordoni, P. 2012. Climate change and the conservation of plant genetic resources. *Journal of Crop Improvement* 26: 329–345.

Hoover, S.M. 2006. *Religion in the Media Age*. London: Routledge.

Huntington, J.L., K.C. Hegewisch, B. Daudert, C.G. Morton, J.T. Abatzoglou, D.J. McEvoy and T. Erickson 2017. Climate engine: Cloud computing and visualization of climate and remote sensing data for advanced natural resource monitoring and process understanding. *Bull. Am. Meteorol. Soc.*, 98, 2397–2409. https://doi.org/10.1175/BAMS-D-15-00324.1.

Intergovernmental Panel on Climate Change—IPCC. 2001.

Iizumi, T. et al. 2018. Crop production losses associated with anthropogenic climate change for 1981–2010 compared with preindustrial levels. *Int. J. Climatol.* 38: 5405–5417. https://doi.org/10.1002/joc.5818.

IPCC. 2019: IPCC Special Report on the Ocean and Cryosphere in a Changing Climate [H.-O. Pörtner, D.C. Roberts, V. Masson-Delmotte, P. Zhai, M. Tignor, E. Poloczanska, K. Mintenbeck, A. Alegría, M. Nicolai, A. Okem, J. Petzold, B. Rama, N.M. Weyer (eds.)]. Cambridge University Press, Cambridge, UK and New York, NY, USA, 755 pp. https://doi.org/10.1017/9781009157964.

Jarosz, L. 2014. Comparing Food Security and Food Sovereignty Discourses. *Dialogues in Human Geography* 4: 168–181. https://doi.org/10.1177/2043820614537161.

Ketiem, P., P.M. Makeni, E.K. Maranga, and P.A. Omondi. 2017. Integration of climate change information into drylands crop production practices for enhanced food security: A case study of Lower Tana Basin in Kenya. *African J. Agric. Res.*, 12: 1763–1771.

Levasseur, T. 2021. *Climate Change, Religion, and our Bodily Future*. Maryland: Rowman and Littlefield.

Lin, B. 2011. Resilience in agriculture through crop diversification: Adaptive management for environmental change. *Bioscience* 61: 183–193.

Lobell, D.B., Schenkler, W. and Costa-Roberts, J. 2011. Climate Trends and Global Crop Production since 1980. *Science*, 333 (6042): 616–620

Lobell, D.B. and Burke, M. 2010. 'Introduction'. In: Lobell, D.B., Burke, M. (eds) *Climate Change and Food Security. Advances in Global Change Research*, 37. Springer, Dordrecht. https://doi.org/10.1007/978-90-481-2953-9_1

Magesa, L. 2013. *What is not Sacred? African Spirituality*. Maryknoll, NY: Orbis Books.

Maseno, L. and Mamati, K. 2022. The nexus between indigenous beliefs on environment and climate change adaptation amongst the Sengwer in Embobut Forest, Kenya. *African Perspectives on Religion and Climate Change*, edited by E. Chitando, E. Conradie and S. Kilonzo. New York: Routledge, pp. 50–64.

Marshall, K. et al. 2018. Engaging religious actors in addressing famine emergencies. *Economics: The Open-Access, Open-Assessment E-Journal*. Discussion Paper No. 2018–10.

Matsa, M. 2021. *Climate Change and Agriculture in Zimbabwe: Sustainability in Minority Farming Communities*. New York: Palgrave Macmillan.

Mbiti, J.S. 1969. *African Religions and Philosophy*. London: Heinemann.

Olsson, L. & M. Opondo, & Tschakert, Petra & Agrawal, Arun & Eriksen, Siri & Ma, S. & Perch, Leisa & Zakieldeen, S.A. (2014). Livelihoods and Poverty. https://doi.org/10.13140/RG.2.1.4508.0406.

Polzin, M. 2018. Food Security Vs. Food Sovereignty. https://wp.nyu.edu/gallatin_human_rights_fellows/2018/09/26/food-security-vs-food-sovereignty/. Accessed 12.7.2023.

Porter, J.R., Howden, S.M., Smith P. 2017. Considering agriculture in IPCC assessments. *Nature Climate Change*, 7, 680–683.

Raj, S., Roodbar, S., Brinkley, C. and Wolfe, D.W. 2022. Food Security and Climate Change: Differences in Impacts and Adaptation Strategies for Rural Communities in the Global South and North. *Front. Sustain. Food Syst.* 5:691191. https://doi.org/10.3389/fsufs.2021.691191

Sanni, J.S. 2016. Religion: A new struggle for African identity. *Phronimon* 17(1): 71–83.

Schmidtko, S., L. Stramma, and M. Visbeck, 2017. Decline in global oceanic oxygen content during the past five decades. *Nature*, 542, 335. https://doi.org/10.1038/nature21399

Searchinger, T., R. Waite, C. Hanson, J. Ranganathan & E. Matthews. 2019. Creating a Sustainable Food Future: A Menu of Solutions to Feed Nearly 10 Billion People by 2050. Digital Report. World Resources Institute. https://www.wri.org/research/creating-sustainable-food-future.

Serventi, L. (ed). 2022. *Sustainable Food Innovation*. New York: Palgrave Macmillan.

Sirba, H.Y. and T.B. Chimdessa. 2021. Review of impact of climate change and food security in Africa. *International Journal of Research and Innovations in Earth Science* 8(3): 40–66.

Smart, N. 1969. *The Religious Experience of Mankind*. New York: Charles Scribner's Sons.

Soudini, S. 2022. *Religious Transnationalism and Climate Change*. Cham: Palgrave Macmillan.

Tarusarira, J. 2021. The grass versus the people: sacred roots of environmental conflict in the Chilonga communal lands in Zimbabwe. *Journal for Religion, Society and Politics*. https://doi.org/10.1007/s41682-021-00090-0.

Umuziranenge, G. 2016. The role of churches in addressing climate change: Case study of Rwanda. In D. Werner and E. Jeglitzka, eds., *Eco-Theology, Climate Justice and Food Security: Theological Education and Christian Leadership Development*. Geneva: Globethics.net: 209–215.

Wiesmann, D. & Biesalski, H. & Grebmer, K. & Bernstein, J. 2015. Methodological Review and Revision of the Global Hunger Index. *SSRN Electronic Journal*. https://doi.org/10.2139/ssrn.2673491.

World Vision Canada. 2020. gendernutritionframework.org. https://www.gendernutritionframework.org/_files/ugd/c632d7_a7d415dcfd8b483288c8de14fa3d4744.pdf. Accessed 10.7.2023.

Wolfe, D.W. 2013. Climate change solutions from the agronomy perspective. In: Hillel D and C Rosenzweig (eds), *Handbook of Climate Change and Agroecosystems: Global and Regional Aspects and Implications*. Imperial College Press. London.

Wlokas, H.L. 2008. The impacts of climate change on food security and health in Southern Africa. *Journal of Energy in Southern Africa* 19(4): 12–20.

World Bank. 2016. Poverty and shared prosperity Report: Taking on inequality. https://www.worldbank.org/en/publication/poverty-and-shared-prosperity-2016.

Yadav, S., V.S. Hegde, A. Habibi, M. Dia and S. Verma. 2019. "Climate Change, Agriculture and Food Security," in S.Yadav, V.S. Hegde, A. Habibi, M. Dia and S. Verma, eds., *Climate Change, Agriculture and Food Security*. Hoboken, NJ: John Wiley & Sons, pp. 1–24.

CHAPTER 2

Prophetic Action, Climate Change, Food Security, and SDG 2 in Africa

David Andrew Omona

Introduction

Prophetic action, climate change, and food security are important issues in Africa, as the continent is one of the most vulnerable regions to the effects of climate change. Prophetic action, that is, the actions taken by religious leaders and communities to address social, economic, and environmental issues through the lens of their faith, becomes strategic in addressing the concerns of climate change and food security in the light of SDG 2 that aims to end hunger, achieve food security and improved nutrition, and promote sustainable agriculture.

Climate change has been identified as one of the greatest challenges of the twenty-first century, with its impacts being felt across the globe. Of the different parts of the globe, Africa, in particular, is highly vulnerable to the effects of climate change due to its heavy reliance on rain-fed agriculture and natural resources. The 2020 State of the Climate in Africa report

D. A. Omona (✉)
School of Social Sciences, Uganda Christian University, Mukono, Uganda

L. Maseno et al. (eds.), *Religion, Climate Change, and Food Security in Africa*, Sustainable Development Goals Series,
https://doi.org/10.1007/978-3-031-50392-4_2

provides a snapshot of climate change trends and impacts (WMO 2021). In the foreword of the 2021 World Meteorological Organization (WMO) report, the Secretary General, Petteri Taalas (WMO 2021), notes,

> During 2020, the climate indicators in Africa were characterized by continued warming temperatures, accelerating sea-level rise, extreme weather and climate events, such as floods, landslides and droughts, and associated devastating impacts. The rapid shrinking of the last remaining glaciers in eastern Africa, which are expected to melt entirely in the near future, signals the threat of imminent and irreversible change to the Earth system.

The observation by Taalas is a true reflection of the state in Africa to date. The International Partnership on Climate Change (IPCC) publications and many other documents on climate change in Africa do affirm this notion. The report of the 2021 WMO did not only highlight the rise of the sea level and the melting of the continent's iconic glaciers; it also highlights Africa's disproportionate vulnerability and shows how the potential benefits of investments in climate adaptation, weather and climate services, and early warning systems far outweigh the costs (WMO 2021).

Like the observation in the WMO report, the IPCC's 2023 Synthesis Report of the IPCC Sixth Assessment Report (AR6) on climate change in Africa indicates "widespread and rapid changes in the atmosphere, ocean, cryosphere and biosphere have occurred" (IPCC 2022). Most of these have been attributed to human-caused climate change which is already affecting many weather and climate extremes in every region across the globe. Invariably, this has led to widespread adverse impacts and related losses and damages to nature and people. Yet in most cases, vulnerable communities have historically contributed the least to current climate change although they are disproportionately affected.

Two factors have been identified as responsible for climate change, namely human activities and natural occurrences (Antwi 2013). The human activities responsible for climate change are those activities that release carbon dioxide and other greenhouse gasses into the air. Consequently, at present, this has led to a lot of carbon dioxide to circulate in the atmosphere than ever before in at least the past 2 million years (Omona 2022b). This has been to the extent that the twentieth and twenty-first centuries have witnessed a 40% increase in the level of atmospheric carbon dioxide. The increase in greenhouse gasses is caused by the following human activities, including, but not limited to:

- Burning fossil fuels such as oil, gas, and coal that contain carbon dioxide, which initially have been "locked away" in the ground for thousands of years, pump a great deal of carbon dioxide and methane into the atmosphere. The increase in the number of automobiles, factories, and burning other items that produce greenhouse gasses increases the concentration of carbon dioxide in the atmosphere "that act like a blanket wrapped around the Earth, trapping the sun's heat and raising temperatures" (UN Climate Action n.d.). Furthermore, damp sites for garbage, energy, industry, transport, buildings, agriculture, and land use are among the main emitters of greenhouse gasses into the atmosphere with direct impact on climate. While this phenomenon is a reality globally, but the major culprits are the developed countries from the global north.
- Deforestation is another contributor to climate change. For instance, when trees are cut, they emit carbon dioxide into the atmosphere. But when forests are not cut, trees remove and store carbon dioxide from the atmosphere. Hence continual cutting down of trees implies that a lot of carbon dioxide is made to build in the atmosphere since there are no trees to absorb it. Since such emitted carbon dioxide into the atmosphere keeps circulating for some time, its presence has direct effects on the climate (Omona 2022b, p. 110). The research by the Global Forest Research Assessment shows that in 2010 alone deforestation resulted to emission of a billion carbon dioxide into the atmosphere (Rinkesh n.d.). Accordingly, FAO's (2021) latest Global Forest Resources Assessment shows that 420 million hectares of forest has been lost since 1990. And almost 90% of such cutting has been driven by agricultural expansion, thus impacting the climate much greater than previously thought.
- Agriculture practice of planting crops and rearing animals that leads to releases many different types of greenhouse gasses into the air also has direct effect on the climate. For example, animals produce methane, which is 30 times more powerful than carbon dioxide as a greenhouse gas (UN Climate Action n.d.). The nitrous oxide used for fertilizers is 10 times worse and is nearly 300 times more potent than carbon dioxide! All these combined in the atmosphere lead to the depletion of the ozone layer, thus affecting the climate greatly.

Whereas human activities are the major causes of climate change, natural occurrences are also responsible for climate change. The natural occurrences that are responsible for climate change are seen to be: Milankovitch cycles, El Nino and La Nina, solar irradiance, and volcanic eruption, explained as follows:

- The Milankovitch cycles—The cycles lead to climate change in that as Earth travels around the sun, its path and the tilt of its axis can change slightly (Bius 2020). These changes, called Milankovitch cycles, affect the quantity of sunbeams that fall on Earth. Thus causing the temperature of Earth to change. Whereas these cycles take place over tens or hundreds of thousands of years and are unlikely to be causing the changes to the climate that is seen today, however, its occurrence over the year could explain what is currently taking place around the globe (Omona 2022a).
- El Niño Southern Oscillation (ENSO)—ENSO is a pattern of changing water temperatures in oceans. In an "El Niño" year, the global temperature warms up, and in a "La Niña" year, it cools down. These patterns have often affected the global temperature greatly. The frequent dry spell and floods around the world are a direct result of the El Nino and La Nina operations. For example, in 2021, scorching temperatures did cause deadly heatwaves in Canada and Pakistan and set off wildfires in Greece and Siberia. While in Madagascar, a particularly prolonged and intense drought has driven 1 million people to the brink of what is being described as the world's first "climate change-induced famine" (Amnesty International n.d.; Capper 2021). On the other hand, there was severe flooding in Germany, China, United Kingdom, and in 2022 in Pakistan.
- Solar irradiance—The rate of energy coming from the Sun, while it seldom occurs, however, affects the climate directly by changing the rate of solar heating of the Earth and the atmosphere, thus indirectly changing the cloud forming processes (Bhargawa and Singh 2019). Therefore, such increase in solar energy would make the entire atmosphere of Earth warm, thus posing a potential danger to the climate of an area.
- Volcanic eruptions—Volcanoes have a mixed effect on our climate. Eruptions produce aerosol particles that cool Earth, but they also release carbon dioxide, which warms it. While volcanoes produce 50 times less carbon dioxide than humans do, it also makes significant

contribution to climate change when it occurs (UN Climate Action n.d.).

From the foregoing, it is unequivocal that human influence has warmed the atmosphere, ocean, and land, to the extent that it has made the global mean sea level to increase by 0.20 [0.15–0.25] mm yr^{-1} between 1901 and 2018. While the average rate of sea level rise was 1.3 [0.6–2.1] mm yr^{-1} between 1901 and 1971, it has made to increase to 1.9 [0.8–2.9] mm yr^{-1} between 1971 and 2006, and further increasing to 3.7 [3.2–4.2] mm yr^{-1} between 2006 and 2018 (IPCC 2022). Human influence is likely seen to be the main driver of these increases since at least 1971. Evidence of observed changes in extremes such as heatwaves, heavy precipitation, droughts, and tropical cyclones, and, in particular, their attribution to human influence, has further strengthened since AR5. Human influence has likely increased the chance of compound extreme events since the 1950s, including increases in the frequency of concurrent heatwaves and droughts (IPCC 2022).

Looking at the global context, approximately 3.3–3.6 billion people live in contexts that are highly vulnerable to climate change. In most cases, the regions and people with substantial development restraints have high vulnerability to climatic threats. For, increasing weather and climate extreme events have exposed millions of people to acute food insecurity and reduced water security, with the largest adverse impacts observed in many locations and/or communities in Africa, Asia, Central and South America, and or generally in Low Developed Countries, Small Islands, and the Arctic (IPCC 2022). In these locations, Indigenous Peoples, small-scale food producers, and low-income households are bearing the brunt of food insecurity since they depend on rain for agriculture. Due to climate change, for example, between 2010 and 2020, human mortality from floods, droughts, and storms was 15 times higher in highly vulnerable regions compared to regions with very low vulnerability (IPCC 2022).

Furthermore, climate change has caused substantial damages, and increasingly irreversible losses, in terrestrial, freshwater, cryospheric, and coastal and open ocean ecosystems. For, due to increases in the magnitude of heat extremes, hundreds of local losses of species with mass mortality events have been recorded on land and in the ocean. The impacts on some ecosystems are approaching irreversibility such as the impacts of hydrological changes resulting from the retreat of glaciers, or the changes in some mountain and Arctic ecosystems driven by permafrost thaw (IPCC 2022).

Climate change has reduced food security in terms of food production, accessibility, availability, and utilization by the population (Omona 2022b) and also affected water security, thus hindering efforts to meet Sustainable Development Goals (IPCC 2022). Although overall agricultural productivity has increased, climate change has slowed this growth over the past 50 years in some high latitude regions. Ocean warming and ocean acidification have adversely affected food production from fisheries and shellfish aquaculture in some oceanic regions, thus directly affecting the physical, social, and "economic access to sufficient, safe, affordable and nutritious food to meet dietary needs and food preferences for an active and healthy life" (Bazzaz and Samboek 1996, p. 4; Oyiga et al. 2011, p. 3). Consequently, "roughly half of the world's population currently experience severe water scarcity for at least part of the year due to a combination of climatic and non-climatic drivers" (IPCC 2022).

Whereas in all regions of the world increases in extreme heat events have resulted in human mortality and morbidity, the situation in Africa is on the higher side. The occurrence of climate-related food-borne and water-borne diseases and the incidence of vector-borne diseases have increased substantially in Africa. This has led to some mental health challenges that are associated with increasing temperatures, trauma from extreme events, and loss of livelihoods and culture (IPCC 2022). Climate and weather extremes are increasingly driving displacement in Africa, Asia, North America, and Central and South America, with small island states in the Caribbean and South Pacific being disproportionately affected relative to their small population size.

Climate change has also caused widespread adverse impacts and related losses and damages to nature and people who are unequally distributed across systems, regions, and sectors. In Africa, as elsewhere, economic damages from climate change have been detected in climate-exposed sectors, such as agriculture, forestry, fishery, energy, and tourism. Invariably, this affected individual livelihoods through, for instance, destruction of homes and infrastructure, and loss of property and income, human health and food security, with adverse effects on gender and social equity (IPCC 2022).

In many urban areas in Africa, observed climate change has caused adverse impacts on human health, livelihoods, and key infrastructure. Given hot extremes have intensified in cities, urban infrastructure, including transportation, water, sanitation, and energy systems, has been compromised by extreme and slow-onset events, with resulting economic losses, disruptions of services, and negative impacts to well-being (Li et al. 2022). Unfortunately, the observed adverse impacts as these are concentrated among the economically and socially marginalized urban dwellers.

While other continents also experience climate change, the level of vulnerability in Africa is at another level. To some extent, Africa has become highly vulnerable to the impacts of climate change due to its heavy reliance on rain-fed agriculture and natural resources (Moss et al. 2019). Consequently, climate change has the potentially exacerbated existing challenges in the region, such as poverty, food insecurity, and inequality (IPCC 2018). Furthermore, it has also led to conflicts over natural resources, which in turn can exacerbate poverty and food insecurity (Homer-Dixon 1991). Therefore, given this, the need for prophetic action on climate change and food security in Africa so as to realize SDG 2 becomes imminent. The following sections provide recourse on prophetic action on climate change, food security, and achieving SDG 2 in Africa.

Prophetic Action and Climate Change in Africa

Climate change is a global phenomenon. In the context of Africa, as elsewhere, the warming up of the planet has resulted in rising sea levels, increased frequency and intensity of storms, and loss of biodiversity. Although it is a key threat to human survival that requires immediate prophetic action, human activity and natural factors that cause change in climate are still going unabated.

Prophetic action in this chapter involves speaking the truth, confronting the status quo, and challenging power structures. This is an ingredient for religious leaders to add their voices to combating the ills of climate change in Africa. On many occasions, religious leaders have been able to confront the powerful with a demand to address concerns of environmental stress, gone public in their address, and were able to center their concerns on justice. While being prophetic can throw one into a thorny thicket, and a briar patch, being prophetic is a long-distance race (United Church of Christ 2022). In the case of climate change, prophetic action means recognizing the science behind it and challenging those who deny its existence. It also means addressing the root causes of climate change, such as the continued use of fossil fuels, deforestation, and unsustainable agriculture (Akinpelu 2020; Ebegbulem 2017).

Consequently, one prophetic action which religious leaders adopt to combat climate change in Africa is through advocacy. In many occasions, religious leaders engaged in speaking to elected officials and urging them to take action on this issue. Through this, religious leaders have been able to mobilize others to do the same in their different locations. Since advocacy can take many forms, from writing letters or making phone calls to organizing protests or lobbying, some concerned people have done that. In a way of advocacy, as Ezra Chitando observes,

> On 7–8 June 2011, African faith leaders met in Nairobi, Kenya, to discuss climate change and how it needed to be approached during the United Nations climate change conference (COP 17) in South Africa. They challenged exploitative interpretations of development; undertook to lead by example, to publicize the threat of climate change, and to draw on their spiritual resources to foster positive ecological values; and acknowledged the negative impact of climate change on already existing vulnerable groups. Further, they undertook to plant indigenous trees and promote ecological restoration. They also addressed world leaders and African political leaders. For their part, South African church leaders acknowledged the challenges posed by climate change. (Chitando 2017, p. 427)

While their action might not have attracted the urgency they envisioned in many parts of Africa, the advocacy role played by the faith leaders as exposed by Chitando has gone a long way to energizing local religious leaders on their prophetic action at local levels. On 11 March 2021 during a discourse organized by Anglican Alliance on "Prophetic Indigenous Voices on the Planetary Crisis," Reverend Doctor Kapya Kaoma from Zambia, using the African worldview as a framework of analysis, argues,

> The African world view sees humanity in relationship to the entire created order. Unlike in the West, the African world view has repeatedly pointed to the interconnectedness of humanity to the world of nature. Humans are part of nature. ... I want us to see how the African world view can help reform our thinking, because until our thinking is reformed, this crisis will continue to haunt us. The most important thing we need to realise is what Africans realised centuries ago: that we live on this earth, but it does not belong to us. (Anglican Alliance 2021)

The above assertion by Kaoma is indicative that when African religious leaders talk about the climate, they look at such a symbiotic relationship between human beings and their environment, hence take a clear stand knowing that the depletion of the climate has direct impact on human livelihood. To show the seriousness of the advocacy during the discourse, a delegation headed by the Anglican Communion Office at the United Nations (ACOUN) was able to lift up those voices at the United Nations Environmental Assembly (UNEA), to which the Anglican Communion is accredited (Anglican Alliance 2021).

Another prophetic action by religious leaders to address climate change in Africa is by calling on people to adopt sustainable practices in our daily lives. Such practices include reducing energy consumption, eating a plant-based diet, and using public transportation or walking instead of driving. It also means supporting companies and governments that prioritize sustainability. Like church leaders, religious leaders from African Traditional Religions and Islamic Faiths have played a pivotal role in actualizing this. As Chitando observes,

> [L]eaders from African Traditional Religions and Islam continue to play a strategic role in their communities. They provide a guide to belief and action, and their proclamations are taken seriously by their followers. Although some religious leaders have abused the trust, they are often regarded as people of integrity who are dedicated to the improvement of their communities. (Chitando 2017, p. 430)

Notwithstanding the abuse that some religious leaders have caused, as noted in Chitando's submission above, in most cases the plea from the diverse religious leaders is heeded by their followers because they act as custodians of community morals. To the extent, when they tell people to take action on something that is beneficial to the masses, and teaching them on issues of climate change, the community would heed to it. To this end, Simone Schuman, Jon-Vegard Dokken, Dewald van Niekerk, and Ruth A. Loubser (2018) report a view from a respondent who asserts,

> [I]f we get taught every time about climate change, like there's projects where people tell people if we do something like this which is wrong that it will affect climate change in a certain way, like at schools, maybe at church, at the house, maybe do some projects just to show people that something that we do can affect climate change in a certain way and how we will be affected also as people. (p. 7)

Given that prophetic action on climate change requires addressing social justice issues and that climate change disproportionately affects marginalized communities, such as low-income, Indigenous, and communities of color, religious leaders have taken it upon themselves to draw people's attention to that. Since climate justice "recognizes the intersectionality of social, economic, and environmental issues" and "takes into account the inequalities of climate change cause and effect" (Grady-Benson 2014, p. 13) "by understanding that historically marginalized communities are least responsible for climate change and yet suffer disproportionately from its impacts" (Zuckerman 2016, p. 11), religious leaders in Africa have ably done so. Since taking action on climate change means addressing the root causes of social injustice and inequality, religious leaders in Africa have been actively involved in the same.

Given taking prophetic action on climate change requires courage and commitment, recognizing the truth about the issue, challenging entrenched power structures, and working to create a more just and sustainable world, African religious leaders have done what it takes to do so. Since such acts are a call to action that everyone must answer if people hope to create a future where humans can thrive in harmony with nature, religious leaders ought not to waiver. And since, as Chitando (2017) observes, "[t]he potential of African religious leaders to motivate their followers to become change agents in the face of climate change is significant," there is no need to tire doing so. This is because, as Chitando (2017, p. 430) further says, "[r]eligious leaders have the capacity to influence their followers in a positive way," especially "[w]hen they possess the right knowledge and information, they are more effective than fly-by-night politicians." In order to keep the fire burning, "more dialogue and interaction between African religious leaders and African theologians and scholars writing on religion and climate change" (Chitando 2017, p 430) is required. Holding such dialogue "can result in the emergence of more religious leaders who are well grounded in eco-theology and sensitive to climate change" (Chitando 2017, p 430). In turn, such leaders will articulate issues relating to climate change with greater clarity and conviction.

In Africa, as noted above, prophetic action has played a significant role in addressing climate change. Religious leaders have been at the forefront of efforts to raise awareness about the impacts of climate change and to advocate for action to address it. They have used their moral authority to mobilize their communities and to call on governments and international organizations to take action. For example, in 2015, Pope Francis released an encyclical on the environment, in which he called for urgent action to address climate change (Francis 2015). The encyclical was widely welcomed by African religious leaders, who saw it as a powerful call to action.

Many religious organizations in Africa have since taken concrete steps to reduce their carbon footprint and to advocate for policies that address climate change (UN Environment Program 2015).

Besides religious leaders, grassroots organizations and communities have also played a critical role in addressing climate change in Africa. These groups have been involved in activities such as reforestation, sustainable agriculture, and the promotion of renewable energy sources. They have also advocated for policies that support the transition to a low-carbon economy.

Despite these efforts, much more needs to be done to address climate change in Africa. The continent is already experiencing the impacts of climate change, including droughts, floods, and rising sea levels. These impacts are having a devastating effect on communities, particularly those who rely on agriculture and natural resources for their livelihoods.

Prophetic Action and Food Security in Africa

Food security is a major challenge in Africa, with millions of people suffering from malnutrition and hunger. The root causes of food insecurity in Africa are complex and multifaceted, including poverty, climate change, conflict, and lack of access to resources. Prophetic action can be used to address these root causes and promote food security in Africa.

One way in which prophetic action has been used to promote food security is by advocating for policies and practices that promote sustainable agriculture. Prophets and other spiritual leaders have used their influence to call for policies that support small-scale farmers, promote sustainable farming practices, and protect natural resources (Nzwili 2020). Once heeded to, such actions can help to address the root causes of food insecurity in Africa by promoting sustainable agriculture and protecting the environment. In this regard, recent calls for increased action against hunger by church leaders, faith-based humanitarian agencies, and development leaders are finding significance as a new report warns of serious levels of under-nutrition in sub-Saharan Africa (Nzwili 2020).

Another way in which prophetic action that has been used to promote food security is by advocating for social justice. Prophets and other spiritual leaders can use their influence to call for an end to the inequalities that contribute to food insecurity. This can include advocating for policies that address poverty, promote gender equality, and protect the rights of marginalized communities. Accordingly, Nzwili (2020) notes that in 2017, as part to call for action, church groups and agencies singled out responding to climate change, resolving conflicts, and promoting good governance as some areas they needed to engage vigorously to tackle the hunger. As a way of advocacy, Bread for the World, an American, non-partisan, and Christian movement to end hunger, has focused on advocacy for policy change by engaging individuals and groups in annual campaigns with a specific focus on matters related to food and agriculture (Costa 2020). It organizes fundraisers for nutrition programs and encourages activists and members to write letters to their members of Congress, urging them to take action on the issue of hunger and food security. Furthermore, the organization has also participated in raising awareness among Christian communities and conducts training in congregations, campuses, and community groups, while working globally with communities hit by natural disasters to tackle post-disaster food insecurity.

Prophetic action in Africa has also been used to address the immediate needs of people suffering from food insecurity. For example, in a wake to uplift the standard of the food insecure people, the Aga Khan Foundation (AKG) focuses on boosting agricultural yields in food-scarce areas to "effectively reduce poverty, ensure tangible food security, and improve the livelihoods of smallholder farmers and their families" (Costa 2020). In Africa, religious and other spiritual leaders have mobilized their communities to provide food and other forms of assistance to those in need. Invariably, this has included organizing food drives, establishing community gardens, and providing education on nutrition and healthy eating.

Prophetic Action and SDG 2 in Africa

The United Nations' Sustainable Development Goal 2 draws people's attention to ending hunger, achieving food security and improved nutrition, and promoting sustainable agriculture. To achieve this, specific goals have been set to be achieved at different periods of time. While moving toward 2030, some of the requirements are far-fetched, religious leaders have directed their prophetic role to ensuring faith groups agitate for the realization of the same. Many religious leaders take this upon themselves because religious traditions have always played a central role in supporting those experiencing poverty and marginalization through service delivery as well as the provision of spiritual resources that provide mechanisms for resilience at both the individual and the community level (Tomalin et al. 2019). While in years past sometimes these interventions were limited to those within one's own religion, as people who are concerned about the human race, many religious leaders extend support to those from other religions or understood it as an essential part of their religious commitment to combine their religious outreach with relief and development efforts. This was a particularly marked phenomenon within the Christian missionary movement which accompanied European colonialism from the nineteenth century onward, from early abolitionist activism and the slogan of "Christianity and commerce" to the "civilising" ideology and the provision of essential services in health and education as colonialism took root (Tomalin et al. 2019). Colonial and missionary interests did not always align and were at times even marked by conflict, but the Christian "civilizing mission" formed a pathway for colonialism and provided it with an important ideological justification in Europe by casting colonial efforts as a service in development.

To this end, in achieving SDG 2, prophetic actions that religious leaders in Africa take make them to participate in several activities so as to end hunger, achieve food security and improved nutrition, and promote sustainable agriculture by 2030. The actions involve taking bold and visionary steps to address systemic issues that contribute to poverty, hunger, and malnutrition. Such actions require a deep understanding of the root causes of these problems and a willingness to challenge unjust policies and practices that perpetuate them. While some foodstuffs are disliked because of peoples' attitudes toward them as being inferior, as we have already seen, other foods are taboos because of religious symbolism attached to them. Eating such foods would be considered as impure, polluting, or spiritually corrupting. In many societies diet and dietary restrictions are often used as a metaphor for purity or virtue (Byarugaba 2017, p. 15). Regardless of the existence of such attitude toward food, given food security is defined contextually, religious leaders have committed to ensure that people's interest are taken care of.

In Africa, where many countries struggle with high levels of poverty and hunger, prophetic action has been particularly effective in bringing about change. Through such actions religious leaders are working with local communities to build sustainable food systems, advocating for policies that promote equitable access to land and resources, and engaging with governments and international organizations to promote food security and sustainable agriculture. Furthermore, some religious leaders, in the context of SDG 2, involve advocating for policies that support smallholder farmers, promoting sustainable agriculture practices, and calling for an end to the unequal distribution of resources that contribute to food insecurity. In Uganda, for example, the Ministry of Agriculture, Animal Industries and Fisheries, through the National Agriculture Advisory Services (NAADS), has signed a Memorandum of Understanding with the Interreligious Council of Uganda (IRCU) that requires the IRCU to mobilize religious leaders and religious groups for sensitization and agricultural skills training, which will be supplemented by NAADS' obligation to liaise with other government actors to ensure smooth and coordinated implementation (MAIF 2020).

Conclusion

As seen from the foregone discussion, climate change, food security, and achieving SDG 2 are major challenges in Africa. Although nation-states and non-governmental organizations have been long in the struggle to fight the ills that come with them and the faith-based organizations have also joined in the fight, a lot still is yet to be done. To address these challenges, there needs to be greater investment in renewable energy, sustainable agriculture, and climate-resilient infrastructure. Governments, international organizations, and the private sector all have a role to play in this effort.

However, it is also important to recognize the critical role that religious leaders and communities can play in addressing climate change through prophetic action. Consequently, to achieve this requires a comprehensive approach to address the root causes. Hence, prophetic action is one approach that can be used to promote food security, address climate change, and realize SDG 2 goals, albeit in part, in Africa. Prophetic actions in their diversity across Africa advocate for policies and practices that promote sustainable agriculture, seek to secure social justice, and address the immediate needs of people suffering from food insecurity. In this way, prophets and other spiritual leaders continue to play an important role in promoting food security in Africa.

Ultimately, addressing climate change, food security, and achieving SDG 2 in Africa require a multifaceted approach that combines prophetic action with innovative solutions, cross-sectoral collaboration, and a deep commitment to addressing the root causes of poverty and hunger.

References

Akinpelu, A.O. 2020. Food security and the prophetic imagination: Towards a theological perspective. *HTS Theological Studies*, 76(1): 1–10.

Amnesty International. n.d. Climate Change, available from https://www.amnesty.org/en/what-we-do/climate-change/?utm_source=google&utm_medium=cpc&gclid=cj0kcqjw_viwbhd8arisah1mcd5tvriciezre3wvybo1cmvrdtt0u-y2-beouo53axftunqab0s1b_uaasxcealw_wcb, accessed 2/8/2022.

Anglican Alliance. 2021. Prophetic Indigenous Voices on the Planetary Crisis—Update, available from https://anglicanalliance.org/prophetic-indigenous-voices-on-the-planetary-crisis-update/, accessed 4/5/2023.

Antwi, A. 2013. Climate change and food security: An overview about the issue, Accra: Friedrich-Elbert-Stiftung.

Bazzaz, F. and Samboek, W. Eds. 1996. Global climate change and agricultural production: Direct and indirect effects of changing hydrological, pedological and plant physiological processes, Rome: FAO.

Bhargawa, A. and Singh, A. K. 2019. Solar irradiance, climatic indicators and climate change—An empirical analysis. *Advances in Space Research* 64(1): 271–277. https://doi.org/10.1016/j.asr.2019.03.018.

Bius, A. 2020, February 27. Milankovitch (Orbital) Cycles and Their Role in Earth's Climate, in NASA News- Global Climate Change: Vital signs of the planet, available from https://climate.nasa.gov/news/2948/milankovitch-orbital-cycles-and-their-role-in-earths-climate/#:~:text=The%20Milankovitch%20cycles%20include%3A,is%20pointed%2C%20known%20as%20precession, accessed 7/8/2022.

Byarugaba, G. 2017. The influence of food symbolism on food insecurity in South Africa: how relevant is the Eucharistic celebration? *Scriptura*, 116(1): 1–19. https://doi.org/10.7833/116-1-1335.

Capper, L. 2021. The Impact of Climate Change on Food Systems in Southeast Madagascar, available from https://madagascar.co.uk/blog/2021/11/impact-climate-change-food-systems-southeast-madagascar?gclid=Cj0KCQjworiXBhDJARIsAMuzAuxklg_1pTUuJ_1j_jpiY2sTxdxKb5iPC0pZZWMbIL4gnWGg3IrwTw4aAr9-EALw_wcB, accessed 6/8/2022.

Costa, S. 2020. 24 Faith-Based Organizations Promoting Food Security, available from https://foodtank.com/news/2019/07/24-faith-based-organizations-promoting-food-security/, accessed 6/5/2023.

Ebegbulem, J.C. 2017. The prophetic vocation in Africa: Towards a theology of social transformation. *HTS Theological Studies* 73(3): 1–7.

Chitando, E. 2017. Praying for Courage: African Religious Leaders and Climate Change. *The Ecumenical Review*, 69(3), available from https://doi.org/10.1111/erev.12304, accessed 4/5/2023.

FAO. 2021. COP26: Agricultural expansion drives almost 90 percent of global deforestation, available from https://www.fao.org/newsroom/detail/cop26-agricultural-expansion-drives-almost-90-percent-of-global-deforestation/en, accessed 05/8/2022.

Francis. 2015. Encyclical Letter *Laudato Si'* of the Holy Father Francis, on Care for Our Common Home, Vatican: Vatican Press.

Grady-Benson, J. 2014. *Fossil Fuel Divestment: The Power and Promise of a Student Movement for Climate Justice*. Pitzer Senior Theses.

Homer-Dicson, T. F. 1991. On the threshold: Environmental changes as a causes of acute conflict. *International Security,* 16(2), 76–116.

IPCC. 2022. Summary for Policymakers. In: Climate Change 2022: Mitigation of Climate Change. Contribution of Working Group III to the Sixth Assessment Report of the Intergovernmental Panel on Climate Change [P.R. Shukla, J. Skea, R. Slade, A. Al Khourdajie, R. van Diemen, D. McCollum, M. Pathak, S. Some, P. Vyas, R. Fradera, M. Belkacemi, A. Hasija, G. Lisboa, S. Luz, J. Malley, eds.]. Cambridge, UK and New York, USA: Cambridge University Press. https://doi.org/10.1017/9781009157926.001.

IPCC. 2018. Global Warming Of 1.5°C: An IPCC Special Report on the Impacts of Global Warming of 1.5°C above Pre-Industrial Levels and Related Global Greenhouse Gas Emission Pathways. Cambridge: Cambridge University Press.

Li, X. Stringer, C.L. and Dallimer, M. 2022. The Impacts of Urbanisation and Climate Change on the Urban Thermal Environment in Africa, in Climate, Volume 10, Issue 11. https://doi.org/10.3390/cli10110164, accessed 7/5/2023.

MAIF. 2020. NAADS, IRCU MoU to fight food insecurity and poverty, available from https://www.agriculture.go.ug/2020/12/11/naads-ircu-mou-to-fight-food-insecurity-and-poverty/, accessed 6/5/2023.

Moss, R. H., Avery, S., Baja, K., Burkett, M., Chischilly, A. M., Dell, J., ... & Zimmerman, R. 2019. Evaluating knowledge to support climate action: A framework for sustained assessment. Report of an independent advisory committee on applied climate assessment. *Weather, Climate, and Society,* 11(3), 465–487.

Nzwili, F. 2020. Church action against hunger in Africa reemphasized as report shows under-nutrition, available from https://www.oikoumene.org/news/church-action-against-hunger-in-africa-reemphasized-as-report-shows-under-nutrition, accessed 6/5/2023.

Omona, A.D. 2022a. Effects of climate change and food security in Uganda, in R. Baikady, et al., eds., The Palgrave Handbook of Global Social Change, Switzerland: Springer Nature, https://doi.org/10.1007/978-3-030-87624-1_339-1.

Omona, A.D. 2022b. The mainline churches and climate change in Uganda, in Chitando, E. Conradie, M.E., and Kilonzo, S.M., eds., *African Perspectives on*

Religion and Climate Change, London, New York: Rutledge. https://doi.org/10.4324/9781003147909-8.

Oyiga, C., Heile, M. and Waschkeit, C. 2011. Implication of climate change on crop yield and food accessibility in Sub-Saharan Africa, https://www.zef.de/fileadmin/downloads/forum/docprog/Termpapers/2011_1_Oyiga__Haile_Waschkeit.pdf, accessed 6/5/2023.

Rinkesh. n.d. What is deforestation? www.concerve-energy-future.com/how-deforestation-affects-climate-change-human-animals.php, accessed 19/2/2022.

Schuman, S., Dokken, J.-V., Van Niekerk, D. and Loubser, R.A. 2018. 'Religious beliefs and climate change adaptation: A study of three rural South African communities', *Jàmbá: Journal of Disaster Risk Studies* 10(1), a509. https://doi.org/10.4102/jamba.v10i1.509.

Tomalin, E., Haustein, J. and Kidy, S. 2019. Religion and the Sustainable Development Goals. Review of Faith & International Affairs, 17 (2). pp. 102–118. ISSN 1557-0274. https://doi.org/10.1080/15570274.2019.1608664.

UN Climate Action. n.d. What is climate change? Available from https://www.un.org/en/climatechange/what-is-climate-change, accessed 1/8/2022.

UN Environment Program. 2015, June 18. UN Leaders React to Pope Francis' Release of Encyclical on Climate and Environment, Press Release, Sustainable Development Goals, available from https://www.unep.org/news-and-stories/press-release/un-leaders-react-pope-francis-release-encyclical-climate-and#:~:text=In%20a%20papal%20encyclical%20letter,Earth%20and%20everyone%20on%20it, accessed 4/5/2025.

United Church of Christ. 2022. Ten "Naked Truths": Prophetic Action and the Climate Crisis, available from https://www.ucc.org/ten-naked-truths-prophetic-action-and-the-climate-crisis/, accessed 3/5/2023.

WMO. 2021. Climate change triggers food insecurity, poverty and displacement in Africa: Melting of iconic African glaciers symbolizes changes to Earth system. https://www.un.org/africarenewal/magazine/november-2021/climate-change-triggers-food-insecurity-poverty-and-displacement-africa, accessed 10/1/2022.

Zuckerman, M. 2016. "Faith in a Changing Planet: The Role of Religious Leaders in the Fight for a Livable Climate", Pitzer Senior Theses, Paper 71. http://scholarship.claremont.edu/pitzer_theses/71.

CHAPTER 3

Interfacing Gender, Religion, Food Security and Climate Change in Africa: An African Ecofeminist Perspective

Mercy Shumbamhini and Sophia Chirongoma

Introduction

We are living in an era of global warming and environmental degradation. Damage to our Mother Earth threatens to disrupt the world's food production. Climate change and its impact on food security are increasingly affecting different parts of the world, including Africa (Masipa 2017). It is widely recognized that climate change already has and will have a devastating impact throughout the African continent. While in Africa agriculture is the mainstay for 70% of the continent's population (WEF 2016), climate change is posing a major threat to agriculture in Africa, and it will reduce crop yields (The Conversation 2019). Since women play a vital role

M. Shumbamhini
Mary Ward Sisters, Harare, Zimbabwe

S. Chirongoma (✉)
Research Institute for Theology and Religion, University of South Africa, Pretoria, South Africa

L. Maseno et al. (eds.), *Religion, Climate Change, and Food Security in Africa*, Sustainable Development Goals Series,
https://doi.org/10.1007/978-3-031-50392-4_3

in food security; the agency of rural female farmers is essential for enhancing agricultural productivity and realizing the Sustainable Development Goals (SDGs), including ensuring food security (SDG 2) and addressing the perils of climate change (SDG 13).

Whilst a lot has already been accomplished in terms of bridging the gender gap in most African countries, most rural women have remained on the bottom rung of the pile. Since subsistence agriculture is the main source of their livelihood, they are the ones who inadvertently bear the brunt of climate change and food insecurity. It is apparent that African women play a vital role in the three components of security, namely; food production, food distribution and food utilization (Mutume 2005; Zvingowanisei, Chitando and Chirongoma 2022). They also undertake a range of community-level activities that support agricultural development such as soil and water conservation, afforestation and crop domestication (UNDP 2012). Indeed, agriculture is central to women's livelihoods. Climate stresses on agricultural production is making women particularly vulnerable to food insecurity. Women often face greater burden in responding and adapting to climate stresses due to their lack of access to land, financial services, social capital and technology (Chirongoma 2021, 2022). Although over the years, women have developed various adaptation strategies to cushion the effects of climate change, they are still caught up in the vicious cycle of poverty and increased vulnerability (Chirongoma and Zvingowanisei 2022).

Cognizant of the challenges posed by climate change on African communities and its far-reaching impact on human livelihoods, the chapter contends that there is an urgent need for religious responses to the climate change crisis. Hence, the central question guiding this discussion is: "How can religion in Africa be a resource for countering climate change and its effects (SDG 13), eradicate poverty (SDG 1), achieve zero hunger and promote food security (SDG 2), as well as ensuring the attainment of gender equality (SDG 5)"? In an endeavour to answer this question, the chapter employs literature review as a research methodology. In this undertaking, we engage in a discussion of the interplay of gender, food security and the climate change crisis in Africa, while probing the philosophical and practical responses of religion. The chapter is structured as follows: introduction, research methodology, ecofeminist framework, African ecofeminist approach, gender, climate change and food security, religion and climate change adaptation and conclusion.

Research Methodology

The study relies on literature review as a research methodology. According to Tranfield et al. (2003), a literature review can broadly be described as a systematic way of collecting and synthesizing previous research. An effective and well-conducted review as a research method creates a firm foundation for advancing knowledge and facilitating theory development (Webster and Watson 2002). By integrating findings and perspectives from many empirical findings, a literature review can address research questions with a power that no single study has. It can also help to provide a comprehensive overview of literature related to a theme/theory/method and synthesizes prior studies to strengthen the foundation of knowledge.

In this chapter, literature review brings the reader up to date with current literature on gender, religion, food security and climate change in Africa. Literature review enables us as researchers to expand current expertise on the topic, improve on what already exists and allows us to respond to the question: "How can religion in Africa be a resource for countering climate change and its effects (SDG 13), eradicate poverty (SDG 1), achieve zero hunger and promote food security (SDG 2), as well as ensuring the attainment of gender equality (SDG 5)"? With the goal of unpacking this overarching question, our chapter also draws some insights from the ecofeminist framework and the African ecofeminist approach. The ensuing sections summarize these two interrelated frameworks.

Ecofeminist Framework

As noted by Ruether (2012: 22), "Ecofeminism or ecological feminism examines the interconnections between the domination of women and the domination of nature." This examination is done from an empirical and conceptual (cultural/symbolic) perspective (Eaton 2009). As articulated by Eaton (2009), the empirical claim explores how the structural restrictions that include the socio-political and economic injustices reduce women's lives to poverty, ecological deprivation and economic incapacity. On the other hand, the conceptual (cultural/symbolic) claim emerges from diverse cultural perceptions that highlight the historical and symbolic connection between women and nature. The conceptual claim exposes the hierarchical and dualistic connection of women and nature's domination at times promoting anthropocentric environmentalism (Eaton 2009). In the same light, Chisale (2021) tenders that ecofeminism invites all to engage in welcoming attitudes and behaviours that enhance all lives, including natural resources. Ecofeminists highlight that due to patriarchal structures, a societal imbalance that typically favours men, the effects of environmental degradation, including climate change, are more likely to have a detrimental impact on women (Evans 2023). It is against this backdrop that the ecofeminist approach deconstructs hierarchies that were created by patriarchy in the past, replacing them with theologies that promote egalitarian communities.

The main goal of ecofeminism is to halt environmental degradation, which disproportionately affects women and eventually the entire human race. When humans pollute and disregard the earth, women and children's health are contaminated primarily. In fact, the entire community is contaminated since women literally birth and nurture communities. Resultantly, the abuse of the earth exacerbates the abuse of girls and women (Ruether 1993). For instance, deforestation impacts more on women's lives since it results in them having to walk twice as far each day to gather firewood (Ruether 1996). Furthermore, ecofeminism exposes the sacredness and integrity that lie in the purpose of creation by God, such as to live in harmony with each other and be stewards of all creation. It argues that religions should be in the forefront to ensure the health and well-being of each member of the community including the non-humans. Ecofeminism therefore challenges all religions to take a strong stance in stopping the destruction of our Mother Earth and to ensure that women and girls have equitable access to justice and the enactment of inclusive and effective climate laws and policies (Berman et al. 2021).

Gebara (1996) calls for a reconsidering and rediscovering of Christian virtues that affirm human connectivity to the ecological system. It is imperative for us all then to recognize the potential, power, and contributions of the earth and nature to facilitate the sustainment of life and as a resource to improve the quality of our life. Hence, ecofeminists believe that Christianity can be a resource for liberating women and non-human nature from the effects of patriarchy. Ruether and McFague find bases for ethical responses to the ecological crisis. For Ruether (1998) it is an ethics of "compassionate solidarity" and for McFague (2008) a "community care." For both, it is a love commitment to an eco-justice of right relations with all creatures. We are all equal, human and non-human, therefore, religion and all of us should see all creation as radically interrelated and interdependent as well as having a clear understanding of our special responsibility for the women and planet's well-being.

Ecofeminism is interpreted differently depending on context, it particularly differs based on race, class, gender and culture. Siwila (2014) highlights this diversity as she argues that third-world feminists have developed their own concepts of ecofeminism, which reflect ways in which women and nature are connected. To illustrate this diversity, the following subsection explores the African ecofeminist approach.

African Ecofeminism

Like other ecofeminists, African ecofeminism highlights the interplay of the oppression of women, other marginalized groups and the degradation of nature (Chirongoma 2005; Phiri 1996; Shumbamhini 2022; Siwila 2014). They actively respond to the call of an earth-centred approach for sustainable development. The Sustainable Development Goals Agenda 2030 includes a call to everyone to become responsible stewards of Mother Earth. For African ecofeminists, the struggle of nature and African women's domination is collective, they are dominated and silenced because of their gender—this is against the backdrop that the earth is perceived as feminine, as denoted in the designation, "Mother Earth" (Chirongoma and Mombo 2021).

In most families in Africa, especially in rural communities where the greater percentage of the population resides, women are responsible for the sustenance of the entire household. They are usually the ones who shoulder the heavier burden of most domestic chores, which are mainly dependent on nature's bounty. The brunt of burden posed by environmental injustice is therefore more intense on women than on their male counterparts. Despite achievements and progress made in most African countries concerning gender equality, African women still face major challenges and obstacles. They remain at the bottom rung of the social hierarchy, with limited access to land, markets, and credit even though they do most of the work. In Africa, women are among the most vulnerable to climate change, partly because as pointed out above, in many countries they make up the larger share of the agricultural workforce and partly because they tend to have access to fewer income-earning opportunities. The African ecofeminists see environmental injustice in the way humans have neglected the mandate given by God to care for the natural world which has resulted in the rampant damaging of natural environmental resources.

Human activity has caused climatic changes which have affected times and seasons resulting in the loss of lives and the destruction of Mother Earth. These changes exacerbate the already entrenched gender-based inequalities, making women's lives more uncomfortable. As such, African ecofeminists pay particular attention to the intersection of food insecurity, climate change, and gender inequality, with the main goal of easing women and girls' burdens wrought by the ensuing ecological crisis. They reiterate the pertinent need for interfacing sustainable development with gender equity, ecological care, ecological justice as well as gender inclusive disaster risk reduction policies and programmes. More importantly, they implore the church to give both men and women access and opportunities to flourish and realize their potential (Oduyoye 1995; Njoroge 1997; Dube 2015; Gabaitse 2011; Chirongoma 2018; Amoah 2007).

In concurrence with fellow African ecofeminists, the chapter restates the need for religion to include women in decisions and operations that affect them, this includes their participation in the crafting of policies to counter climate change and its effects (SDG 13), eradicate poverty (SDG 1); achieve zero hunger and promote food security (SDG 2) and ensuring gender equality (SDG 5). Climate change disproportionately affects those who mainly depend on natural resources and whose livelihoods are climate sensitive, many of which are marginalized communities, including women in agriculture. If religion is to be the panacea for the ecological crisis, then, religious communities must be at the forefront, spearheading support for women's initiatives on sustainable development that help them to earn an income, especially during the tumultuous climate change crisis. With adequate support and equitable access to resources and opportunities, women's central role in sustaining families, building community resilience, and responding to climate-related disasters can be enhanced. The following section reflects on the intersection of gender, climate change and food security.

Gender, Climate Change and Food Security

According to the UNDP (2017), the interlinked challenges of climate change and food security are most evident in the agricultural sector in Africa and women are hardest hit by climate shocks and food insecurity. Ortiz (2015), commenting on the interplay between gender, climate change and food security argued that women are vital to food production around the world and yet they are left out of decision-making forums deliberating on climate change. Women and girls, especially in rural communities, continue to carry the heavier yoke of the effects of the climate change crisis. This intensifies the pre-existing gender-based inequalities and inadvertently jeopardizes their food security (WFP 2022). At the same time, climatic stresses on agriculture and food systems present formidable food security and livelihood challenges to many African women who are the primary land tillers. Their stories are connected to the land where they live and where they produce food, fetch firewood and dig wells. For women, the land is their sacred home.

The effects of the climate change crisis in agriculture requires integrated approaches that increase productivity, enhance adaptive capacity and cut back net emissions UNDP (2017). The agency of rural female farmers is essential for enhancing agricultural productivity and realizing the Sustainable Development Goals (SDGs), including ensuring food security (SDG 2) and addressing the perils of climate change (SDG 13). The ability of both women and men to make decisions about their own lives and influence those of their communities is fundamental to gender equality and food security, as expressed in various international instruments. For example, Target 5.5 of the 2030 Agenda for Sustainable Development is to "ensure women's full and effective participation and equal opportunities for leadership at all levels of decision-making in climate change, political, economic development, and public life" (WFP 2020).

Since the rural women are particularly vulnerable to current and future climate change and food insecurity, a sustainable future is only possible when women and girls have what they need to adapt to the changing climate (WFP 2022). Religion's response to climate change within the agricultural sector should therefore consider gender dynamics and its gender-responsiveness should ensure that technologies and extension services cater to women's needs. Religion should address the rights of women, for instance, hiding behind the banner of culture or sacred texts to justify gender-discriminatory access to resources, power, leadership, and decision-making should be a thing of the past. It is also vital to be cognizant of the fact that while women may be more vulnerable to climate change, seeing them as victims would be severely underestimating their strength and capacity as actors and agents of change. Women possess a strong body of knowledge and expertise (Ortiz 2015). These abilities and strength should be used in climate change mitigation, disaster reduction and adaptation strategies. For instance, in most rural communities, women are the reservoirs of indigenous knowledge systems, including taboos which protect and preserve Mother Earth as a sacred entity, indigenous methods of detecting the weather patterns and indigenous ways of engaging in organic farming which is ecologically friendly (Chirongoma and Zvingowanisei 2022). Such wealth of indigenous wisdom can be utilized for the betterment of our interaction with Mother Earth. This links us to the next subsection on religion and climate change mitigation and adaptation.

Religion and Climate Change Mitigation and Adaptation

In Africa and in other parts of the world, religion plays a pivotal role in shaping human behaviour and in providing basic services and social cohesion. According to Skirbekk et al. (2020), previous research has found that religion influences many aspects of lifestyle that affect the environment. It is, therefore, imperative to make climate change and environmental responsibility an integral part of contemporary religious life and practice. All major religious traditions, spanning from Buddhism; Christianity; Hinduism; Islam, as well as African indigenous religions, acknowledge the need for environmental stewardship. Similarly, all the sacred texts urge adherents to be responsible caretakers of the earth and its biodiversity (UNEP 2019). Steffen (2017) acknowledges that religious faith and spiritual commitments are today providing the energizing fuel to sustain involvement with issues like climate change and environmental responsibility. Religion provides invaluable moral lessons in times of crisis, they spread the awareness of climate change, and, through community ritual, religious institutions can nurture a sense of collective responsibility in climate adaptation and mitigation (Stephens 2019). Therefore, climate change is acknowledged to be both a moral and ethical issue (Tarusarira 2017).

According to Hitzhusen and Tucker (2013), religion has the power to bring about socio-ecological change by promoting the principles of environmental stewardship to enhance environment resilience and a positive response to climate change. It is noteworthy that two important basic responses to climate change are mitigation and adaptation. Jerneck (2018: 4) defined mitigation as, “the reduction of future climate change” and adaptation as “adjusting to actual or expected climate effects.” Foregrounding the interlocking relationship between socio-cultural factors, gender, and climate change, the WHO (2014: 3), tendered that “the effects of climate on our world, including Africa, and our ability to mitigate and adapt to them, are mediated by social factors, including gender." Women working in agriculture bear the brunt of extreme weather events, they are hardest hit by climate shocks and food insecurity. It is, therefore, imperious that mitigation and adaptation processes provide opportunities to promote gender equality and women's empowerment. By tackling the challenges posed by the changing climate from the perspective of gender, religion would help to curb the unforeseen outcomes on poverty eradication and food security (SDGs 1 and 2) (UNDP 2012).

Churches promote ecological stewardship ethics through the belief that God gave human beings the mandate of caring for everything that exists on Mother Earth (Norgaard 2002). Some leaders of religious faiths such as Pope Francis, Pope Benedict XVI, the Dalai Lama and Archbishops of Canterbury have pressed for environmental stewardship (Mcleod and Palmer 2015). Whilst the impact of these religious traditions is quite central, it is equally important that climate change approaches in Africa be intentional about acknowledging the influence of African indigenous religions and the religious significance attached to environmental sites. This will encourage the people to be open to new concepts, ideas, philosophies and processes. The integration of African indigenous religion in climate change efforts shows respect and acknowledgement of the religious beliefs of the people of the land. This will help to increase the success of climate change education and campaigns (Tarusarira 2017).

In sum, it can be said that religion plays a major role in climate change mitigation and adaptation. Hence, the intensity of our current climate crisis, makes it imperative that there be some meaningful collaborations among religions and other domains of society, including science, economics, education and public policy (Hitzhusen and Tucker 2013). Pope Francis (2015) calls on all people of good will to care for a creation entrusted to them by God. By making the world more humane, we will make the environment a "common good" for which all people must accept responsibility (Pope Francis 2015). It is therefore crucial for all of us to work together; religious and nonreligious, government departments, NGOs, international organizations and lobbyists, men and women as well as our children. Such multidimensional collaborations will bear abundant fruits for the attainment of the goal to counter climate change and its effects (SDG 13), eradicate poverty (SDG 1), achieve zero hunger, and promote food security (SGD 2) and ensure gender equality (SDG 5). The ensuing subsection will discuss the relationship between religion and gender equality.

Religion and Gender Equality

The relationship between religion and gender equality is a complex one. Since time immemorial, humanity has been grappling with the task of flattening the curve of gender inequality. African countries have not achieved gender equality despite the legislative, social and economic gains for women. Gender-based violence and gender injustice are major obstacles to the fulfilment of women's and girls' human rights and to the achievement of the 2030 Agenda for Sustainable Development and its 17 Sustainable Development Goals (UNDP 2019).

Religion plays a vital role in shaping cultural, social, economic, norms and gender roles. It is deeply rooted in people's experiences and influences the social, economic and political direction of societies (Stump 2008). Hence, gender roles and the status of women and men in society are closely related to the way religious texts have been interpreted over the years by those in positions of authority, positions that are held primarily by men. This is supported by Klingorová and Havlíček (2015) who argue that the status of women in society is an outcome of the interpretation of religious texts and of the cultural and institutional set-up of religious communities. Religion, which serves as a reference point for gender norms and expectations, is often seen as institutionalizing and perpetuating patriarchy, thus frustrating many advocates for women's rights and equality (Berkley Forum 2016). Despite the widespread engagement of women in their religious communities, the processes of the production of religious knowledge and all decision-making concerning the church life largely exclude women (Mhuru 2023; Sibanda 2011). Yet, within the family and the church community, women teach and implement religious wisdom and traditions, and pass them on to future generations.

Cognizant of how religion is immensely influential in the African society, our chapter exhorts religious leaders to support and advocate for gender equality. Different religions can also create gender sensitive and gender inclusive policies, strategies and programmes that promote gender equality within their faith communities. It will also be important for all religio-cultural communities to embark on some introspection and soul searching such that we can begin to acknowledge that there are some religio-cultural norms and practices that promote gender inequality. This would hopefully take all religious leaders and religious practitioners to a place where they can start to not only challenge, but to become agents of deconstructing these life sapping and life denying traditions. The process of deconstructing these harmful and retrogressive traditions should not be a lone journey, rather it calls for meaningful collaborations with other stakeholders such as government departments, civil society, ecofeminists, NGOs and any other like-minded stakeholders in their respective countries, who are committed towards attaining SDG 5, which is aimed towards achieving gender equality. It is therefore apparent that since religion is amongst the most powerful forces for societal change, it must be in the forefront to ensure that women and girls have equal access to justice and the enactment of inclusive and effective climate laws and policies. As has been noted above, all religious traditions express a commitment towards ecological stewardship, they also acknowledge the importance of addressing gender inequality, promoting food security as well as eradicating poverty. It, therefore, makes practical sense to engage in multi-pronged collaborations, e.g., inter-faith and religious-nonreligious collaboration, in an endeavour to address climate change and its effects (SDG 13), eradicate poverty (SDG 1); achieve zero hunger and promote food security (SDG 2) (Mcleod and Palmer 2015; Tarusarira 2017).

Conclusion

This chapter commenced with acknowledging the influential position of religion in Africa. Drawing insights from ecofeminism, and African ecofeminism, the chapter proceeded to proffer how religious values and traditions can be tapped into as pivotal resources to counter climate change and its effects (SDG 13), eradicate poverty (SDG 1), achieve zero hunger, and promote food security (SDG 2), as well as ensuring gender equality (SDG 5). It is against this backdrop that our chapter also recommends that different religions and culture must seriously consider collaborating with other like-minded stakeholders such as the government, NGOs, civil society, policy makers, ecofeminists, and all people of good will, committed towards attaining the targets of SDGs 1, 2, 5 and 13, which are the chapter's focus of discussion. Religions can serve as a ray of hope in turning the tidal waves of destruction wrought by climate change. It can also serve as a catalyst for attaining gender equality, eradicating poverty and achieving socio-economic development in twenty-first-century Africa. Through collaborations between the secular community committed towards ecological conservation and the religious community in Africa and worldwide, we can gradually stop the destruction of our beloved Mother Earth and create a future where respect and care for the environment encourages responsible choices. In this way, all people will have access to resources without discrimination based on gender, race, culture or nationality. Partnerships between scientists and religious groups have a higher chance of being successful when it supports both environmental solutions and action on the promotion of food security and ensuring gender equality. Hence, the intertwinement of religion, gender, poverty, food security, and the climate change crisis can be utilized as a springboard for achieving the goals enshrined in SDGs 1, 2, 5 and 13 in Africa and world over.

References

Amoah, E., 2007, Reflections on the Impact of "The Teaching for Change" Conference, *Journal of Feminist Studies in Religion,* Vol. 23, No. 2, pp. 95–98.

Berkley Forum, 2016, Religion and Women's Equality (georgetown.edu).

Berman, S.K., Leshota, P.L, Dunbar, E.S., Dube, M.W. & Kgalemang, M. (eds.). 2021. Mother Earth, Mother Africa and Biblical Studies: Interpretations in the Context of Climate Change. Bamberg: University of Bamberg Press. https://doi.org/10.20378/irb-49839.

Chisale, S.S. 2021. When women and Earth connect: African ecofeminist or ecowomanist theology? In: S.S. Chisale & R. Robson Bosch (eds.), Mother Earth, Mother Africa and Theology. HTS Religion & Society Series, Volume 10. Cape Town: AOSIS. 9–17 https://doi.org/10.4102/aosis.2021.BK237.01.

Chirongoma, S., 2005, 'Motherhood and ecological conversation of Mother Earth', *Women in God's Image* 10&11, 8–12.

Chirongoma, S., 2018, *Gleaning for Gender Justice in Laudato Si: Envisioning a radial eco-feminist conversion.* Conference on the radical ecological conversion after Laudato Si: Discovering the Intrinsic Value of all Creatures, Human and Non-human. Rome: The Pontifical Gregorian University. 2.

Chirongoma, S., 2021, "Voices from the Margins: Religio-Cultural Perspectives of Women, Children and the Elderly amidst the Tokwe-Mukosi Dam Displacements in Zimbabwe" in Sophia Chirongoma and Esther Mombo (eds.) *Mother Earth, Postcolonial and Liberation Theologies*, Lexington Publishers, 151–170.

Chirongoma, S., Mombo, E., 2021, "Introducing Mother Earth, Postcolonial and Liberation Theologies" in Sophia Chirongoma and Esther Mombo (eds.) *Mother Earth, Postcolonial and Liberation Theologies*, Lexington Publishers, 1–12.

Chirongoma, S., 2022, "Where Earth and Water Meet: Exploring the impact of Tokwe-Mukosi Dam in light of African Spirituality and Religion in Zimbabwe" (eds.) Lily Mendoza and George Zachariah *Decolonizing Eco-theology: Indigenous and Subaltern Challenges*, Wipf & Stock, 146–161.

Chirongoma, S., and Zvingowanisei, S. 2022, "Karanga Women's Utilization of Indigenous Knowledge Systems on Climate Change Adaptation and Mitigation in Zimbabwe: Towards the Attainment of Sustainable Development Goal 13 (SDG 13)," in Sophia Chirongoma and Ven. Scholar Wayua Kiilu (eds.) *Mother Earth, Mother Africa: World Religions and Environmental Imagination*, African Sun Media Press

Dube, Musa W., 2015, "And God Saw that it was Very Good": An Earth-Friendly Theatrical Reading of Genesis 1." Black Theology, Vol. 13 No. 3, 230–246.

Eaton, H., 2009, 'Women, nature, earth', in S. Bergmann & Y.-B. Kim (eds.), *Religion, ecology and gender: East-West perspectives*, pp. 7–22, LIT Verlag, Münster-Hamburg-Berlin-Wien-London-Zürich.

Evans, O. G., 2023, Ecofeminism Feminism: Intersection of Gender & The Environment (simplysociology.com).

Gabaitse, R., 2011, Reading Beyond the Veil: A feminist reading of Luke 23: 45, *Journal of Gender and Religion in Africa*, Vol. 17, No. 1, 2011, pp. 154–169.

Gebara, I., 1996, 'Ecofeminism', in L.M. Russell & S.J. Clarkson (eds.), *Dictionary of feminist theologies*, pp. 76–78, Westminster John Knox Press, Louisville, KY.

Hitzhusen, G.E. & Tucker, M.E., 2013, The potential of religion for Earth Stewardship https://www.undp.org/sites/g/files/zskgke326/files/publications/TM1_Africa_GenderClimateChange_Overview.pdf.

Jerneck, A., 2018, 'What about gender in climate change? Twelve feminist lessons from development', *Sustainability* 10(3), 1–28. https://www.mdpi.com/2071-1050/10/3/627.

Kamila Klingorová, K. Tomáš Havlíček, T., 2015, Religion and gender inequality: The status of women in the societies of world religions (sciendo.com).

Masipa, T., 2017, https://www.ncbi.nlm.nih.gov/pmc/articles/PMC6014268/#CIT0005.

McFague, S., 2008, A *new climate for theology: God, the world, and global warming*, Fortress Press, Minneapolis, MN.

Mcleod, E., & Palmer, M., 2015, (PDF) Why Conservation Needs Religion (researchgate.net).

Mhuru, L., 2023, Gender justice, law and religion in Zimbabwe: An evaluation of the role of sacred texts | HTS Teologiese Studies / Theological Studies (ajol.info).

Mutume, G., 2005, https://www.un.org/africarenewal/magazine/july-2005/african-women-battle-equality.

Norgaard, R.B., 2002, 'Can science and religion better save nature together?', *BioScience* 52, 842–846.

Njoroge, N., 1997, The Missing Voice: African Women Doing Theology, *Journal of Theology for Southern Africa*, Vol. 99, 1997, pp. 77–83.

Phiri, I.A., 1996, 'The Chisumphi cult: The role of women in preserving the environment', in R.R. Ruether (ed.), *Women healing the Earth: Third world women on ecology, feminism, and religion*, pp. 63–105, Orbis, Maryknoll, NY.

Pope Francis, 2015, *Laudato si'*. Encyclical Letter, 24 May.

Oduyoye, M. A., 1995, *Daughters of Anowa: African women and patriarchy*, Orbis Books, Maryknoll, New York.

Ortiz, M., 2015, Gender, climate change and food security by Monica Ortiz (sheffield.ac.uk).

Ruether, R.R., 2012, 'Ecofeminism—The challenge to theology', *Deportate, Esuli, Profughe* 20(1993), 22–33.

Ruether, R.R., 1998, *Women and redemption: A theological history*, Fortress Press, Minneapolis, MN.

Ruether, R.R., ed. 1996, *Women healing earth: Third World women on ecology, feminism, and religion*. Orbis Books.

Ruether, R.R., 1993, 'Ecofeminism symbolic and social connections of the oppression of women and the domination of nature', in C.J. Adams (ed.), *Ecofeminism and the sacred*, pp. 13–23, Continuum, New York, NY.

Shumbamhini, M., 2022, Mother Earth, our life support system: Perspectives from a Catholic nun. In Chirongoma, S. & Kiilu, W. (eds.) *Mother Earth, Mother Africa: World Religions and Environmental Imagination*. Stellenbosch: South Africa.

Siwila, L.C., 2014, 'Ecohuman by Siwila.pdf', *Studia Historiae Ecclesiasticae* 40(2), 131–147.

Sibanda, M., 2011, Married-Too-Soon-Child-Marriage-in-Zimbabwe.pdf (zgc.co.zw)

Skirbekk, V., de Sherbinin, A., Adamo, S., 2020, How Religion Influences Our Relationship With the Environment (columbia.edu).

Stump, R., 2008, *The geography of religion: Faith, place, and space*, Rowman and Littlefield Publishers, Maryland.

Stephens, J. H., 2019, [PDF] Acts of God: Continuities and change in Christian responses to extreme weather events from early modernity to the present | Semantic Scholar

Steffen, L., 2017, Religion and Climate Change | HuffPost Religion.

Tarusarira, J. 2017. https://pure.rug.nl/ws/files/49079994/Tarusarira_2017_The_Ecumenical_ Review.pdf.

The Conversation, 2019, Climate change is affecting crop yields and reducing global food supplies (theconversation.com).

Tranfield, D., Denyer, D., & Smart, P., 2003 Towards a methodology for developing evidence-informed management knowledge by means of systematic review. British Journal of Management, 14, 207–222. https://www.cebma.org/wp-content/uploads/Tranfield-et-al-Towards-a-Methodology-for-Developing-Evidence-Informed-Management.pdf.

WFP, 2020, The power of gender equality for food security | World Food Programme (wfp.org).

WFP, 2022, Climate crisis threatens food security for women and girls | World Food Programme (wfp.org).

UNEP, 2019, How all religious faiths advocate for environmental protection (unep.org).

UNDP, 2019, Ending Gender Based Violence to achieve Sustainable Development Goals (SDGs) | United Nations Development Programme (undp.org).

UNDP, 2017, Gender, climate change and food security | United Nations Development Programme (undp.org).

UNDP, 2012, https://www.undp.org/sites/g/files/zskgke326/files/publications/TM1_Africa_ GenderClimateChange_Overview.pdf.

Webster, J., & Watson, R. T., 2002, Analyzing the past to prepare for the future: Writing a literature review. Management Information Systems Quarterly, 26, 3.

WEF. 2016, 70% of Africans make a living through agriculture, and technology could transform their world | World Economic Forum (weforum.org).

WHO, 2014, https://www.who.int/publications/i/item/9789241508186.

Zvingowanisei., S., Chitando. E., and Chirongoma, S., 2022. "Karanga Women and Indigenous Knowledge Systems (IKS): Towards enhancing Agricultural Production and Food Security in Zimbabwe" in Sophie Chirongoma, Ezra Chitando and Mazvita Machinga (eds.) *African Thought: A Journal of Afro-centric Knowledge, Special Issue on African Women and Indigenous Knowledge Systems* Volume 2, No.1, BIBLIOTIOS Publications, 202–259.

CHAPTER 4

Food Availability, Accessibility, and Climate Change in Africa: Religion as a Transforming Structure-Process

Sokfa F. John

Introduction

> *In October [2022], Nadifa Abdi Isak brought her malnourished daughters to hospital in Mogadishu. That day, a nurse said, 42 other children had already been checked into the emergency unit, ravaged by hunger. There were 57 the day before that.*

This opening story in a Reuters' report (Bavier et al. 2022) brings home the reality of the food crisis that currently defines the lives of millions of people in Africa and around the world, and climate change is one of the key phenomena behind the situation. This chapter probes the potential role that religion in Africa can play in addressing this food crises, with a focus on food availability and accessibility as key dimensions of the food security environment. The chapter will focus specifically on climate change as it affects food availability and accessibility and the extent to which

S. F. John (✉)
Centre for Mediation in Africa, University of Pretoria, Pretoria, South Africa

L. Maseno et al. (eds.), *Religion, Climate Change, and Food Security in Africa*, Sustainable Development Goals Series,
https://doi.org/10.1007/978-3-031-50392-4_4

religion in Africa can be assessed as a resource or impediment to food availability and accessibility in this regard. Climate change does not operate in isolation to affect food security. There are significant mutual influences and overlap between, for example, climate change, poverty, conflicts, and other creating the hunger situation. With this in mind, I use the Sustainable Livelihoods Framework (SLF), particularly the idea of Transforming Structures and Processes (TSP), as an effective framework for reimagining religion in order to best understand its role in food availability and accessibility.

Food availability and accessibility refer, respectively, to adequate food supply and the ability to conveniently access such supply (Shearer et al. 2015). They are also core components of food security. There is food security when there is adequate nutritious, quality, and safe food in a society, and members are able to produce and acquire such food to sustainably live a healthy life (Hlophe-Ginindza and Mpandeli 2021). An understanding of these essential food security dimensions is often shaped by the need to define them in empirically measurable ways. Thus, food availability is typically used to mean the number of locations where food can be obtained/purchased in a given area or adequate supply of nutritious food in specific locations, and obtainable in desired or needed quantities. Food accessibility then refers to the ability to obtain such food at such locations with ease. In other words, the distance to the food location, how easy it is to get to such locations with reference to travel, time, safety, cost of transportation, amongst other things (Carson and Boege 2020; Caspi et al. 2012).

Food availability and accessibility, thus, depend on such factors as food production, livestock ownership, distribution, behaviours of the international and local markets, security, government decisions, and several other social, political, and environmental factors (Connolly-Boutin and Smit 2016). The Food and Agriculture Organization of the United Nations (FAO) affirms that food security has multiple dimensions involving complex interaction of political, environmental, social, economic, and other pressures (FAO 1996). This is important, considering the tendency to think about food accessibility and availability primarily in measurable economic terms. It also opens up the space for meaningful engagement of religion. Most of these factors are further impacted by climate change.

There is growing confidence in the ability of religion to impactfully act to address the climate crisis. As Veldman et al. (2013) noted, from the personal to the transnational, the influence of religion in society is felt at every level, and religions' voices and action on climate change decisively shape society's response. This applies to almost every other issue, including food access and availability. Thus, this chapter argues that religion can exponentially contribute to food accessibility and availability if approached as a transforming structure and process. This places it at the same level with other structures, institutions, and processes—governmental and non-governmental—which have the power to shape or undermine livelihoods, including food accessibility and availability.

The State of Food Availability and Accessibility in Africa

The world committed to Zero hunger by 2030 (SDG 2), and the African Union's target in this regard is to eliminate hunger and end food insecurity by 2025 as articulated in Agenda 2063 and other AU Summit decisions (African Union 2014, 2015; Hedden et al. 2021). Despite all the efforts put into achieving these goals, the hunger situation appears to only get worse overall. Food appears to be less available and less accessible to more people, making significant achievements in addressing hunger almost invisible. The United Nations noted that more than 720 million people were suffering from hunger as of 2020, an increase of 161 million from the figures of the previous year (Martin n.d.). According to the World Food Program (2023), the current number of people facing 'acute' food insecurity rose from 287 million in 2021 to 349 million—a 200-million-people increase from pre-COVID-19 figures. The WFP further notes that the number of people suffering from acute hunger is rising at a pace that is unlikely to be matched in financial responses, amidst an all-time high cost of delivering food support (World Food Programme 2023). The food crisis in Africa is distressful. About 140 million Africans are experiencing acute food insecurity (IFRC 2022). Indicators point to a dire lack of sufficient healthy food, especially amongst those with challenging livelihoods, such as poor and/or low-income earning households. According to Hedden et al. (2021) about one in five people in Africa is hungry.

Poverty, where populations are unable to afford sufficient food, and food shortages, which could be due to inadequate food production and distribution resource and infrastructure to make food available and at rates that are affordable, are some challenges to food availability and access in Africa. Hlophe-Ginindza and Mpandeli (2021) argue that poverty and poor distribution of food are the root of the hunger situation in Africa. They further argue that agricultural policy and challenges, such as land reforms and expropriation, are also a major contributor and affect poor people who mostly depend on small-holder agriculture for their food. Income inequality, stagnant agriculture, poor governance, rapid population growth, lack of land, poor policy, dependence on agricultural imports, poor economic incentive for investment in agriculture implementation, corruption, and others are also often cited challenges (Giller 2020). However, amongst the most impactful challenges are conflicts and violence resulting in displacement and disruption of agricultural activities and the needed stability for sustained food production and distribution, as well as climate change. The next section focuses on how climate change impacts food accessibility and availability.

How Climate Change Contributes to Food Availability and Accessibility Issues in Africa

There is a general agreement in the literature that climate change dangerously affects food security, through accessibility, affordability, and availability, and will continue to do so unless urgent and radical measures are taken (Alvi et al. 2021; Barnett 2011; Kabubo-Mariara and Kabara 2018; Masipa 2017; Wheeler and von Braun 2013; Zewdie 2014). This is primarily presented in terms of how climate change causes weather and climate variability, loss of biodiversity, and destruction of natural resources. Consequently, agricultural production and natural sources of food are negatively impacted, while aggravating other conditions that make food unavailable or inaccessible.

Climate change creates weather extremes, natural disasters, droughts, soil degradation, desertification, increase in pests and disease which thrive in hotter conditions, loss of biodiversity, shortage of water, catastrophic floods, rising sea levels, negative shifts, and increases in human diseases, amongst several other things (Ahmadalipour et al. 2019; Awazi et al. 2021; Coates and Norton 2021; Mcleman and Smit 2006; Salih et al. 2020; Schilling et al. 2020; Tadesse et al. 2019). The links to food availability and accessibility are immediate. Droughts, for example, make less water available for crops and irrigation; floods make it impossible to access land for agricultural purposes, and often cause serious damage to crops; soil erosion due to extreme rains and drought-induced desertification mean that crops can hardly grow; increased pests and diseases destroy crops and animals; destruction of biodiversity means that less wild food sources are available to people who depend on them. All of these lead to poor crop yields, destruction of agriculture, and other sources of livelihoods and food.

For four consecutive years, the Horn of Africa, for example, has experienced droughts and catastrophic flooding; 19 countries in West and Central Africa were affected by severe floods as of 2022, affecting five million people including the destruction of crops, loss of lives, and displacements (Madjiangar 2022; Wilson 2023). These have also resulted in high prices of food and fertilizer, worsened by the Russia-Ukraine war. For example, the price of wheat in Mauritania rose 49 percent higher than the five-year average, imported rice price in Sierra Leone rose by 87 percent, and staple sorghum price in Burkina Faso increased by 85 percent (Madjiangar 2022). Similar situations are observed in several other African countries, such as Nigeria, Mali, Zimbabwe, Madagascar, and Angola due to severe droughts and its consequences. It is projected that around 52 million people in Djibouti, Ethiopia, Kenya, Somalia, South Sudan, Uganda, and Sudan will face more tragic levels of food insecurity (IFRC 2022).

Shortages of important resources such as water, grazing lands, as well as the destruction of sources of food and livelihoods due to climate change lead to forced migration of people and conflicts. This is already a major source of conflicts in Africa, where large populations depend on agriculture for their livelihoods and where pasturing is an integral aspect of the lifestyle and cultures of many people, beyond being about livelihood (McGuirk and Nunn 2020a, 2020b; Mwiturubani and Van Wyk 2010; Nordås and Gleditsch 2015). The overlapping of economic factors, climate change, and conflict, for example, has caused displacements in the Sahel Region, with 30 million people needing urgent food relief as of 2022 compared to 7 million a few years prior (IFRC 2022). To locate the issue of food accessibility and availability within a broader conceptual context of development and to pave the way for my construction of the role of religion in addressing these issues, I discuss the Sustainable Livelihoods Framework in the following section.

The Sustainable Livelihoods Framework

The Sustainable Livelihoods Framework is a way of thinking about poverty and vulnerability, often in regard to livelihoods. It aims to help scholars and practitioners to understand the complex factors that intertwine to determine the livelihoods of people, especially the poor. It is a useful tool in identifying and analysing the factors that drive poverty and to design appropriate and sustainable interventions (Levine 2014; Natarajan et al. 2022). This approach has received considerable criticism. Some of these, as outlined by Natarajan et al. (2022), include concerns about the theoretical basis of its claims of being people-centred, holistic, and empowering, also linked to limited attention to the role of institution; excessive focus on the household and the local, without sufficient attention to the web of relationships that influence livelihoods beyond the local; inadequate attention to politics; how the framework is responsive, given the discursive context from which it emerged, to the decoloniality discourse. Nonetheless, the framework continuously proves useful and adaptable in addressing issues from a people-centric, community, and, to some degree, institutional perspectives. Food security through availability and access is an integral aspect of the outcomes targeted through livelihood strategies (Connolly-Boutin and Smit 2016). The livelihoods approach makes possible a thorough understanding of the forces that influence the unavailability and lack of access to food by individuals and households, as well as those that can transform the situation (Connolly-Boutin and Smit 2016).

Livelihood is the capabilities, activities, and resources that provide an individual or household with a means of living. Livelihoods are sustainable if they are able to confront a crises and shocks, cope in such situations, and recover while maintaining their capabilities and activities at the time of stress and in the future without having to exploit or undermine their natural resource base (Serrat and Serrat 2017). The sustainable livelihoods approach offers a tool to understand the factors that affect (enhance and constrain) livelihoods and subsistence opportunities, and how these factors interact with each other. The premise is that livelihoods are composed of assets or capital which are used to create and sustain livelihoods, and a sustainable response will seek to expand such capital which poor and vulnerable people make choices and trade-offs on. These assets include human capital, such as education, skills, health, nutrition, experience; social capital, like social or communal networks, formal and informal groups, religious belonging; natural capital, such as land, water, wildlife, biodiversity, trees, forest products; physical capital, including infrastructure and technology, such as roads, transport, water supply, sanitation, energy, communications, seeds, fertilizers, and various technologies; and financial capital, such as savings, credit, wages, pensions, and remittances (Natarajan et al. 2022).

With reference to food availability and accessibility, the central question would be, what livelihoods capital do affected people have to help them sustainably have and access food? Religious network is often one of the most important social capital of people in crisis and other support-needing situations (Fukuyama 2001). This usually depends on contexts and situations (Halman and Pettersson 2003); in difficult times, many people often find spiritual, emotional, and material support from within their religious networks. Religious entities also sometimes provide support in terms of other forms of capital. In fact, charity, or philanthropy, is a central tenet of most known religious doctrines and theologies. Religion inspires charitable behaviour and supportive behaviour that are instrumental to survival and livelihoods (Abdelkader 2017; Herzog et al. 2020). But beyond charity and social support in times of need, religious stakeholders, and organizations can contribute more sustainably to food accessibility and availability, and I suggest that we begin by more explicitly framing religion as a transforming structure and process.

Transforming structures and processes are integral to sustainable livelihoods. The structures include institutions (different levels of government, agencies, private sector, development organizations, industries, etc.), and processes (legislation, policies, regulations), local cultures (relations among agricultural groups, stokvels, conflict prevention strategies, employment opportunities), services and available facilities, and labour market (Mondal et al. 2012). They have the power to shape poor people's livelihoods (DfID. 2007). They can change both the livelihood assets and vulnerability contexts of the people. Vulnerability contexts are shocks, trends, or seasonality that affect livelihoods, such as the COVID-19 shock in 2020, for instance. The vulnerability context also begins with understanding the pre-existing vulnerabilities of the context, such as prevailing food insecurity, persistent droughts, conflicts, and political instability, among others (DfID. 2007). All of these determine the extent to which a shock or trend can impact livelihoods, the sensitivity of a people's livelihoods, and their ability to cope with such challenges. The COVID-19 shock, for example, made life extremely difficult for millions of people around the world, and the most affected were people who did not already have any strong capital, people with no sustainable source of income, people whose access to food depended on their ability to work each day, communities whose subsistence agriculture has been affected by poor yields due to droughts or shortage of water, and people in conflict zones (Hamadani et al. 2020; John et al. 2022; Workie et al. 2020).

The power of transforming structures was also demonstrated during the pandemic. For example, lock down regulations in many places further entrenched starvation and suffering, requiring interventions by governments, religious organizations and networks, and other entities, through food parcels and several others. Transforming structures and processes are powerful because they enable choices in people, grant or deny access to capital assets, enable the transformation of one type of asset into another, among other things, thereby denying or promoting livelihoods (Levine 2014; Serrat and Serrat 2017). They shape the livelihoods' strategy of people, and weak structures and processes often result in weak livelihoods. They operate between the vulnerability context and livelihood outcomes, which include food security through availability and access, wellbeing, sustainable assets, and other outcomes.

Religion as a Transforming Structure and Process

Religion in many parts of the world holds as much power and influence over the realities of people as government institutions and other processes—negatively or positively. Herein lies the manifestation of religion as a transforming structure and process. As structure, I refer to religious institutions, authorities, organizations, networks, and other entities of decisive direct or indirect influence on people's realities, including their livelihoods assets, behaviours, and choices. Process includes religious doctrines, laws, faith, belief, moral principles, and so on that also limit or enable decisions and activities that impact livelihood resources and their sustainability. This influence in some cases operates independently on people's personal morality and their relationship with other people and society, shaping important decisions and behaviours. In other cases, it is intertwined with other institutions such as government, politics, culture, and education, and co-creates both the spiritual and material experiences of members of society, including issues around food security.

On the more individual and relational level, religious beliefs, doctrines, and belonging shape behaviour and practices, as noted earlier, such as charity, voting decisions, moral justification of decisions to participate in violence, or choose non-violence, care, and social capital—all of which affect the questions of food security, and other issues, directly or indirectly. On the more institutional level, religion has a strong relation to politics and the state in some countries. Taking Nigeria as an example, religion has played a defining role in the formation of the state, how it is currently structured, how it functions, and its power dynamics (Kukah 1993; Olayiwola 2012; Salau 2020; Vaughan 2016). Religious interests shaped the constitution and continues to influence choices of political leadership at both federal and state levels (Ilesanmi 2001; Rufai 2011; Usman 1987; Yesufu 2016). For example, political parties boost their chances of winning presidential elections by pairing a Muslim with a Christian as presidential and vice-presidential candidates; religious interests are crucial parts of political campaigns and pursuits, and key policies or bills can succeed or fail based on the extent to which they receive a Christian or Muslim buy-in. Religion is also central to the vulnerability context of the country, including its implication in the emergence and resolution of violent conflicts, poverty, corruption, insecurity, employment, and access to national and state resources. Religion and religious networks and collaborations significantly contribute to development, sometimes intervening to meet needs through provision of services that are typically the responsibility of governments, such as food, healthcare, security, among others (Hoffmann and Patel 2021). All of these affirm religion as a crucial transforming structure in Africa.

To be sure, my acknowledgement of the influence of religion in Africa is not an affirmation of the narratives that represent Africa as notoriously religious or that every aspect of its personal and public life at individual and other levels is religiously determined, or that African thinking and cultures are inherently religious. These representations of Africa are reductionist, political, have limited understanding of the dynamics of religion and religiosity on the continent, and in some cases implicitly advance logics of coloniality (Green 2006; Paas 2019; Platvoet and van Rinsum 2003; Togarasei 2015). This critique will not be addressed in further detail, as that would not serve the purpose of this chapter. It suffices to note that while I reject the kinds of religious determinism implied in such representations of Africa, I acknowledge that religion is one of the most powerful institutions on the continent and influences lived reality in the private and public lives of many people as well as in state and private institutions and processes. It is equally important to note that such religious influences are also widely contested and rejected in many spaces. Furthermore, it is constantly negotiated, rather than merely accepted, in private and public life. While, for example, many politicians persistently appeal to religious grievances, sentiments, and belonging to win support, there are often many voices, including those of religious leaders, asking citizens to consider other more important factors in their political decisions or to think more critically about their religious interests and the injunctions of their faith towards the good of humanity beyond their immediate interests and needs.

Responding to Food Availability and Accessibility in Africa

There are several implications of thinking about religion as a transforming structure and process in relation to food accessibility and availability in Africa. Most importantly, viewing religion this way could expand our level of engagement with its role from a reactive interventionist approach to a more macro and institutional level transformation of underlying systems and structures that undermine food availability and accessibility. It means that a religious intervention would begin with a thorough analysis of the vulnerability context of the situation, including an assessment of the role that religion may have played in creating such vulnerabilities. This would be followed by a further analysis of the vulnerability context, including the role of religion that makes people's livelihoods, including food security, particularly challenging and unsustainable.

Thinking of religion as a transforming structure in this regard may include analysis of what religious organizations, networks, and infrastructures are doing or not doing that makes food availability or accessibility worse in a given context. In light of the drought, floods, and other crises in the Sahel region and the Horn of Africa, for example, what omissions or inactivity or commissions have made religious structures complicit or implicated in how these disasters are impacting people? Such analysis can move religious response towards more reflexivity and sustainability. It would go beyond interventions such as supplying food, to include issues around whether or to what extent are religious structures working with government to develop and accelerate the implementation of relevant policies and strategies to rapidly reverse and mitigate the adverse impact of climate change due to human activities. Or what macro-level response can produce lasting results beyond recognition of the problem, speaking out against it or calling for action.

As a process, religious institutions can assess how their teachings, doctrines, and regulations contribute towards beliefs and behaviours that undermine or encourage sustainable livelihoods, including food availability and accessibility. Evidence shows that, for example, the religious beliefs of many individuals affect their attitude towards climate change (Jenkins et al. 2018). While such level of critical reflections may be common among scholars and theologians, such critical self-reflection is not a practice that has been widely encouraged in the religious culture of the continent. Rather, self-preservation and competition in several African environments appear to be the more prevailing religious culture. The pockets of critical or reflexive practices that exist in religious spaces needs to be expanded so that religious processes are more relevant to everyday material struggles of people. With this in mind, there are specific areas where religion can serve as a transforming resource for food availability and accessibility. However, I primarily look at climate change, believing that any major progress in the mitigation of the negative impacts of climate change will significantly bring progress in other interconnected issues, such as conflicts, reduction in poverty, and the availability and accessibility of food. Climate change, as earlier acknowledged, is indeed one of numerous interconnected factors—a crucial one nonetheless.

That religion has a role to play in addressing climate change is an increasingly popular position. Hulme (2017) argues that religions' engagement in this regard is not only necessary, but also inevitable. Reasons vary for this belief, but at the centre of it is the influence and ubiquitous presence of religions around the world; the types and networks of relationships that religion fosters; and how religion shapes behaviour. Veldman et al. (2013) outlines four reasons why religions can be crucial to how society responds to climate change. These are 1) their influence on the cosmologies or worldviews of believers, through narratives that provide meaning, purpose, ethics, and the notions of the sacred, which helps define that which deserves protection, preservation, and respect. 2) The influence of religion is profound, and religious leaders have access to a large audience—the majority of the world's population, which respects their leadership and moral authority, positioning them to potentially mobilize large populations for climate change. 3) A significant amount of the world's institutional and economic resources belongs to religion, and in some places, considerable political power as well as interfaith collaborations, and symbolic resources that can be drawn upon to pursue climate change. And, 4) their potential to foster the achievement of collective goals through connectivity and social capital, and these are extensively involved in humanitarian, emotional, and other forms of support and intervention. These reasons demonstrate the extensive potential of religion to bring about massive change in the climate crisis. It may not easily translate into practice due to the many varied dynamics entrenched in these religious resources and relationships. They, however, further affirm the view of religion as transforming structure and process that significantly impact issues.

Veldman et al. (2013) further observed that religion has been responding to climate change for over three decades—including issuing powerful declarations on the moral urgency of addressing climate change, partnerships with religious and secular institutions, and others. Jenkins et al. (2018) observed that, in fact, statements on climate change by religious authorities have existed for as long as calls for action by scientists. The World Council of Churches, the author notes, already had an existing climate change programme, with the Intergovernmental Panel on Climate Change was established in 1988.

These responses are also visible across Africa, where some religious organizations and leaders are attuned to the dangers of climate change and call for it to be treated as an urgent matter for human action rather than something to left to divine will. Most of the contribution from religious stakeholders have been on the conversation and debates on climate change. The recent edited volume by Chitando et al. (2022) gives an excellent sense of the state and nature of such contribution to the climate change debate. Most importantly, it highlights African perspectives on these conversations, accentuating important African indigenous religious and knowledge resources, which aligns with the growing literature reminding us of the fact that African religious and indigenous systems fostered environmentally friendly attitudes and practices that we could still draw on to mitigate the impact of climate change irrespective of the current religious ecosystem of Africa.

In his introduction to the volume, Chitando (2022) made the indispensable observation that the African ethic of Ubuntu, in the sense of solidarity, and indigenous knowledge systems inform African perspectives on religious role in climate change and remain important resource for mitigating the impact of climate change. Contributors to the volume and other African scholars have also highlighted how indigenous systems and religions advanced respect, preservation, and protection of the environment in ways that could help prevent or reduce the impact of climate change. For example, Shoko (2022) examined how the environment, from a Zimbabwean indigenous perspective, is held in high esteem because of the strong connection between God, the spiritual, and the environment; and the strong connection between religion, land, and people. Thus, entities like totems, sacred waters, sacred trees, sacred forests and mountains were respected and protected because they, as well as animals, are important sites for the operation of spirits, have religious significance, and are necessary for the wellbeing of the people.

Scholars have identified the role that religious actors have played in aggravating the impact of climate change in African communities through their aggression towards indigenous religions and knowledge systems. The demonization and mocking of tradition and indigenous religions by colonialists and Christian missionaries; the conflict between tradition and Christianity or Islam in some communities, sometimes involving the profaning and even destruction of sacred forests and other habitats of the spirits for several purposes (Cherif and Greenberg 2013; Sarfo-Mensah and Awuah-Nyamekye 2013).

Such contributions to the climate crisis debate are invaluable and point to essential religious and cultural resources that can considerably change religious response to climate change. Yet, Chitando (2022) challenges that much more practical and innovative intervention needs to come from the religious sector, including extensive educational intervention, mobilizing local action, and multi-religious meetings on climate change (Chitando 2017).

As a transforming structure, religious institutions and collaboration need to also think about their response in terms of high-level policies with clear strategies for implementation, as well as the practical translation of such policies at all levels—from the national to the most local grassroots level. For example, can such powerful and binding religious legislative systems, as the canon law, also expand to address such issues as climate change and food security, as integral to how religious organizations imagine and express themselves?

While successful strategies in response to climate change will also impact food availability and accessibility, these strategies can also work directly on efforts to ensure that food is available. Religious institutions can make food, as a basic human need, one of their priorities and direct their resources towards achieving this goal. Religious institutions can be more involved in agricultural programmes that empower communities to learn and develop sustainable and smart agricultural practices that enable them grow their own food with much ease. Religious institutions can invest in food production and distribution, as well as participating in the food market, to ensure more affordable access to food. They could explore opportunities for partnerships with governments and private sector entities to support communities with subsidized fertilizers, for instance; address problems with access to land, markets, or transportation; address corrupt practices that hinder food availability; and contribute to the appropriate regulation of costs of healthy and nutritious foods. These can be done from a religious perspective without interfering or replacing government's responsibilities. They could also be explored in partnership with government institutions responsible for such responses where the situation allows. Charitable activities, which are already an integral aspect of religious life, could be reimagined to support food availability and accessibility. Tithes, offerings, and other donations could significantly go into relevant investments that can support availability of food for all. 'Giving to God' and popular Christian metaphors like 'sowing of seed', which often translates into bringing money to the church, could be reinterpreted in terms of empowering each other and communities, and such contributions can be mobilized and invested in practical projects that can help address food related challenges in Africa.

Conclusion

I echo Hulme's (2017) and Chitando's (2017) sentiments, in noting that the role of religion in such issues as climate change and food security is a matter of necessity and urgency. Religious entities can expand their ongoing contribution by thinking more innovatively and practically about their response. Pronunciations, statements, conversations, and even prayers are extremely important, but the urgency of the problem demands, additionally, a more radical shift towards practical intervention. Religion as a transforming institution has the moral and material authority, the financial and infrastructural resources, the networks and responsive constituency, and, in some quarters, the theological and material motivation to do so with dramatic impact. A major aspect of the work will also constitute a critical examination of the consequence of non-action or existing attitudes, and considerable work in achieving appropriate buy-in and prioritization from within religious organizations. Thinking of religion as transforming structure and process may sometimes appear as an abstraction. Thus, it is necessary to highlight that religious influence in this regard resides in people, especially, religious authorities. Thus, the invitation to think this way about religious potential to positively transform food availability and accessibility is an invitation to religious leaders to deeply reflect on their role and to religious believers to demand more from themselves, their religious institutions, networks, resources, and leaders—knowing how much transformation they are capable of realizing.

Acknowledgement The author acknowledges the funding and support of the National Institute of Social Sciences and Humanities (NIHSS) through the Centre for Mediation in Africa, University of Pretoria.

References

Abdelkader, E. 2017. Humanitarian Islam. *Pace Int'l L. Rev.*, *30*, 175.

African Union. 2014. *Malabo Declaration on Accelerated Agricultural Growth and Transformation for Shared Prosperity and Improved Livelihoods Doc. Assembly/Au/2 (Xxiii).*

African Union. 2015. *Agenda 2063 report of the commission on the African Union Agenda 2063 The Africa we want in 2063.*

Ahmadalipour, A., Moradkhani, H., Castelletti, A., & Magliocca, N. 2019. Future drought risk in Africa: Integrating vulnerability, climate change, and population growth. *Science of the Total Environment*, *662*, 672–686.

Alvi, S., Roson, R., Sartori, M., & Jamil, F. 2021. An integrated assessment model for food security under climate change for South Asia. *Heliyon*, *7*(4), e06707. https://doi.org/10.1016/j.heliyon.2021.e06707.

Awazi, N. P., Tchamba, M. N., Temgoua, L. F., & Tientcheu-Avana, M.-L. 2021. Farmers' adaptive capacity to climate change in Africa: Small-scale farmers in Cameroon. In *African handbook of climate change adaptation* (pp. 87–115). Springer.

Barnett, J. 2011. Dangerous climate change in the Pacific Islands: Food production and food security. *Regional Environmental Change*, *11*(S1), 229–237. https://doi.org/10.1007/s10113-010-0160-2.

Bavier, J., Sheikh, A., Ovaska, M., & Bhandari, A. 2022, December 14. Africa's food crisis is the biggest yet—Five reasons why. *Reuters*. https://www.reuters.com/graphics/AFRICA-HUNGER/lgpdkknwlvo/.

Carson, J., & Boege, S. 2020. *The Intersection of Food Availability, Access, & Affordability with Food Security and Health.* New Hampshire Children's Health Foundation.

Caspi, C. E., Sorensen, G., Subramanian, S. V., & Kawachi, I. 2012. The local food environment and diet: A systematic review. *Health & Place*, *18*(5), 1172–1187. https://doi.org/10.1016/j.healthplace.2012.05.006.

Cherif, S., & Greenberg, J. H. 2013. Religious perspectives on climate change in the west ivoirian mountainous region. In *How the World's Religions are Responding to Climate Change* (pp. 142–154). Routledge.

Chitando, E. 2017. Praying for Courage: African Religious Leaders and Climate Change. *The Ecumenical Review*, *69*(3), 425–435. https://doi.org/10.1111/erev.12304.

Chitando, E. 2022. Introduction: African perspectives on religion and climate change. In *African Perspectives on Religion and Climate Change* (pp. 1–21). Routledge.

Chitando, E., Conradie, E. M., & Kilonzo, S. M. 2022. *African perspectives on religion and climate change*. Taylor & Francis.

Coates, S. J., & Norton, S. A. 2021. The effects of climate change on infectious diseases with cutaneous manifestations. *International Journal of Women's Dermatology*, *7*(1), 8–16.

Connolly-Boutin, L., & Smit, B. 2016. Climate change, food security, and livelihoods in sub-Saharan Africa. *Regional Environmental Change*, *16*, 385–399.

DfID. 2007. *Sustainable livelihoods guidance sheets. UK DFID Department for International Development: London.*

FAO. 1996. *World Food Summit—Final Report—Part 1*. Food and Agriculture Organization of the United Nations. https://www.fao.org/3/w3548e/w3548e00.htm.

Fukuyama, F. 2001. Social capital, civil society and development. *Third World Quarterly*, *22*(1), 7–20. https://doi.org/10.1080/713701144.

Giller, K. E. 2020. The food security conundrum of sub-Saharan Africa. *Global Food Security*, *26*, 100431.

Green, M. 2006. Confronting Categorical Assumptions About the Power of Religion in Africa. *Review of African Political Economy*, *33*(110), 635–650. https://doi.org/10.1080/03056240601119018.

Halman, L., & Pettersson, T. 2003. *Religion and Social Capital Revisited* (pp. 162–184). Brill. https://doi.org/10.1163/9789004496354_011.

Hamadani, J. D., Hasan, M. I., Baldi, A. J., Hossain, S. J., Shiraji, S., Bhuiyan, M. S. A., Mehrin, S. F., Fisher, J., Tofail, F., & Tipu, S. M. U. 2020. Immediate impact of stay-at-home orders to control COVID-19 transmission on socioeconomic conditions, food insecurity, mental health, and intimate partner violence in Bangladeshi women and their families: An interrupted time series. *The Lancet Global Health*, *8*(11), e1380–e1389.

Hedden, S., Hughes, B., Rothman, D. S., Markle, A. J., & Maweni, J. 2021. Ending Hunger in Africa: The Elimination of Hunger and Food Insecurity on the African Continent by 2025-Conditions for Success. *Available at SSRN 3941433.*

Herzog, P. S., Strohmeier, A., King, D. P., Khader, R. A., Williams, A. L., Goodwin, J. L., Doan, D. R., & Moyo, B. 2020. Religiosity and generosity: Multi-level approaches to studying the religiousness of prosocial actions. *Religions, 11*(9), 446.

Hlophe-Ginindza, S. N., & Mpandeli, N. S. 2021. The Role of Small-Scale Farmers in Ensuring Food Security in Africa. In B. Mahmoud (Ed.), *Food Security in Africa*. BoD—Books on Demand.

Hoffmann, L. K., & Patel, R. N. 2021. *Context: Religion in Nigeria*. https://www.chathamhouse.org/2021/03/collective-action-corruption-nigeria/context-religion-nigeria.

Hulme, M. 2017). Climate change and the significance of religion. *Economic and Political Weekly, 52*(28), 14–17.

IFRC. (2022). *Africa: Hunger crisis | IFRC*. International Federation of Red Cross and Red Crescent. https://www.ifrc.org/emergency/africa-hunger-crisis.

Ilesanmi, S. O. (2001). Constitutional treatment of religion and the politics of human rights in Nigeria. *African Affairs, 100*(401), 529–554.

Jenkins, W., Berry, E., & Kreider, L. B. 2018. Religion and Climate Change. *Annual Review of Environment and Resources, 43*(1), 85–108. https://doi.org/10.1146/annurev-environ-102017-025855.

John, S. F., Okem, A. E., Mubangizi, B. C., Adekanla, N., Ngubane, L. P., & Barry, I. 2022. Rural vulnerability and institutional dynamics in the context of COVID-19: A scoping review. *Jàmbá: Journal of Disaster Risk Studies, 14*(1), 11.

Kabubo-Mariara, J., & Kabara, M. 2018. Climate change and food security in Kenya. In *Agricultural Adaptation to Climate Change in Africa* (pp. 55–80). Routledge.

Kukah, M. H. 1993. *Religion, Politics and Power in Northern Nigeria*. Spectrum Books.

Levine, S. 2014. How to study livelihoods: Bringing a sustainable livelihoods framework to life. *Secure Livelihoods Research Consortium: Researching Livelihoods and Services Affected by Conflict, 22*.

Madjiangar, D. (2022, October 7). *THE ROAD TO COP: Deadly floods deepen food crisis in West and Central Africa amid climate crisis | World Food Programme*. https://www.wfp.org/stories/road-cop-deadly-floods-deepen-food-crisis-west-and-central-africa-amid-climate-crisis.

Martin. (n.d.). Goal 2: Zero Hunger. *United Nations Sustainable Development*. Retrieved January 26, 2023, from https://www.un.org/sustainabledevelopment/hunger/.

Masipa, T. S. 2017. The impact of climate change on food security in South Africa: Current realities and challenges ahead. *Jàmbá: Journal of Disaster Risk Studies, 9*(1). https://doi.org/10.4102/jamba.v9i1.411.

McGuirk, E. F., & Nunn, N. 2020a. *Nomadic pastoralism, climate change, and conflict in Africa*. National Bureau of Economic Research Cambridge, MA.

McGuirk, E. F., & Nunn, N. 2020b. *Transhumant pastoralism, climate change, and conflict in Africa*. National Bureau of Economic Research.

Mcleman, R., & Smit, B. 2006. Vulnerability to climate change hazards and risks: Crop and flood insurance. *Canadian Geographer/Le Géographe Canadien, 50*(2), 217–226.

Mondal, M. A. H., Ali, M. M., Sarma, P. K., & Alam, M. K. 2012. Assessment of aquaculture as a means of sustainable livelihood development in Fulpur upazila under Mymensingh district. *Journal of the Bangladesh Agricultural University, 10*(452-2016–35668), 391–402.

Mwiturubani, D. A., & Van Wyk, J.-A. 2010. Climate change and natural resources conflicts in Africa. *Institute for Security Studies Monographs, 2010*(170), 261.

Natarajan, N., Newsham, A., Rigg, J., & Suhardiman, D. 2022. A sustainable livelihoods framework for the 21st century. *World Development, 155*, 105898. https://doi.org/10.1016/j.worlddev.2022.105898.

Nordås, R., & Gleditsch, N. P. 2015. *Climate change and conflict*. Springer.

Olayiwola, A.-R. O. (2012). Federalism, ethnic politics, state and religion in Nigeria. *Journal of Arts and Contemporary Society, 4*, 1–22.

Paas, S. 2019. 'Notoriously Religious' or Secularising? Revival and Secularisation in Sub-Saharan Africa. *Exchange, 48*(1), 26–50.

Platvoet, J., & van Rinsum, H. 2003. Is Africa incurably religious? *Exchange, 32*(2), 123–153.

Rufai, S. A. 2011. The interplay of power and religion in Nigeria from colonization to democratization. *World Journal of Islamic History and Civilization, 1*(3), 168–177.

Salau, M. B. 2020. Religion and politics in Africa: Three studies on Nigeria. *Journal of Law and Religion, 35*(1), 165–177.

Salih, A. A., Baraibar, M., Mwangi, K. K., & Artan, G. 2020. Climate change and locust outbreak in East Africa. *Nature Climate Change, 10*(7), 584–585.

Sarfo-Mensah, P., & Awuah-Nyamekye, S. 2013. 10 Climate change and indigenous African religion. *How the World's Religions Are Responding to Climate Change: Social Scientific Investigations*, 139.

Schilling, J., Hertig, E., Tramblay, Y., & Scheffran, J. 2020. Climate change vulnerability, water resources and social implications in North Africa. *Regional Environmental Change, 20*, 1–12.

Serrat, O., & Serrat, O. 2017. The sustainable livelihoods approach. *Knowledge Solutions: Tools, Methods, and Approaches to Drive Organizational Performance*, 21–26.

Shearer, C., Rainham, D., Blanchard, C., Dummer, T., Lyons, R., & Kirk, S. 2015. Measuring food availability and accessibility among adolescents: Moving beyond the neighbourhood boundary. *Social Science & Medicine, 133*, 322–330. https://doi.org/10.1016/j.socscimed.2014.11.019.

Shoko, T. 2022. African Traditional Religion and climate change: Perspectives from Zimbabwe. In *African Perspectives on Religion and Climate Change* (pp. 22–33). Routledge.

Tadesse, W., Bishaw, Z., & Assefa, S. 2019. Wheat production and breeding in Sub-Saharan Africa: Challenges and opportunities in the face of climate change. *International Journal of Climate Change Strategies and Management*, *11*(5), 696–715.

Togarasei, L. 2015. Modern/charismatic Pentecostalism as a form of 'Religious' Secularisation in Africa. *Studia Historiae Ecclesiasticae*, *41*(1), 56–66.

Usman, Y. B. 1987. *The manipulation of religion in Nigeria 1977–1987.*

Vaughan, O. 2016. *Religion and the Making of Nigeria*. Duke University Press.

Veldman, R. G., Szasz, A., & Haluza-DeLay, R. 2013. Social science, religions, and climate change. In *How the World's Religions are Responding to Climate Change* (pp. 19–36). Routledge.

Wheeler, T., & von Braun, J. 2013. Climate Change Impacts on Global Food Security. *Science*, *341*(6145), 508–513. https://doi.org/10.1126/science.1239402.

Wilson, J. (2023, January 13). *Climate action: What's new and what's next in 2023 | World Food Programme*. https://www.wfp.org/stories/climate-action-whats-new-and-whats-next-2023.

Workie, E., Mackolil, J., Nyika, J., & Ramadas, S. 2020. Deciphering the impact of COVID-19 pandemic on food security, agriculture, and livelihoods: A review of the evidence from developing countries. *Current Research in Environmental Sustainability*, *2*, 100014.

World Food Programme. 2023. *A global food crisis.* https://www.wfp.org/global-hunger-crisis.

Yesufu, M. L. 2016. The impact of religion on a secular state: The Nigerian experience. *Studia Historiae Ecclesiasticae*, *42*(1), 1–11.

Zewdie, A. 2014. Impacts of Climate Change on Food Security: A Literature Review in Sub Saharan Africa. *Journal of Earth Science & Climatic Change*, *05*(08). https://doi.org/10.4172/2157-7617.1000225.

CHAPTER 5

Religious Perspectives on Climate Change and Food Security in Ghana

Seyram B. Amenyedzi and Rose Omari

Introduction

The United Nations (UN) aspires, in the Agenda 2030, to end poverty, inequalities, protect and conserve the earth. In the agenda are the Sustainable Development Goals (SDGs). The world was on course toward achieving these goals but for the impact of the COVID-19 pandemic, climate change and wars. SDG 1: No Poverty, SDG 2: Zero Hunger and SDG 13: Climate Action are directly connected to food security. Food security is a situation that exists when all people, at all times, have physical, social, and economic access to sufficient, safe, and nutritious food that

S. B. Amenyedzi (✉)
Department of Practical Theology and Missiology, Stellenbosch University, Stellenbosch, South Africa
e-mail: seyram@sats.ac.za

R. Omari
Science and Technology Policy Research Institute, Council for Scientific and Industrial Research, Accra, Ghana
e-mail: romari@csir-stepri.org

L. Maseno et al. (eds.), *Religion, Climate Change, and Food Security in Africa*, Sustainable Development Goals Series,
https://doi.org/10.1007/978-3-031-50392-4_5

meets their dietary needs and food preferences for an active and healthy life (FAO 2002). Food production and preservation in many ways depend on the climate hence the current climate catastrophe impacts hugely on food security and Ghana is not excused. There are socio-cultural, economic, political, and religious perspectives on climate change and food security. This chapter is an ecotheological reflection on climate change and food security in Ghana. It posits that God created and cares for the earth and demands from us to care for the environment. Hunger is not God's plan; therefore, there is provision for creation to be sustained through food production. The church in Ghana is encouraged to consider taking measures in the creation of awareness, training, and helping in addressing climate change and food insecurity.

Methodology

The paper employs a qualitative literature review approach through an ecotheological lens. Ecotheology is a kind of constructive theology that addresses the issues relating to the environment. It draws a relationship between religion, nature, and spirituality especially ecological values, sustainability, and degradation (cf. Conradie 2020; Environment and Ecology n.d.). Ecotheology considers how the earth is suffering from threats of biodiversity, global pollution of the air, water, and the land, and climate change. There is therefore the need to understand this ecological crisis while making the efforts to bring healing to the earth. While this is an interdisciplinary venture, theology is vital in the process. Hence, ecotheology provides a theological framework for understanding and addressing the ecological crisis. Ecotheology has stewardship and ecojustice as core elements (Jorgenson and Padgett 2020).

Conradie (2020) emphasizes Christian ecotheology while Brunner et al. (2014) present an evangelical dimension to the discourse. Kebaneilwe & Kgomotso (2021: 13) toe a Pentecostal line indicating that "the Earth is currently in a state of emergency, needing urgent rescue from possible collapse. We will argue that there is need for a shift from anthropocentric practices evident in Pentecostal missions to joining hands with other stakeholders in efforts to save our environment". We seek to do a Christian ecotheological reflection on the climate and the impact on food security in Ghana. We argue that there are spiritual/religious dimensions to food security which very much depends on the environment, especially the climate.

SDGs, Climate Change and Food Security

The SDGs are global aspirations for the UN to achieve by 2030. Member states are to work toward eradicating poverty, while creating a dignified life and opportunities for all, withing a save and sustainable environment (UN 2022a). Among the 17 goals are SDG 1: No Poverty, SDG 2: Zero Hunger and SDG 13: Climate Action which are linked unswervingly to food security. Poverty reflects on access to healthy food while hunger simply means inaccessibility to food which all point to food insecurity. Climate action undoubtedly affects food security as food production and preservation depends a great deal on the climate. The recent report on the SDGs shows a reversal on the actualization of the goals due to the impact of COVID-19 pandemic and wars (UN 2022b).

Some of the report's findings include:

SDG 1: Between 657 and 676 million people are currently projected to live in extreme poverty in 2022, compared to the pre-pandemic projection of 581 million;

SDG 2: About one in ten people are suffering from hunger worldwide, with 161 million additional people having slid into chronic hunger in 2020 alone.

SDG 13: Energy-related carbon dioxide (CO2) emissions increased by 6% in 2021 reaching their highest level ever (UN 2022b)

The world is on the brink of a climate catastrophe, and the window to avert it is closing rapidly. Increased heatwaves, droughts, and floods caused by climate change are already affecting billions of people around the world and causing potentially irreversible changes in global ecosystems (UN 2022a).

The World Food Programme (WFP) reports the increasing hunger rates due to the effects of the COVID-19 pandemic and wars on escalating food cost. One out of every ten people are not able to access nutritious food and children are the most vulnerable and malnourished. 'The world is hungrier than ever' with 828 million people uncertain of their next meal, and about one billion people going to bed without food every day. The call is to act quickly to save lives and ensure food security (WFP n.d.-a, b; Footsteps 2023; World Vision 2012).

The above reports show that the world is very far from achieving the SDGs: the current state of global poverty is devastating, there is chronic hunger and catastrophic climate which all impact on food security. Chilongozi (2020) observes that, the religious dimension to the SDGs is missing but needful since the sacred must not be separated from the secular. While there are several dimensions to food security the focal point would be the religious dimensions.

Climate Change and Food Security in Ghana

The Intergovernmental Panel on Climate Change (2001) refers to climate change as any change in climate over time, whether due to natural variability or as a result of human activity, which alters the composition of the global atmosphere. Thus, climate change can be defined as the slow change in the composition of the global atmosphere, which is caused directly and indirectly by various human activities in addition to natural climate variability over time (Antwi 2013). There are two main causes of climate change. Both natural and human activities are known to change earth's climate. Many studies and observations proved that before the industrialization era, human activities were at a very minimal level to cause climatic changes, and changes in climate resulted entirely from natural processes and forces such as changes in earth's orbit, changes in solar activity, and volcanic eruptions that causes climate variations. After industrialization era, human activities are increasingly accounting for about 80% of climate change and it is mainly due to burning of fossil fuels for industries and transport, and deforestation for agriculture purposes, urbanization, or infrastructural development (Antwi 2013). These have increased heat trapping greenhouse gases (GHG) in the atmosphere causing warming of earth's atmosphere, land, and oceans. The warming produces dramatic changes in climatic conditions with numerous risks such as more severe extreme events, heat waves, droughts, precipitation events as floods, increased health risks, rising sea levels, and other negative impacts.

Food security has four components namely, availability (having sufficient quantities of safe and nutritious food available), access (having adequate income or other resources to access food), and utilization or consumption (having adequate dietary intake and the ability to absorb and use nutrients in the body) and stability of all the first three components. According to the 2020 CFSVA report, food insecurity in Ghana stands at 11.7%, implying a food insecure population of 3.6 million people (MOFA et al. 2020). Of these food insecure people, 78% are in rural areas, and 22% in urban areas. Food insecurity in Ghana is largely because of low production which is caused by several challenges in the agriculture. The challenges include over-reliance on rain-fed agriculture (which has become erratic because of climate change), limited irrigation facilities to support all-year round food production, declining soil fertility, increasing incidences of crop and livestock pests and diseases as well as issues related to climate change. Even though close to 70% of Ghana's land area is used for farming, the country remains a net importer of agricultural products, including rice and wheat (Oxford Business Group n.d.). In 2015 the US Foreign Agriculture service (FAS) estimated that only around half of all cereal and meat consumed in Ghana is produced locally.

Nutrition and food safety are an integral part of food security and in the last decade, Ghana's food and nutrition situation has improved, however, there are concerns about increased intake of energy-dense and nutrient-poor foods which is characteristic of the nutrition transition being experienced in many developing countries. Micronutrient deficiency notably, vitamin A, iron and zinc deficiencies are high in children and women of reproductive age. This is due to the low intake of micronutrient-rich foods, poor dietary diversity, and impaired absorption or utilization of micronutrients due to infections, worm infestation, diarrhea and other diseases. Factors that contribute to the low intake of micronutrient-rich foods are high cost of food, limited production, post-harvest losses, food safety concerns, abundance of cheaper energy-dense and low nutrient processed products, limited nutrition knowledge, quest for convenience, and low micronutrient content of major staple crops (Omari et al. 2017).

These notwithstanding, Ghana has achieved significant improvements in reducing undernutrition among young children and women of reproductive age. For example, for children under five years, the rate of stunting has declined from 28% in 2008 to 19% in 2014 (GSS et al. 2015). But more work is still needed to reach the Malabo Declaration target of 10% by 2025. Among women of reproductive age, anemia, which is indicative of micronutrient status, has declined from about 58% in 2008 to 42% in 2014 and further to 22% in the 2017 Micronutrient survey. Nevertheless, the rate of anemia in pregnant women (45%) remains above the severe public health threshold. At the same time, overweight and obesity rate among women has increased from 30% in 2008 to 40% in 2014 (GSS et al. 2015) with the rate exceeding 50% in the Greater Accra Region, an urbanized area. Thus, a triple burden of malnutrition (undernutrition, overnutrition, and micronutrient deficiency) is still commonly observed among the Ghanaian population. As a result of the changes in dietary patterns, incidences of non-communicable diseases such as diabetes and hypertension are now increasing.

Furthermore, Ghana experiences several issues related to food safety, plant health and animal health that impact food security. The major concern with plants especially horticulture are pests and diseases, which lead to increased use of pesticides and subsequent deposits of pesticide residues above acceptable levels. In relation to animal health, the major issues of concern are zoonotic diseases such as Avian Influenza (bird flu) and misuse of antimicrobials resulting in antimicrobial resistance in livestock including cultured fish. Outbreaks of bird flu have occurred in 2007, 2015, 2016, 2018, 2021 and 2022 with significant economic losses and in 2021 a total of US$2.7 million was paid by the government in compensation to poultry farms affected by the avian influenza outbreak that year (MOFA 2021). Other causes of microbial and chemical (including heavy metals) contamination are improper disposal of solid waste and effluents and illegal mining activities which end up polluting soils and water bodies used for crops cultivation and irrigation, respectively. Although the Environmental Sanitation section (that is, Part 5) of Ghana's Public Health Act, 2012- Act 851 prohibits the sale or serving of unwholesome food; sale of food under insanitary conditions, and sale of food unfit for human consumption, these practices persist in both rural and urban centers largely due to inadequate enforcement of the law. These unsanitary practices increase the risk of bacterial contamination resulting in foodborne illnesses such as diarrhea, cholera, typhoid fever, and others (Omari and Frempong 2016). The worst cholera epidemic in recent times occurred in 2014 where 17,000 people were affected with over 150 deaths (Omari et al. 2018). Another major food safety hazard of public health significance is aflatoxin, which contaminates food and feed and subsequently affects human and animal health and poses a great threat to trade.

Climate change and variability are among the multiple factors that can provoke changes in food insecurity including the nature and occurrence of food safety hazards. Effects of climate change such as increase in temperature, irregular or erratic rainfall patterns, extreme weather conditions in the form of floods and droughts have aggravated the challenges in the food and agriculture sector. Climate change effects of higher temperatures, water scarcity, droughts, floods, and greater carbon dioxide (CO2) concentrations in the atmosphere affect food production. Production of staples such as maize has suffered recently in Ghana due to the extreme weather events and plant pests notably the fall army worm. As a result of poor yields and low production, the amount of food accessible to the population has reduced and causing inflation. This also happened during the COVID-19 pandemic when people resorted to panic buying, and prices of some foods sharply increased. For example, one *olonka* (about 2.5 kg) of *gari* (cassava grits) which used to be sold at GH¢8 was sold at GH¢18 due to inflation. In 2020, Ghana's rice self-sufficiency ratio was about 43% therefore the deficit was imported at a cost of $391 M, making Ghana the 20th largest importer of Rice in the world (MOFA 2021; OEC n.d.). As a result of economic instability, devaluation of the Ghanaian currency and inflation, the price of imported rice has skyrocketed making affordability difficult.

Climate change has compromised the quality of food in terms of nutritional value and safety. For example, studies show that higher concentrations of carbon dioxide in plants reduce their protein, zinc, and iron content and it is estimated that by 2050, 175 million people could develop zinc deficiencies while 122 million people may be protein-deficient (Concern 2022). Additionally, climate change increases food waste, post-harvest losses, and food safety problems. Food waste also contributes to greenhouse gases and thus adds to the climate crisis. Temperature increases and changes in rainfall patterns have an impact on the persistence and patterns of occurrence of bacteria, viruses, parasites, and fungi as well as their corresponding foodborne diseases. Thus, there are increasing incidences of plant and animal diseases, pest infestations, all of which could increase the spread of zoonotic pathogen and diseases from animas to humans, for example, Ebola and bird flu (FAO 2008). The more pests and diseases there are, the greater the challenges it poses to their control, which may result in misuse of pesticides and antimicrobials and subsequently the presence of their residues in the food chain.

Extreme weather conditions such as excessive floods and droughts may lead to contamination of soil, agricultural lands, water and food and animal feed with pathogens, chemicals and other hazardous substances originating from sewage and polluted lands and industrial effluents (Tirado et al. 2010). Emergency situations after natural disasters often pose water and sanitation problems resulting in food and water borne diseases such as cholera. Climate change also results in ocean warming and acidification which causes changes in ocean salinity and precipitation. These conditions of the ocean cause harmful algal blooms (HABs) (for example, phycotoxins in shellfish) and increase harmful methyl mercury production which gets absorbed by fish thus making them unsafe for human consumption (Tirado et al. 2010).

Essentially, climate change negatively impacts food safety, which is integral to attaining the SDGs as stated above and other SDGs which are not the main focus of this paper. Thus SDG 3: Good health and wellbeing has a significant influence on attaining goals SDG 6: Clean, water and sanitation, SDG 1: No poverty, SDG 2: Zero Hunger (sustainable agriculture). Ecotheologians highlight the God factor in caring for and sustaining the environment as imperative to the ecological discourse.

Religion and Food Security in Ghana

Raman (2014) asserts "food is an everyday need a *sine qua non* for living" (p959); and that there are many aspects to food including religious aspects. Similarly, Schanbacher and Gray (2021) in their article titled 'Religion and Food Insecurity in the Time of COVID-19: Food Sovereignty for a Healthier Future' highlight how the COVID-19 pandemic has heightened the level of food insecurity in "marginalized communities of color". They also identify the religious dimensions to food security and sovereignty and the role religion, mainly faith-based organizations play in mitigating the situation. According to Alonso (2015), culture, religion, and traditional knowledge are drivers of food and nutrition security in developing countries so they must be rendered the needed attention in research and policy making.

The religious dimensions to food are reflected in various religious practices and rituals. Christians observe the Eucharist/communion, lent, fasting and feastings during celebrations like Christmas. Prayers are offered before and after meals for different religions, there are food taboos, totems, and rituals associated with certain foods for instance pork as a taboo for Muslims and rice for celebrations and meat as a taboo for Hindus. *Halal* and *Haram* among Muslims are significant. It is both ethical and sacred for almost all religions, at least the popular ones to feed the poor, and guard against overindulgence in food (Chilongozi 2020; Raman 2014; Schanbacher and Gray 2021).

The lack of food causes famine and compels various religions to perform rituals, this is common in African traditional religions. In Ghana, there are significant events such as the *Homowo* festival (hooting at hunger) among the Ga people, the *Yam* (a starchy root tuber) festival among the Ewes and Akan people. The *Homowo* festival historically was a celebration to hoot or jeer at hunger marking the end of a famine experienced by the Ga people, this has now become an annual event. The celebration is centered around food: the planting of maize and performance of rituals with *Koipoi/Kpekple* (food prepared with maize and usually eaten with palm nut soup)(Visit Ghana n.d.). During the yam festivals, *Te Za* in the Volta Region and *Fofie* Yam Festival in the Brong Ahafo Region, there are celebrations to thank God and the gods for the new harvest and prayers are made for sustenance and blessings. Farmers are also honored in the process (UVOLUNTEER 2016; Xinhua 2022).

There are other food related festivals and rituals, even if the festivals are not directly linked to food, the rituals and the feastings involve food, hence, food plays an integral role in the religious livelihood of the people which means that food security has a huge impact on the religious lives of Ghanaians. The availability and utilization will definitely enhance fulfillment of religious rituals and celebrations while the lack of it will deprive people from performing their religious obligations which in a way affects their wellbeing through their fear of incurring some punishments or curse from the gods; this can also affect the people psychologically. The downside of the religious aspects of food security is perhaps that the focus may be more on the availability and utilization where prayers are made in terms of lack and jubilations when there is bumper harvest. Religion may not be concerned with the accessibility, and the nutritious factor as some of the food taboos may lead to malnourishment. Alonso (2015) opines that there are cultural and religious practices regarding food security which may be either good or harmful to the health of the people, for example, food taboos and dietary prescriptions during pregnancy and traditional healing processes. It is also worth noting that some Christians advise pregnant women and others, whose health conditions may not support food abstinence, not to fast during fasting and prayer seasons.

A Focus on Christianity and Food Security

Jude Collins (2023), the editor for *Footsteps* magazine in her editorial for the edition on Food and Nutrition indicates that hunger is not God's plan for humanity, however global inequalities, climate change and conflicts account for the hunger and malnutrition we face today.

According to World Vision (2012) "food should not be understood only as part of our physical sustenance. It is part of a story that calls us to act as grateful recipients of God's bounty. Food is a life-sustaining gift of

God intended for all people". Some biblical reflections are helpful to throw more light on Collins's statement above.

The creation account in the Book of Genesis in the initial chapters records how God created the garden of Eden and placed humankind in it. God provided food for Adam (food security) but at the same time Adam was supposed to work and produce his food, he was to be fruitful (food sovereignty) (SPEAK Network n.d.; World Vision 2012). Adam was required to care for and cultivate the land which is a reflection of the concerns of ecotheology. Furthermore, there are narratives of famine situations in the Bible where families had to relocate to places where they could be sustained as in the cases of Abraham moving to Egypt (Gen 12), Isaac to Philistine (Gen 26), Jacob and his entire family to Egypt (Gen 43) and the story of Ruth is centered around famine. There are several accounts of famine in the history of the Israelites where God had to intervene. Famine has been normally associated with the divine wrath of God while abundance is a sign of God's blessings and open heavens (Baden 2021).

Notable are other significant narratives of food security in the Bible such as God raining down manna, quails, and providing water for the Israelites in the wilderness in the historical books. The widow of Zarephath and Elijah had their miraculous supply during famine, Elijah fed by ravens, Samaria's deliverance from famine within a day (1Kings 17; 2 Kings 7). The stranger and the poor were all catered for in the law. In the Gospels of the New Testament, one cannot ignore the story of Jesus feeding the multitudes with few loaves of bread and fishes. In the book of Acts, the early church ensured that no member lacked anything. The above narratives all point to the truth that God ensures food security for His creation. God even fends for the plants and birds, and much more for us humans as indicated by Jesus Christ (Matt. 6:25–34). In the Lord's prayer, Jesus taught us to pray for our daily bread, meaning what we eat comes from God (Matt. 6:11; Lk 11:3).

The Role of Christianity in Ensuring Food Security and Climate Action in Ghana

In essence, SDG 1: No Poverty, SDG 2: Zero Hunger and SDG 13: Climate Action are all reflections of God's will and plan for humanity and the world. As indicated early on, the religious factor though missing in the SDG discourse is vital. It is established that there is a religious dimension to food security; now, it is imperative to highlight the role of Christianity in this regard.

In World Vision's (2012) Theological Foundations on Food and Livelihood Security the following proposed *Call to Action* made are relevant to the issues raised in this conversation on Christianity, climate change and food security:

1. Care for the Poor/Hungry: Since there will never cease to be some in need on the earth, I therefore command you, "Open your hand to the poor and needy neighbour in your land"' (Deuteronomy 15:11).
2. Care for the Environment: 'God took the man and put him in the Garden of Eden to till it and keep it' (Genesis 2:15)
3. Achieve Sustainability: 'The good leave an inheritance to their children's children' (Proverbs 13:22). Through our Economic Development programs, we work toward building up the financial resources (savings) of households to help in times of weather shocks, and we encourage communities to be connected so that they help one another through difficult times (social capital)

Expounding on the above action plan, as Christians we are expected to care for the poor and needy. Some Christian humanitarian organizations as the one just mentioned are involved in many activities to alleviate poverty. Some churches organize outreach programs from time to time where they distribute food and essentials to people in need. Others have institutions in place such as orphanages and homes mainly for children. During the COVID-19 pandemic, some churches distributed food to households who were in need. The *Kristo Asafo* Mission is a church that has engaged in farming and production of agricultural mechanical tools and vehicles. This church continues to provide various forms of food especially grains to schools and those in need. (Kristo Asafo Mission n.d.). All these efforts by Christians must be applauded while at the same time identifying the gaps.

Many churches seem to exert financial pressure on members in the form of fund raisings, pledges, dues, various forms of seeds and offerings instead of caring for their needs, as basic as food. This is a call for the Church and Christian organizations to look beyond the spiritual to find ways and means to promote food security among their members and clients. It does not necessarily mean to cook for them but to empower them in the form of trainings and equipping them to be financially able to provide for themselves. Training on food security can enhance not only availability and access to any kind of food but to nutritious food which would boost their holistic wellbeing.

From a Christian ecotheological stance, Christians have the responsibility to take care of God's creation as a responsibility as mandated in the Book of Genesis. For example, to help curb the environmental and health effects of agro-waste, sustainable renewable energy sources are being explored to produce pellets and briquettes to be used as fuel for cooking (Zubairu and Gana 2014). Thus, promoting the production and use of briquettes from agro-waste will reduce our dependency on wood fuel and this will lead to the conservation and restoration of trees, reduction in greenhouse emissions from agro-waste, and increase sequestration of carbon by the trees for a healthy and sustainable environment. The church provides a good platform for carrying out education and awareness-raising on climate change mitigation, adaptation, impact reduction and early warning as indicated in the Climate action; target 13.3. This will foster a behavioral change and build a sense of responsibility and active contribution toward the protection of the food and agricultural sector, the environment and human and animals lives against the effects of climate change. By protecting the food and agricultural sector against the impacts of climate change, we will also be contributing to achieving the SDG 2, 'End hunger, achieve food security and improved nutrition and promote sustainable agriculture', as well as the Africa Union's Malabo declaration that commits African governments to End hunger by 2025. These commitments specially require African governments to increase access to agriculture inputs and technologies, increase agricultural productivity, reduce post-harvest loss, and increase food and nutrition security.

The church and Christian organizations can also support research and the application of climate smart technologies some of which have been developed by the Council for Scientific and Industrial Research (CSIR) in Ghana. Examples are (1) drought tolerant soybean variety, known as Quarshie, developed by CSIR-Savannah Agricultural Research Institute (SARI), (2) disease tolerant and pests resistant groundnut variety named Crop Research Institute (CRI)-Adepa developed by CSIR-CRI between 2001–2006, which are also resistant to rosette and Cercospora leaf spot, and (3) wide adaptability soybean variety such as Songda developed by CSIR-SARI, which is broadly adapted to various ecologies. The Catholic Relief Services in Ghana is an example of Christian Organizations that works on projects to improve child and maternal health, increase access to clean water and sanitation, and scale up farm production.

Conclusion

The chapter was an ecotheological reflection on the religious dimensions to food security and climate change in Ghana. It is approached from the premise that God created and cares for the earth, and requires us to care for the environment. Moreover, it is not God's plan for humankind to be hungry hence He made provision for human beings to be sustained and for them to produce more food. Various religious rituals and taboos reflect the religious dimensions to food. We argue that although Christian organizations and churches are in a way involved in various activities to help the poor and needy people, they are still called upon to be deliberate about this in the form of training and facilitating empowerment programs. The church can use its platform to create awareness about the dangers of climate change, early warnings, mitigation, and adaptation measures. Everybody has the responsibility to ensure the availability of adequate amount of safe and nutritious food that can support healthy and active lives. Apart from education and awareness creation, the church can support the needy with quality food especially in times of crises such as floods, extreme droughts conditions and pandemics such as the COVID-19.

References

English Standard Version Bible. (2001). ESV Online. https://esv.literalword.com/.

Alonso, Elena Briones. 2015. The Impact of Culture, Religion and Traditional Knowledge on Food and Nutrition Security in Developing Countries. FOODSECURE Working paper no. 30. https://lirias.kuleuven.be/retrieve/315294.

Antwi, Anna. 2013. Climate Change and Food Security: An overview about the issue. https://library.fes.de/pdf-files/bueros/ghana/10518.pdf.

Baden, Joel. 2021. "Famine in the Bible Is More than a Curse: It Is a Signal of Change and a Chance for a New Beginning." The Conversation. April 21, 2021. https://theconversation.com/famine-in-the-bible-is-more-than-a-curse-it-is-a-signal-of-change-and-a-chance-for-a-new-beginning-152288.

Brunner, Daniel, Jennifer L. Butler, and A. J. Swoboda 2014. Introducing Evangelical Ecotheology ([edition unavailable]). Baker Publishing Group. https://www.perlego.com/book/2039600/introducing-evangelical-ecotheology-pdf.

Chilongozi, Mwawi Nyirenda. 2020. "The Role of Religion in Sustainable Development: Theological Reflections on Sustainable Development Goals and Mother Earth." In *Mother Earth, Mother Africa & African Indigenous Religions*, 163–77. African Sun Media. https://doi.org/10.18820/9781928480730/10.

Concern. 2022. How climate change increases hunger — and why we're all at risk. https://www.concern.net/news/climate-change-and-hunger.

Conradie, Ernst. 2020. "The Four Tasks of Christian Ecotheology: Revisiting The Current Debate." *Scriptura* 119 (1). https://doi.org/10.7833/119-1-1566.

Collins, Jude. 2023. *Footsteps* "2023_tearfund_footsteps_119_en," no. 119.

Environment and Ecology. n.d. "Ecotheology." Accessed January 17, 2023. http://environment-ecology.com/religion-and-ecology/321-ecotheology.html.

FAO 2008. Food safety and climate change. FAO conference on food security and the challenges of climate change and bioenergy. http://www.fao.org/ag/agn/agns/files/HLC1_Climate_Change_and_Food_Safety.pdf.

FAO. 2002. The State of Food Insecurity in the World 2001. Rome. https://www.fao.org/3/Y1500E/y1500e00.htm#TopOfPage.

Footsteps. 2023. "2023_tearfund_footsteps_119_en," no. 119.

Ghana Statistical Service, Ghana Health Service, and ICF Macro (2015). Ghana Demographic and Health Survey (2014): Key Indicators. http://www.statsghana.gov.gh/docfiles/publications/Ghana%20DHS%202014%20-%20KIR%20-%206%20April%202015.pdf.

Jorgenson, Kiara A., and Padgett, Alan G. 2020. Ecotheology: A Christian Conversation. Wm. B. Eerdmans Publishing Co. https://www.perlego.com/book/3044728/ecotheology-pdf.

Kebaneilwe, Mmapula & Scotch, Kgomotso. 2021. Re-Imagining Mission through Inculturation: Pentecostalism and Sustainable Environment. 10.52779/9781991201317/01.

Kristo Asafo Mission. n.d. "Kristo Asafo Mission—Service to Humankind and Nation Is Service to God." Accessed February 23, 2023. https://kristoasafomission.org/.

Ministry of Food and Agriculture. 2021. Concerted efforts needed against Bird Flu. https://mofa.gov.gh/site/media-centre/agricultural-articles/389-concerted-efforts-needed-against-bird-flu.

MOFA, GSS, WFP, FAO. 2020. Comprehensive Food Security and Vulnerability Analysis (CFSVA) for Ghana. https://ghana.un.org/sites/default/files/2022-03/WFP-0000137744_%20Comprehensive%20Food%20Security%20and%20Vulnerability%202020.pdf.

Observatory of Economic Complexity (OEC). n.d. Rice in Ghana. Accessed January 12, 2023 https://oec.world/en/profile/bilateral-product/rice/reporter/gha.

Omari, Rose, Quorantsen, Kojo Egbert, and Omari, Paul Kwadwo. 2017. Nutrition knowledge and food consumption practices and barriers in rural Ghana: The case of foods for preventing vitamin A and iron deficiencies. Afr. J. Food Agric. Nutr. Dev. 17(1): 11639-11656 DOI: https://doi.org/10.18697/ajfand.77.15815.

Omari, R., Frempong, G.K., and Arthur, W. 2018. Public perceptions and worry about food safety hazards and risks in Ghana. *Food Control* 93: 76–82.

Omari, Rose, and Frempong, Godfred Kwasi. 2016. Food safety concerns of fast food consumers in urban Ghana. Appetite 98 (2016) 49–54

Raman, Varadaraja V. 2014. "Food: Its Many Aspects In Science, Religion, And Culture." *Zygon: Journal of Religion and Science* 49 (4): 958–76.

Schanbacher, William, and Heewon L. Gray. 2021. "Religion and Food Insecurity in the Time of COVID-19: Food Sovereignty for a Healthier Future." *Ecology of Food and Nutrition* 60 (5): 612–31. https://doi.org/10.1080/03670244.2021.1946689.

SPEAK Network. n.d. "Does the Bible Promote Food Sovereignty or Food Security?" Accessed February 23, 2023. https://www.speak.org.uk/blog/does-bible-promote-food-sovereignty-or-food-security.

Tirado, M.C, Clarke, R., Jaykus, L.A., McQuatters-Gollop, A., and Frank, J.M. 2010. Climate change and food safety: A review. Food Research International 43 (2010) 1745–1765.

United Nations. 2022a. "— SDG Indicators." 2022. https://unstats.un.org/sdgs/report/2022/Goal-05/.

———. 2022b. "SDGs Report 2022 Delivers 'Reality Check' on Reversal of Progress | News | SDG Knowledge Hub | IISD." 2022. https://sdg.iisd.org/news/sdgs-report-2022-delivers-reality-check-on-reversal-of-progress/.

UVOLUNTEER. 2016. "Fofie Yam Festival: Ghanaian Culture and Cuisine Combine | UVolunteer." October 30, 2016. https://www.uvolunteer.net/blog/fofie-yam-festival-ghanaian-culture-and-cuisine-combine/.

Visit Ghana. n.d. "Visit Ghana—Homowo Festival." Accessed February 22, 2023. https://visitghana.com/attractions/homowo-festival/.

WFP. n.d.-a "A Global Food Crisis | World Food Programme." Accessed February 24, 2023a. https://www.wfp.org/global-hunger-crisis.

———. n.d.-b "Global Report on Food Crises—2022 | World Food Programme." Accessed February 24, 2023b. https://www.wfp.org/publications/global-report-food-crises-2022.

World Vision. 2012. "Theological Foundations on Food and Livelihood Security." https://www.wvcentral.org/cwb/Pages/Overview.aspx.

Xinhua. 2022. "The People of Asogli Celebrate Traditional Yam Festival | News Ghana." Https://Newsghana.Com.Gh. September 12, 2022. https://newsghana.com.gh/the-people-of-asogli-celebrate-traditional-yam-festival/.

Zubairu, Abdu, and Gana, Sadiq Abba. 2014. Production and characterization of briquette charcoal by carbonization of agro-waste. Energy Power 4(2):41–47.

PART II

Religion, Climate Change and Aspects of Food Insecurity in Africa

CHAPTER 6

Rastafari, Climate Change and Crop Failure in Africa

Fortune Sibanda

Introduction

Only after the last tree has been cut down;
Only after the last river has been poisoned;
Only after the last fish has been caught;
Only then will you find that money cannot be eaten. (Cree Indian Prophecy cited in Van Reenen 2011: 215)

There is enough to satisfy every man's need but never enough to satisfy no man's greed. (Joseph Hill cited in Savage 2018; Mahatma Gandhi n.d.)

The first epigraph, attributed to the 'Cree Native American Indians' or to Alanis Obomsawin of the same community, has been employed in various contexts (see, for example, Van Reenen 2011; Speake 2009; Townsend 1995) as a wake-up call for the need to conserve the natural resources for

F. Sibanda (✉)
Department of Theology and Religious Studies, University of Eswatini, Matsapha, Eswatini
e-mail: fsibanda@uniswa.sz

L. Maseno et al. (eds.), *Religion, Climate Change, and Food Security in Africa*, Sustainable Development Goals Series,
https://doi.org/10.1007/978-3-031-50392-4_6

human flourishing. It cautions the 'bird-brained' short-term thinking inspired by greed and narrow consumerist motivation leading to the destruction of humanity's sources of survival. On the basis of this saying, people can be challenged to self-introspect their actions in light of the climate change crisis. The second epigraph shows how Mahatma Gandhi and later the Jamaican reggae musician, Joseph Hill, pronounced the existential reality that the world has adequate food to feed all the people but what is lacking is accessibility by some people due to greed. This chapter draws its parameters for Rastafarian perceptions, responses and engagements on the aspect of food (in)security/crop failure in Africa on these two epigraphs, which tally with the Rastafari claim to be typical 'ancient environmentalists' and '*Italists*' (Sibanda 2020) through the 'ecological ethic' (Sibanda 2012), 'Rastafari green philosophy' (Sibanda 2015) and '*Ital* foodways' (Sibanda 2019).

The climate change emergency and crop failure constitute urgent maladies in the African context. Paradoxically, Africa is the richest continent in terms of natural resources including the possession of vast tracts of land, but continues to be among the poorest and most underdeveloped, suffering from food shortages and hunger (Ngambi 2014; Sibanda 2017a: 243). In addition, Africa has not been spared from the environmental crisis. The impact of the climate emergency continues to exacerbate the livelihoods and the food security of the majority of the citizens in the global South as compared to those of the global North (Chitando 2020; Sibanda 2020). In other words, it is absurd that the western anthropogenic greedy consumerism that endangers the ecosystem through climate change crisis tends to harm the least industrialized African states as compared to their western counterparts (Sibanda 2020: 53, 54), thereby destroying the "last tree" and "last fish".

In recent years, sub-Saharan Africa has been witnessing constant extreme weather patterns including cyclones, heat waves and floods. It is untenable to exonerate the human activities from the destruction that has occurred to Mother Earth resulting in "climactic disaster, global warming, and pollutions of land, air and water, which endanger the ecosystem" (Sibanda 2020: 53). Notably, despite deforestation and over-exploitation of the land, it is a pity that most people in the global South earn their living from land through agricultural practice (Darkoh 2009; Chitando 2020: 4). The external and internal problems associated with climate change affect the levels of production and productivity as already the majority of people are exposed to hunger, famine, poverty and diseases. A holistic approach is urgently called for to address the ecological crisis, which includes political will, techno-fixes, spirituality and morality (Moe-Lobeda and Helmiere 2013; Sibanda 2023). What Moe-Lobeda and Helmiere (2013: 207) point out is instructive to this chapter when they say: "[h]uman societies' relationships to their environments are shaped, often unconsciously, by religiously influenced worldviews, cosmologies, and value systems." Therefore, religion (in this sense Rastafari) can be placed among the agents for transformation to the climate change emergency.

Specifically, this chapter seeks to examine the Nyahbinghi Rastafari perceptions, responses and engagements on the aspect of food (in)security/ crop failure in selected states in Africa, particularly, Zimbabwe, South Africa, Malawi and Ethiopia. The selection of these countries was influenced by a number of factors including the significant presence of Rastafari communities and the lessons that can be drawn from these states in the context of the discussion at hand. For instance, Ethiopia, Malawi and Zimbabwe have records of experiencing intermittent food shortages, hunger and famines. Ethiopia occupies a central place in Rastafari world view. South Africa, being more industrialized and developed, presents a state with a stable food supply but has a relatively heavy carbon foot print on the environment than its African counterparts. The food shortages of South Africa are a result of maldistribution of available resources as testified in epigraph two above. The study also considers how and to what extent Rastafari serves as a resource (or confounding factor) in relation to land, and food systems stability in Africa. The chapter posits that in spite of being marginalized and misunderstood, the Rastafari environmental ethic guided by a 'green philosophy' and dependence on *Ital* (natural) food is consistent with their principle of organic farming, food security and living in harmony with Mother Earth in contrast to the toxic and greed-driven consumerist 'Babylon *shitstem*' (western corrupt and oppressive system). Rastafarian communities perceive that western consumerism cannot be exonerated from the climate change crisis and its ripple effect on food availability and crop failure in Africa. This is partly due to their long-term pollution of the soil, water and air through pesticides and fertilizers as well as genetically modified food crops. Therefore, the chapter will reflect on how far Rastafari could be a resource to counter climate change and its effects as stipulated in Sustainable Development Goal (SDG) 13 and also achieve zero hunger by promoting food security to meet SDG 2. The chapter concludes that Rastafarians serve as a strong resource for supporting organic farming, food security and climate restabilization within a framework of their environmental I-consciousness. After this introductory section, the chapter explores the theoretical framework and the research methodology used in the study. In the subsequent section, the chapter provides a historical overview of Rastafari before interrogating Rastafari perspectives on land use and food systems stability in Africa, which is followed by the Rastafari perceptions on climate change, food (in)security/ crop failure in Africa. The concluding section caps the chapter.

Theoretical Framework and Research Methodology

The study utilized a decolonial theoretical framework. The theory of decoloniality is a specific type of decolonization, which advocates for the dismantling of the legacies of racial, gender and geopolitical inequalities and domination (Tamale 2020). According to Ndlovu-Gatsheni (2015), decoloniality is a way of thinking, knowing and doing. Some of the proponents of the concept include the Peruvian sociologist, Anibal Quijano, the Argentinian semiotician, Walter D. Magnolo and Nelson Maldonado-Torres. For Quijano, there are four principal elements of coloniality, which decoloniality resisted in thought and action, namely, control of the economy; control of authority; control of gender and sexuality; control of knowledge and subjectivity (Ndlovu-Gatsheni 2015). In addition, Ndlovu-Gatsheni (2015: 490), says there are three concepts, on which decoloniality is premised, namely, (1) coloniality of power, which explains the centre-periphery dichotomy; (2) coloniality of knowledge, which looks at the politics of knowledge generation; (3) coloniality of being questioned on the self-pride and self-assertion of the Africans.

The decolonial theory is relevant to the study partly because the decolonial movements assumed various forms such as Ethiopianism, Pan-Africanism, Garveyism, Black Consciousness and African Renaissance. Rastafari is one of these movements of resistance. Furthermore, this theory is also helpful in understanding the power dynamics in terms of the land use and management characterized by chemical-intensive farming and food (in)stability systems that have side-lined Rastafari epistemological treasures on organic farming, ecological ethic, and *Ital* foodways. They have been pushed to the margins on the basis of coloniality of power, coloniality of knowledge and coloniality of being. Therefore, the observation of Vandana Shiva, an Indian environmental activist, is instructive that the USA, during the Vietnam war, coined the phrase "food as a weapon" and is the "real cause of world hunger" (Shiva 2020). Thus, decoloniality can help one to discern that food systems have been weaponized and nutritionally impoverished through Genetically modified foods, which explains the existential experiences of food (in)security and crop failure in Africa. A 'decolonial turn' would decolonize knowledge, power and being from a Rastafari perspective in Africa. In terms of research methodology, this study followed a qualitative research design. Data were gathered through a combination of in-depth interviews, observation, documentary analysis. In addition, a desk top review of relevant works on Rastafari was also utilized.

Rastafari: A Historical Overview

Building up on researches I have published previously, the history of Rastafari is important in order to understand the movement within the broader framework of why it occupies the margins despite the collective transformation consciousness to resolve the climate change crisis. Originating from Jamaica in the 1930s from ex-black slaves, the movement emerged as an anti-society outfit occupying the lowest strata of society that resisted the oppressive white system through the influence of Marcus Garvey (Sibanda 2017b: 413). At the heart of Rastafari is the deification of the Ethiopian King, His Imperial Majesty Emperor Haile Selassie I from which the pillar of Ethiopianism emanated. Ethiopia, regarded as a "promised land" and "Zion" in Rasta diction, is the source of Rastafari consciousness, "natural *livity* and an eschatological land of freedom and liberation from 'Babylon'" (Sibanda 2015: 193). In Rastafari cosmology, Africa and Ethiopia are sometimes conceived interchangeably. Ethiopia symbolizes free and uncolonised African identity. Alongside Garveyism and Ethiopianism, the ideological history of Rastafari was anchored on Pan-Africanism with its decolonization ideals, which were spread partly through Rastafari reggae music, art, poetry and philosophy to Africa and the rest of the world. Rastafari evolved from a combination of spin-offs from the Judeo-Christian tradition, Hinduism and African Indigenous Religions (Sibanda 2020: 60).

Today, there are sizeable Rastafari communities in Zimbabwe, Malawi, South Africa and Ethiopia, which the study focused on. These countries offer helpful insights on the issues pertaining to food (in)security in relation to the chapter. Therefore, as a movement of resistance, Rastafari used 'Jah music' to cascade critical messages on peace, love, living in harmony with nature, promotion of vegetarian diet of *Ital* foodways and the Rastafari green philosophy, meant to reduce the carbon foot print of humanity. The colour green derived from the Ethiopian flag with red, green and gold colours is important in relation to the climate change crisis. However, the role of Rastafari in preserving the environment and food security has been underplayed in scholarship on minority religions and their potential towards socio-economic development as compared to narratives on Christianity, African Indigenous Religions and Islam (see, Atiemo 2020: 170; Kaunda and John 2020: 72). The ecological and sustainable approaches to environmental consciousness can be addressed through Rastafari views on land use and food systems stability in Africa, to which we now turn.

Rastafari Views on Land Use and Food Systems Stability in Africa

The Rastafari theology of land is intricate and cannot be separated from Africa and Ethiopia (Sibanda 2017c: 194). As already alluded to, the Rastas refer to Ethiopia and Africa as "home", "Zion" and "the promised land." However, from their environmentalist perspective, Rastas also aim to preserve the entire planet, which they refer to as "Mother Earth". They emphasize that in principle what constitutes their diet must come from their own land. It is on this basis that Rastas uphold the ideals of natural *livity*, which inform their views on natural land use through organic farming and *Ital* foodways and diet. From a decolonial perspective, Rastas refer to land as a "stolen legacy" mired in a historical error, which resulted in the alienation of ancestral lands from the majority of the black populace under the colonial hegemony. In Zimbabwe, during the Fast Track Land Reform Programme, the majority of Rastas did not benefit from the land allocation but there were some promises, which were largely not fulfilled. The Rastafari desire for land from the government was summarized in song by Man Soul Jah, a Zimbabwean reggae artist through the song "Mr Government Man" released in 2000 (Sibanda 2017c: 196). In this song, Man Soul Jah complains that the majority of the people (including 'Jah people'), are landless in the land of their heritage. He further adds that people were living like squatters and therefore he challenges the government to give land to the people. He then exclaims in part: "Mr Government Man [...] Give us the land right now!", showing the urgency of the matter, which he describes as "revolutionary, visionary and spiritual". He threatens the possibility of a revival of the 16 years of the liberation struggle if social justice on the land question remains elusive. As a song from a Rastafari artiste, the piece is 'Jah Music', which provides interactive maps for understanding the vitality of land to human flourishing. This resonates with the observation that the Land Reform Programme in Zimbabwe was a perpetuation of the struggle against neo-colonialism (Sibanda and Maposa 2014). The land question remains a contentious challenge and unfinished business with implications on food (in)security in Africa. The lack of equitable land distribution is partly explained by what epigraph two contends. The importance of land towards food security cannot be over-emphasized as it is a vital resource for one to produce and eat green in a sustainable way.

Rastafarians work to attain food stability through organic farming. In South Africa in Grassmarry of Gauteng province, organic Rastafari farmers such as Elder Ras Sipho Mantula and Ras Makush are worth mentioning because they constitute success stories in producing food organically (Sibanda 2015: 199). In Zimbabwe, Malawi, Ethiopia and South Africa, a common observation is that although lacking ownership of large tracts of land, the majority of Rastafari households have functional small gardens where organic farming principles are employed. They grow different varieties of vegetables and herbs, which are food-medicines without the use of chemicals and artificial fertilizers in those organic farming practices (Sibanda 2017c: 198). The Nazarene family of Dangwe Arts and Culture Village provides a fascinating Zimbabwean case of typical strict vegetarians who claim to significantly eat and follow an *Ital* food and diet from what they produce. In Ethiopia, the Shashamene community also provides a good example of promoting organic farming and naturality that are environmental-friendly in land use. In Malawi, the Rastafari communities from the Nyahbinghi and Bobo Ashanti strands are still inspired by the historic visit of Emperor Haile Selassie I in 1965 to four districts of Zomba, Blantrye, Cholo and Chiradzulu, where he visited farmers (Sibanda 2019: 371). This is important to Rastafari participation in organic food production that would enhance food stability and food security. The kind of crops that Rastas recommend should be grown by Rastafarians and non-Rastafarian farmers alike, and should include legumes crops such as soya beans as it improves the fertility of the soil, besides contributing positively to food stability. This is contrary to some of the cash crops, such as cotton and tobacco, which cannot thrive without the use of chemicals and artificial fertilizers. The net effect is that the agro-chemicals and artificial fertilizers deplete the soils and pollute the rivers, dams and the air, which is detrimental to aquatic life and human flourishing.

Healthy food systems in Rastafari are *Ital* (natural) and therefore organic. Human flourishing can be realized through Rastafari *Ital* foodways, which are anchored on their "green philosophy" as inspired by Emperor Haile Selassie I. Essentially, Selassie was a good advocate for the greening of the Mother Earth and the natural environment. Rastafarians claim that they have been admonished by His Imperial Majesty to "plant a million trees every year for food (to nourish and feed the hungry), medicine (to heal the sick and partaken as food-medicines) and as a spiritual consciousness for [preserving] the environment" (Sibanda 2019: 374). The Rastafari lifestyles testify the importance of the principles, which they follow in their diet and foodways. The Rastafari culinary heritage is meant to develop sound physical, mental and spiritual states. Thus, the slogans such as "You are what you eat" and "Let your food be your medicine and your medicine be your food" (Sibanda 2019: 374), have become guiding principles. Ideally, Rastafarians follow a vegetarian diet, which shuns the use of Genetically Engineered Foods. It also explains the Rastafari use of herbs as food-medicines. Rastafarians in South Africa, Malawi, Zimbabwe and Ethiopia overwhelmingly agreed on the importance of industrial hemp towards enriching the soil and promoting an environmental-friendly ecosystem, notwithstanding the ongoing legislative processes on legalizing both hemp and medicinal marijuana. From the Bible, the Rastas justify the use of marijuana as food and medicine. For instance, Gen 1:12, which says "and the earth brought forth grass, and herb yielding seed … and God saw that it was good", shows the basis of using it in their lifestyle as divinely sanctioned. It also shows that it is based on nature and what nature provides. This explains the Rastafari argument that humanity's biological make up was designed to follow a vegetarian diet and not to eat flesh, blood and bones. Food stability is attained in Rastafari from this perspective. Apparently, Rastas also make use of some herbs to improve the soil texture and for preparing organic liquid fertilizer as well as controlling insects and rodents (Sibanda 2017c: 200). A 'decolonial turn' is urgent in promoting the use of these environmental crops and herbs in Africa for human health, economic development and the environmentally sustainable use of land.

Rastafari Perceptions on Climate Change, Food (in) Security/ Crop Failure in Africa

This section provides the Rastafari perceptions on climate change and how they are connected to issues of food (in) security or crop failure in Africa.

The nature I-consciousness in Rastafari is part of their ecological ethic, which also entails the "green philosophy". Rastafarians are committed environmentalists and stewards, who are concerned about western consumerism and environmental injustices affecting society (Sibanda 2020: 61). The climate change crisis emanates from the anthropogenic activities and greedy consumerism that endangers the ecosystem. Through climate change crisis the least industrialized African states are affected most as compared to their western counterparts, including the flora, fauna and the food production systems. The use of agro-chemicals and artificial fertilizers negatively impact on the land, which is a basis for organic food production. In Ethiopia, Zimbabwe, Malawi and South Africa, the 'reasonings' among Rastafarian communities reflect the chanting down of Babylon system for the pollution of the water, soil and air, which are detrimental to human flourishing. This resonates with the observation made in Jamaica by Dickerson cited in Sibanda (2020: 62) on how the abuse of land by 'Babylon' had ripple effects on the environment and food supplies of Africans. Thus, through a 'decolonial turn' from a Rastafari perspective, 'Babylon' is complicit in crop failure being experienced on the African continent. The climate change crisis and environmental injustice are a result of Western modes of thinking (see, Tarusarira 2017) with the related toxic carbon foot print hazardous to health and a cause for crop failure.

In line with the above, to further show Rastafarian concern for food security, they 'reason up' on such topics on their social media platforms. The following is a typical example of such reflections among Rastafarians, which is quoted at length, verbatim:

> Greetings and solutations brothers and sisters of the faith. May HIM Haile Selassie I guide and protect the I in his name, Ras Tafari … I give much thanks for the reasoning today about healthy foods. I am an organic farmer and a Rastafarian who is very interested in continuing the reasoning on this topic, but right now I am steady working I cultivation and targeting His Imperial Majesty's 130th Earthstrong [birthday] for I to be at a certain point. I pray that ones and ones will hold firm to this topic as I will be certain to bring forth some of my thoughts and experiences to this table in due time. It is important that I n I have an overstanding of where we came from, where we are today and where we want to be in the future when it comes to food security, food safety and economic growth.

The above excerpt shows that Rastas focus on cutting edge topics to the extent that they no longer leave their future to chance. They actively participate on sharing "orthognosy" (correct knowledge), in their diction, promoted through 'reasoning' for human flourishing, liberated from western consumerism. On this basis, the correct knowledge that Rastas refer to is not similar to a universal mode of thinking advanced by the western knowledge system. It is typically pluriversal in contrast to universal through a decolonial turn in its holistic approach to knowledge (Tarusarira 2017: 409). In celebrating the 130 years Earthstrong (birthday) of Emperor Haile Selassie I, the Rastas in Zimbabwe through Dangwe Arts and Culture Village had a Tree Planting Day event on 22 July 2022 at a school called Ken Zim College in Chaitezvi Village. Among the trees planted were some indigenous fruits trees, which were meant not only to preserve the environment but also to enhance food security and for medicine.

Another observation of Western influence on food insecurity and crop failure in Africa relates to seeds. Research has shown that "food security is dependent on seed security. Before a country becomes food-secure it needs to be seed-secure" (Hlatshwayo 2023: 7). According to Shiva (2020) countries known to be the high victims of hunger can be the highest producers of food. Africa is cited as an example of a continent with large tract of land per human being per capita and yet because of the deliberate neo-colonial policies, it has been turned into a continent of hunger, today. This militates against the attainment of SDG 2 on the need to ensure food security. The Rastafarian perspectives are understood in the same light. In addition, just as the Rastas hold 'Babylon' systems as accountable for the misery of food hunger, the environmental activist, Vandana Shiva (2019) refers to the destructive impact of billionaires. For instance, one billionaire is described as the "Christopher Columbus of the modern world" for forcing GMOs onto communities and sideling the more nutritious natural seeds. Thus, through manipulating seed and food supply, hunger becomes an instrument of war. From a Rasta perspective, crop failure is artificially created through bio-piracy on seed. Mbambo (2021) indicates that GMOs are a threat to Mother Africa and all her children, their health, environment and economic freedom. Along the same lines, Mbambo (2021: 11) writes: "Life starts with seeds. So, let's protect our inheritance by refusing to eat fruits or food with no seeds and refusing to eat Genetically Engineered Foods with seeds that don't reproduce." In other words, seed is important because it "creates, nourishes and nurtures life and the ecosystem required for life" (Mbambo 2021: 11). This resonates with the Rastafari standpoint on the GMOs and Western consumerist lifestyles that are not pro-life.

Conclusion

The chapter demonstrates that in spite of their numerical inferiority and being underplayed in scholarship on minority religions, the role of Rastafari in preserving the environment and promoting food security and healthy lifestyles is exemplary. They teach and live a pragmatic *Ital* lifestyle against the backdrop of western consumerism. The use of agro-chemicals and artificial fertilizers is an anathema in Rastafari organic farming methods, which they recommend. The natural environment is a source of food-medicines and human flourishing. The United Nations SDGs 2 and 13 are critical in understanding the Rastafari contribution towards food security and environmental stewardship. The Rastafari principles and *Ital* lifestyle are vital for teaching humanity to make peace with nature. As the two epigraphs used in this chapter have shown, the greedy aspects of humanity can disturb the ecosystem, whilst egocentric motives create artificial shortages. Therefore, the climate change crisis is perceived to be a result of Western reckless lifestyles, whilst crop failure can be equally blamed on the anthropogenic factors as well. As claimed in Rastafari reasonings on one of the social media platforms, "poverty exists not because we cannot feed the poor but because we cannot satisfy the rich". In addition, from a Rastafari perspective, crop failure is artificially created through bio-piracy on seed such that food insecurity is a result of seed insecurity. A decolonial turn is urgent to avert the climate change crisis and cultivate living seed, living soil as well as a living economy in Ethiopia, Zimbabwe, Malawi and South Africa. The chapter concludes that Rastafarians serve as a strong resource for supporting organic farming, food security and climate restabilization within a framework of their environmental I-consciousness.

References

Atiemo, A.O. (2020) "Religion's Contribution to Development: The Case of African Hindus in Ghana" in: E. Chitando, M.R. Gunda & L. Togarasei (Eds.) *Religion and Development in Africa*. Bamberg: Bamberg University Press, pp. 165–184.

Chitando, E. (2020) "Ecotheology in Africa: An Overview and Preliminary Assessment" in: M.C. Green and M. Haron (Eds.) *Law, Religion and the Environment in Africa.* Stellenbosch: African Sun Media, pp. 3–16.

Darkoh, M.B.K. (2009) "An Overview of Environmental Issues in Southern Africa", *African Journal of Theology*, Vol. 47(1), pp. 93–98.

Gandhi, M. (n.d.) https://www.epa.gov/history/quotations-about-environment

Hlatshwayo, V. (2023) "Fortune Favours the Brave at Smiling Through." *Eswatini Farming*, December–January.

Kaunda, C.J. & John, S.F. (2020) "Religion and Development in Africa—A Critical Analysis" in: E. Chitando, M.R. Gunda & L. Togarasei (Eds.) *Religion and Development in Africa.* Bamberg: Bamberg University Press, pp. 59–74.

Mbambo, S.S. (2021) *Mbambo's African GMO Free Products and African Creativity Catalogue, 2020–2021.* South Africa. [Author's Self Publication].

Moe-Lobeda, C. & Helmiere, F. (2013) "Moral Power at the Religion-Development-Environmental Nexus" in: M. Clarke (Ed.), *Handbook of Research on Development and Religion*, Cheltenham, UK: Edward Elgar Publishing, pp. 201–219.

Ngambi, H. (2014) "RARE Leadership: An Alternative Leadership Approach for Africa" in: K. Kondlo (Ed.) *Perspectives on Thought Leadership for Africa's Renewal.* Pretoria: Africa Institute of South Africa, pp. 110–129.

Ndlovu-Gatsheni, S.J. (2015) "Decoloniality as the Future of Africa" *History Compass*, Vol. 13(10), pp. 485–496.

Savage, J. (2018) "Face to face with the Great Joseph Hill of Culture". https://www.modernghana.com/news/846357/face-to-face-with-the-great-joseph-hill-of-culture.html

Shiva, V. (2020) "On the Real Cause of Hunger" Available at: https://www.youtube.com/watch?v=jEqS6rnoyYc

Shiva, V. (2019) "Protecting the Planet: The Destructive Impact of Billionaires" Available at: https://www.youtube.com/watch?v=MNM833k22LM

Sibanda, F. (2023) "Gender, Climate Change and Food Insecurity: A Zimbabwean Rastafari Perspective", *Pharos Journal of Theology*, Vol. 104(2), pp. 1–15.

Sibanda, F. (2020) "Promoting Environmental Justice through the Prism of Rastafari in Zimbabwe" in: M.C. Green and M. Haron (Eds.) *Law, Religion and the Environment in Africa.* Stellenbosch: African Sun Media, pp. 53–70.

Sibanda, F. (2019) "Promoting Human Flourshing through Rastafari *Foodways in Africa*" in M.C. Green (ed). Law, Religion and Human Flourishing in Africa. *Stellenbosch: African Sun Media*, 363–381.

Sibanda, F. (2017a) "Servant Leadership and the Paradox of Africa's [under-]development Predicament" in: M. Mawere, T.R. Mubaya and J. Mukusha (Eds.) *The African Conundrum: Rethinking the Trajectories of Historical, Cultural, Philosophical and Developmental Experiences of Africa.* Mankon, Bamenda: Langaa Research & Publishing Common Initiative Group, 243–266.

Sibanda, F. (2017b) "Praying for Rain? A Rastafari Perspective from Zimbabwe," *The Ecumenical Review*, Vol. 69(3), pp. 411–424.
Sibanda, F. (2017c) "Rastafari Perspectives on Land Use and Management in Postcolonial Zimbabwe" in: M.C. Green, R.I.J. Hackett, L. Hansen & F. Venter (Eds.) *Religious Pluralism, Heritage and Social Development in Africa*. Stellenbosch: Sun Media Stellenbosch, pp. 189–203.
Sibanda, F. (2015) "Rastafari Green Philosophy for Sustainable Development in Postcolonial Zimbabwe: Harnessing Eco-theology and Eco-justice" in: F.H. Chimhanda, V.M.S. Molobi & I.D. Mothoagae (Eds.) *African Theological Reflections: Critical Voices on Liberation, Leadership, Gender and Eco-justice*. Research Institute for Theology and Religion: UNISA, pp. 187–206.
Sibanda, F. (2012) "The Impact of Rastafari Ecological Ethic in Zimbabwe: A Contemporary Discourse", *The Journal of Pan African Studies*, Vol. 5(3), pp. 59–76.
Sibanda, F. & Maposa, R.S. (2014) "Beyond the Third *Chimurenga*?: Theological Reflections on the Land Reform Programme in Zimbabwe, 2000–2010", *The Journal of Pan African Studies*, Vol. 6(8), pp. 54–74.
Speake, J. (ed.) (2009) *The Oxford Dictionary of Proverbs*. Oxford: Oxford University Press.
Tamale, S. (2020) *Decolonization and Afro-Feminism*. Ottawa: Daraja Press.
Tarusarira, J. (2017) "African Religion, Climate Change and Knowledge Systems", *The Ecumenical Review*, Vol. 69(3), pp. 398–410.
Townsend, J. (1995) "Letter to the Editor: The Last Tree", New York Times, August 17.
Van Reenen, R. (2011) *The Gospel of the Lion of Judah: A Student's Guide*. USA: Roland van Reenen.

CHAPTER 7

Catholicism, Climate Change and Pests in Africa

George C. Nche

Introduction

In its latest report released on the 9th of August 2021, the Intergovernmental Panel on Climate Change (IPCC) notes that human-caused changes in the earth's climate are not only observable in every region but also promise a very bleak future for millions of people across the world if nothing drastic is done (IPCC 2021). Already, climate change poses an existential threat to many societies across the globe. The recent devastating floods that swept through over ten countries (e.g., Germany, US, China, Turkey, Spain, Ethiopia, Afghanistan, Bangladesh, Nigeria.) (Aljazeera 2021); the distortion of century-old rainfall patterns across the globe (United States Environmental Protection Agency 2022a; United Nations 2022); the extreme cases of drought that have been reported in Africa (United Nations 2022); and the extreme weather and wildfires that have engulfed some cities

G. C. Nche (✉)
Department of Religion Studies, University of Johannesburg, Johannesburg, South Africa
e-mail: gcnnche@uj.ac.za

L. Maseno et al. (eds.), *Religion, Climate Change, and Food Security in Africa*, Sustainable Development Goals Series, https://doi.org/10.1007/978-3-031-50392-4_7

in the US and Europe (European Commission 2022; Canon 2022) attest to this.

While these impacts seem to be widespread across the globe, they have disproportionately compounded already existing socio-economic issues in developing regions like Africa. Some of these issues include water scarcity, food insecurity, communal conflicts over scarce resources and insecurity. Of these issues, food insecurity seems to stand out as it is at the root of the survival of the continent. Although the subject of food insecurity in Africa has been a source of concern for many years, it currently appears to be a more pressing issue especially in the context of climate change. This is because climate change significantly contributes to the proliferation of temperature-resilient insect pests that pose enormous threat to food production in the continent (Skendžić et al. 2021; Mafongoya et al. 2019). By feeding on plant tissues and transmitting viral infections, these pests account for about 49% loss in yearly crop production in the continent (Botha et al. 2020). With over 60% of its population engaging in subsistence farming and about 23% of its GDP coming from agriculture (Goedde et al. 2019), this challenge of insect pests has significant social, economic and political implications in Africa.

This raises serious concerns about adaptation capacities in the continent and the prospects of achieving Goal 2 (i.e. Zero Hunger) of the United Nations' Sustainable Development Goals (SDGs) by 2030. Even in the just concluded United Nations' COP27 in Egypt, these concerns took preeminence, resulting in the creation of a special fund for climate change adaptation in Africa and other developing regions (UNEP 2022). The case of Africa seems peculiar because the continent does not only lack the adequate technologies and finance to efficiently address this challenge of pests but also lack the political will to implement relevant and sustainable pests management policies (Nche 2020). It is against this backdrop that the intervention of the Catholic Church in the management of climate change-induced insect pests in Africa is encouraged.

The Catholic Church has a population of almost 200 million (Diseko 2019) spread across 31 countries in Africa. This suggests that Africa has the world's third largest Catholic population, after the Americas and Europe. Nearly 1 out of every 5 Africans—19.2%—is Catholic (Hellweg 2019). The church's influence and contributions to the development of the continent are enormous. For instance, the church protested against the trans-Atlantic slave trade (Hellweg 2019), and has continued to promote Western education (Grace and Wodon 2021), provide healthcare services (Johnston et al. 2018; Onyalla 2020), invest in economic development (Zalot 2002) and intervene in the area of security (Ogunje 2022) in Africa. The Catholic Church has also made tangible contributions to the campaign against climate change and environmental degradation in the continent (Nche 2020, 2022; Kiarie 2020). The roles of the Catholic Church in addressing hunger and food security in Africa have further been documented (Baxter 2019). However, research on the role of the Catholic Church in climate change-induced pest management is still sketchy and skeletal.

Hence, the present study examines the role of the Catholic Church in addressing climate change-induced pest proliferation and its threat to food security in Africa. The study employs a desk or secondary research method involving a complete reliance on online scholarly, organizational and ecclesiastical materials and reports. The chapter proceeds in the following order: the next section discusses how climate change enhances the destructive tendencies of insect pests and the consequences on crop production in Africa. The section that follows discusses the Catholic Social Teaching (CRS) as a moral framework for the Catholic Church's intervention in the challenge of climate change-induced pests in Africa. The penultimate section focuses on the existing interventions the Catholic Church has made in the area of pest management in Africa while highlighting a specific shortcoming in the Church's disposition towards the link between climate change, pest management and food insecurity in the continent. The last section then discusses what the Catholic Church's future interventions should look like as well as the specific contributions that should be made. It should be noted that this study treats both the Church at the Vatican and the local dioceses and parishes in Africa as one and the same. Hence, the term "the Catholic Church" as used throughout this chapter refers to either of them depending on the context.

Climate Change, Insect Pests and Crop Loss in Africa

Insect pests and their role in biodiversity are known to be natural to the ecosystem. Hence, climate change does not manufacture insect pests. However, the high carbon dioxide levels which largely drive climate change and climate change-induced high temperatures, chronic floods and droughts have had (and is still having) both direct and indirect impacts on insect pests. These impacts are multi-dimensional as they affect the feeding patterns, the population dynamics, the physiology (i.e. body function) and metabolism (i.e. the chemical process that converts food into energy) of insect pests with important implications for food security.

For instance, the high concentration of carbon dioxide in the atmosphere has been shown to diminish the nutritional quality of plant tissues (e.g., the reduction of protein and amino acids concentrations) (Johnson et al. 2020; DeLucia et al. 2008).[1] Drought and flooding which are caused by climate change-induced change in precipitation patterns (e.g. prolonged dryness or rainfall) are also having similar effects on plant nutritional quality in addition to other impacts on plant chemistry, defences, palatability and digestibility (Bauerfeind and Fischer 2013). Again, the climate change-related extreme high temperatures are increasing pests' physiological and metabolic activities (Ma et al. 2021; Ngumbi 2021). All these are collectively making insect pests to consume more plant tissues to compensate for the lost nutrients as well as for survival. And because they eat more, they are expected to grow faster, reproduce, flourish and pose more threats to crop yields (Ma et al. 2021; Ngumbi 2021; Johnson et al. 2020; DeLucia et al. 2008).

On the other hand, prolonged rainfall can support fresh vegetation (food for insects) that can support the growth and proliferation of insect pests (Ngumbi 2021). The population of insect pests is further being promoted by a climate change-induced reduction of pests' natural enemies or predators. For instance, due to their sensitivity to increases in temperature and precipitation, some natural predators die out while their prey thrives (Thomson et al. 2010). Also, as pests move to other areas due to climate change-induced pressures, they escape from their natural enemies and as such flourish in the new areas unhindered (Evans et al. 2013; Damien and Tougeron 2019).

In Africa, there are many insect pests that pose great danger to the continent's most important crops on which millions of people rely for survival. Some of these pests include the lepidopteran stem borers, *Busseola fusca* (Fuller) and *Chilo partellus* Swinhoe (Crambidae) that damage cereal crops especially in Eastern Africa with a production loss of about US $450 million (Pratt et al. 2017; Ngumbi 2017); the *Bemisia tabaci*

[1] Research has also shown that the high concentration of carbon dioxide in the atmosphere can cause a reduction in the population of some insect pests such as dung beetles (see Tocco et al. 2021).

(Gennadius) that damage root and tuber crops in East and Central Africa with an estimated yield loss of US $1.9–2.7 billion (Patil and Fauquet 2009); and the bean flies, aphids, thrips, leafhoppers, whitefly and leaf beetles that damage legume crops in Nigeria, Niger and Burkina Faso with a yield loss of up to 80% (Ngumbi 2017). All these collectively contribute to the high rate of food insecurity in Africa.

However, this situation is currently being made worse by climate change-driven outbreak of two major pests: desert locust and fall armyworm. The 2019–2020 desert locust outbreaks, for instance, are reportedly the largest seen in 25 years in Ethiopia and Somalia—and in 70 years in Kenya (FAO 2022; Dunne 2020). Severe climate change-induced cyclones (e.g., Cyclone Mekunu and Cyclone Luban) in 2018 and prolonged wet conditions that followed allowed the locusts to have food, multiply rapidly—increasing their numbers 8000-fold in just a few months and spread (Dunne 2020; Ngumbi 2021). These swarms of locusts devoured tens of thousands of hectares of croplands and pasture across East Africa, endangering the lives of about 15 million people who were already suffering from food scarcity (Dunne 2020; FAO 2020). This locust outbreak is expected to get worse as its breeding conditions persist. This causes serious concern as the locust has an enormous potential for destruction; for instance, a locust swarm of one square kilometre can eat the same amount of food in one day as 35,000 people, assuming that one person consumes around 2.3 kg of food every day (FAO 2020).

The fall armyworm, on the other hand, poses the greatest threat to maize production in Africa, but also to wheat, sorghum, millet, sugarcane, vegetables and cotton productions (FAO 2021). The fall armyworm made its first entry into West Africa from the Americas in 2016 (Rwomushana et al. 2018). Due to prolonged rainfall and the resultant availability of host plants, this worm has multiplied and currently it is found in 44 countries in Africa (Rwomushana et al. 2018; Ngumbi 2021). A survey of the impacts of the fall armyworm in Ghana and Zambia revealed that the average maize loss was 26.6% (i.e. a lost value of US $177 million) and 35% (i.e. a lost value of US $159 million), respectively (Rwomushana et al. 2018). Day et al. (2017) have noted that the fall armyworm has the potential to cause maize yield losses in a range from 8.3 to 20.6 million tonnes per annum, in the absence of any control methods, in just 12 of Africa's maize-producing countries. This, according to these authors, represents a range of 21–53% of the annual production of maize averaged over a three-year period in these countries. The value of these potential losses is estimated at between $2481 million and $6187 million.

There have been some efforts both from the farmers themselves and the governments and from the non-governmental organizations to address these pests and their devastating impacts in Africa. For instance, there have been decades of research to find an ecologically sustainable solution to lepidopteran stem borers and develop crop varieties that are resistant to pests and the plant diseases they spread (Ngumbi 2017). There has also been reported use of pesticides and bio-pesticides by farmers as recommended by governments in many countries in the continent (Rwomushana et al. 2018). However, the problems with these include the inadequacy of these chemicals, unhealthy nature of the chemicals (especially pesticides), development of pesticide resistance, farmers' neglect of safety precautions and a lack of farmers' knowledge on how best to use the chemicals (Rwomushana et al. 2018; Day et al. 2017).

Hence, the Integrated Pest Management (IPM) has been recommended. The IPM is an effective and environmentally sensitive approach to pest management that relies on a combination of biological, chemical, physical and crop-specific (cultural) strategies (United States Environmental Protection Agency 2022b; Food and Agricultural Organization of the United States 2022). However, beyond the IPM, it has been suggested that the war against pests (especially climate change-related ones) can only be won in Africa when the enabling framework is created. And creating this framework requires coordinated action from multiple stakeholders (Day et al. 2017) of which the Catholic Church is one. The tasks of these stakeholders would include awareness creation among farmers, promotion of research on pest management including IPM, provision of necessary resources and equipment for management of pests, etc.

The Catholic Social Teaching Framework

The justifications for the Catholic Church's interventions in social issues are anchored on the principles of its social teaching framework. These principles are articulated through the tradition of papal, conciliar and episcopal documents (United States Conferences of Catholic Bishops [USCCB] 2005) and based on the belief among the Catholic leadership about the need to build a just and peaceful society even while building one's spiritual life. Wright (2017: 11) defines the Catholic Social Teaching (CST) as a "branch of moral theology addressing contemporary issues within the social structures of society: political, economic, and cultural". It "teaches Christians to reflect, analyze, and act on social issues on the basis of the Roman Catholic Church's (RCC) social doctrines" (Ballano 2021: 93). While it is not an economic or political programme, "it offers a powerful way of thinking about what the common good requires, and how structures in society can promote or undermine human well-being and the requirements of justice" (Nichols as cited in Wright 2017: 11).

These are seven principles/themes that make up this framework and they are as follows:

- **Life and Dignity of the Human Person**: The Catholic Church firmly believes that human life is sacrosanct and as such should be protected and promoted by all means possible. It also believes that the dignity of the human person is the foundation of a moral vision for society.
- **Call to Family, Community and Participation**: Through this principle, the Catholic Church encourages the promotion of family and community values as well as the right of every individual to participate in all societal processes aimed at achieving the common good and well-being for all.
- **Rights and Responsibilities**: The Catholic Church teaches that every human person irrespective of race, religion, sex and socio-economic status has fundamental rights (e.g., rights to life, liberty, work, education) that should be protected and responsibilities (e.g., to one another, to our families and to the larger society) that should be met.
- **Preferential Option for the Poor and Vulnerable**: This principle calls on all Catholics to prioritize the weakest and those who are in most need of basic necessities of life and extend love and solidarity to them.
- **The Dignity of Work and the Rights of Workers**: The Catholic Church believes that there is dignity in labour and that whoever labours should be respected and given commensurate wages and conducive working conditions. Labourers or workers also reserve the right to join unions, own private property and engage in economic initiatives.
- **Solidarity**: Through this principle of solidarity, the Catholic Church promotes a sense of unity and oneness among humanity notwithstanding the national, racial, ethnic, economic and ideological

differences. In this light, people are called to love, show kindness and promote peace among one another.

- **Care for God's Creation**: This is a call to be stewards over God's creation amidst the current devastating environmental crisis. This is because the Catholic Church believes that the care for the natural environment is a moral imperative.

These are the principles that guide the Catholic Church in their engagement with society and issues therein. Although Deberri et al. (2003) and Ballano (2021) have argued that the CST principles have been difficult to be implemented by Catholics and Christians in society due to its lack of empirical illustrations from social science research, it is important to note that the leaderships of Catholic Church both at the international and at the local levels have been successfully implementing these principles in different spheres of society and in different countries of the world. The evidence of these implementations are found in the Catholic Bishops' condemnation of the killing of over 300 people in Gambella and Western Wollega regions of Ethiopia (Isenjia 2022); the Pope's criticism of the massacre in Ukraine (McKeown 2022); the South African Jesuits' disapproval of xenophobia and discrimination against migrants in South Africa (Bordoni 2019); and the toughening of the Catholic Church's punishment against sexual abuses (Horowitz 2021).

More examples of this evidence include the Catholic Church's provision of relief materials to victims of earthquakes in Asia and the South Pacific (Carvalho 2022; O'Neill 2009); the Catholic Church's support of 150 million striking Indian workers over poor working conditions and welfare packages (Union of Catholic Asian News 2015); and the use of energy efficient lights, the recycling of water and other materials and the discontinuation of the use of chemical pesticides as well as the planting of trees in the Vatican garden and other Catholic parishes (Brockhaus 2020). It is in the light of the above principles and the corresponding actions of the Catholic Church that this study argues that the Catholic Church has the moral framework as well as the wherewithal to contribute to the campaign against the climate change-driven proliferation of insect pests and their impacts on food production in Africa (Baxter 2019). The next section discusses the current state of this contribute and what future contributions should look like.

The Catholic Church and Pest Management in Africa

Perhaps the best way to commence the conversation about the Catholic Church and pest management is a discussion of the church's concern about the health and environmental implications of chemical pesticides. Pope Francis, through the 2015 encyclical on the care for creation, discouraged the use of chemical products like pesticides (others include herbicides, fertilizers, insecticides, fungicides and agrotoxins in general) in agricultural areas or industries due to their polluting and harmful consequences on human and environmental (e.g., biodiversity loss) health especially that of the poor and vulnerable (Pope Francis 2015). He also raised concerns about the environmental suitability of the technologically driven genetic modification of both plants and animals especially when excessively used. It is important to note that these concerns raised by the Pope reinforce Carson's concern in her popular 1962 book entitled *Silent Spring*. Through this book, Carson, in a rather more compelling way, brought to the fore the dangers of inordinate use of chemical pesticides to both human and environmental health especially in the American context (Carson 1962).

Nevertheless, the Vatican and many Catholic parishes have called on governments and farmers to act on this injunction by ensuring that they engage and promote diversified and innovative systems of food production and pest management which impact less on both human and environmental health (Pope Francis 2015). The Vatican has even gone ahead to discontinue the use of chemical pesticides in the Vatican gardens (Brockhaus 2020). Studies (e.g., Rwomushana et al. 2018; Day et al. 2017) have shown that the most environmentally friendly pest management strategy is the IPM. The IPM innovatively combines biological, chemical, physical and crop-specific (cultural) strategies in the management of pests (Food and Agricultural Organization of the United States 2022). The Catholic Church also approves of this method and has been assiduously encouraging its adoption in Africa. This is because the church appreciates the threat and impacts of pests on food production in the continent.

For instance, in partnership with the Catholic Church, the International Maize and Wheat Improvement Center, through the project entitled "Effective Grain Storage for Sustainable Livelihood of African Farmers," successfully introduced the development and deployment of metal silo technology in Kenya (Embu and Homa Bay districts). This was, among other objectives, aimed at providing protection against major storage pests of staple food crops like maize (George 2011). Also, in collaboration with the US-based Catholic Relief Services and the Ethiopian Catholic Church-Social Development Coordination Office of Harare (ECC-SCDOH), the farmer-to-farmer organization, from 2 to 18 October 2015, trained 163 persons (comprising Farmers, Woreda Agriculture Experts, Zonal Agriculture Experts) on IPM techniques and practices and also provided information on major pests in the Woreda region of Ethiopia and their management techniques (Longland 2015). In 2017, the Ethiopian Catholic Church-Social and Development Coordinating Office of Adigrat (ECCSDCOA) and that of Sodo (ECC-SDCO/S)-Wolayta Zone, respectively, hosted similar collaborated projects for farmers and other stakeholders in their respective regions (see Ringuette 2017; Alfonso 2017).

In fact, for the past 60 years or more, the Catholic Relief Services, in partnership with other organizations (e.g. farmer-to-farmer), has been supporting the needs of rural farmers in many countries (e.g., Ethiopia, Tanzania, Kenya and Uganda) in Africa (Catholic Relief Services 2009). Although not all the agro-based projects executed by the Catholic Relief Services in these countries entirely focused on pest management, it is safe to assume that it is almost impossible to discuss how to improve food production in rural communities in Africa without making direct or indirect reference to pest management. It may not be easy to get information on the total number of farmers that have been successfully trained by the Catholic Relief Services in Africa. However, the FAO on its own part reported that as of 2020 nearly a million and a half African farmers were trained on scouting and monitoring the appearance and spread of the fall armyworm (FAO 2021). Yet, Qu Dongyu, the Director-General of the organization, has noted that more effort is needed to significantly address the impacts and spread of pests generally and the fall armyworm in particular (FAO 2021).

However, it appears that the Catholic Church has not yet attached enough relevance to how climate change is affecting the population and activities of pests and their impact on crop production. For instance, on 18 August 2022, which was a day ahead of the World Humanitarian Day, the leadership of the global confederation of Catholic relief agencies, Caritas Internationalis (CI), reportedly acknowledged the link between climate change and food insecurity in Africa (Chimtom 2022b; Atemanke 2022). However, while acknowledging this link, the attention was largely given to highlight how climate change-driven extreme weather conditions like floods and droughts have pushed people out of their communities and exposed them to food scarcity and hunger. Little or no attention was given to the role of climate change-induced insect pests' proliferation in compounding the challenge of food scarcity in Africa (Chimtom 2022b; Atemanke 2022). The Catholic Church is not alone as Rev Fidon Mwombeki, general secretary of the All Africa Conference of Churches on a different occasion, acknowledged this link between climate change and food insecurity in the continent, but also did not make any reference to the ravaging role of climate change-related pests. He rather talked about how climate change has distorted rainfall patterns and consequently changed farming periods and activities in Africa (Nzwili 2020a). This is despite the fact that church leaders especially those of the Catholic Church have not only experienced the devastations that are caused by the climate change-driven 2019–2020 locust outbreak in Africa, but have also reportedly called for increased urgent action against the outbreak (Nzwili 2020b).

This suggests that either the link between climate change, insect pests and food insecurity in Africa seems yet to be fully understood and acknowledged by churches especially the Catholic Church or the Church considers the link negligible. And this has important implications on the amount of attention, resources and energy that the Church gives to the fight against climate change-induced pests and their impacts on food production in the continent.

Future Interventions

After leading the Angelus prayer at the Vatican's St Peter's Square on 14 August 2022, Pope Francis drew the attention of the global community to "the serious humanitarian crisis" that the people of Somalia and the Horn of Africa nations are facing (Atemanke 2022). Extreme cases of drought and flooding have become so rampant in this region of the continent (especially in Ethiopia, Kenya and Somalia) that about 15 million people currently face forced displacement, acute food scarcity and hunger (International Organization for Migration [IOM] 2022). A recent report has it that this crisis is likely to result in one person dying every 48 seconds across Ethiopia, Kenya and Somalia (Chimtom 2022c). People from other African countries also have similar experiences as one in every five Africans goes to bed hungry (Food Security Information Network [FSIN] 2022) and this is being made worse by factors or events such as COVID-19 and the Russia-Ukraine war. Hence, the Pope called on the international community to, in solidarity, respond effectively to this emergency (Atemanke 2022).

However, the food crisis that is currently ravaging Africa, especially the eastern part of the continent, is not only being worsened by climate change-induced droughts, flooding, wars and COVID-19. It is also being, as this study has shown, directly exacerbated by climate change-driven proliferation of insect pests (see FAO 2021). In fact, apart from the forceful displacement of people, another major way through which drought and flooding contribute to the food crisis is aiding the multiplication of insect pests (Ma et al. 2021; Ngumbi 2021; Johnson et al. 2020; DeLucia et al. 2008). Hence, for the international community to effectively address the challenge of food scarcity in Africa, special attention needs to be given to the management of pests, whose population and destructive tendencies have been significantly enhanced by climate change in the continent. While the responsibility of addressing the challenge of pests statutorily lies with the government, the Catholic Church should on moral basis and principles enshrined in its CST framework contribute in the following areas:

- Invest in more research to understand the impacts of climate change on insect pests and their new impact on crops in Africa.
- Promote awareness of and provide finance (perhaps from USAID grants) for the implementation of IPM techniques (which comprises biological, chemical, physical and crop-specific [cultural] strategies in the management of pests) by farmers and other stakeholders in Africa. The Catholic Church has done a lot in this direction, but a lot is still needed to be done just as the Director-General of the FAO noted (see FAO 2021).
- Enlighten African farmers, extension agents and plant health inspectors on the existence of climate change-induced pests like locust and fall armyworm outbreaks and how climate change is generally changing the population dynamics as well as the destructive tendencies of pests globally and particularly in Africa
- Educate farmers on how to explore available technologies that are designed to fight pests. For instance, the Church could start with teaching farmers how to use the Android app (FAMEWS) that was designed to fight the fall armyworm with the sponsorship of FAO and translated into 29 languages. The app analyses manually entered

data and photos, and uses a mix of artificial and human intelligence, to detect the presence of the worm and offer guidance (FAO 2021).
- Use its credible platforms to attract resources or funds to sponsor effective surveillance and monitoring of pests' outbreak in order to stop invasive species from spreading to new areas. This surveillance should be able to predict when pest invasion is likely to happen as well as the impact of climate change on the distribution of pests (Ngumbi 2017). The successful prediction of the impact of temperature changes on the future distributions of lepidopteran maize stem borers and their natural enemies (see Mwalusepo et al. 2015) is evidence of the practicability of this recommendation.

Conclusion

This study examined the role of the Catholic Church in addressing climate change-induced pests in Africa. It was found that while the church has recorded tangible interventions in the area of pest management, especially in training African farmers on Integrated Pest Management (IPM), it seems not to be aware of the changes climate change has brought to the pest's biology and environment. These changes fortify and enhance the population dynamics and destructive tendencies of pests, and this has serious implications for crop production and food security in Africa. Already, the climate change-aided locust and fall armyworm have caused huge crop loss worth millions of dollars and worsened the lives of millions of vulnerable communities who were already suffering acute food scarcities especially in the Horn of Africa. It is on this basis that the study listed some important areas to which the Catholic Church could make contributions in order to facilitate the elimination of hunger—a goal which aligns with Goal 2 of the United Nations' Sustainable Development Goals (SDGs).

References

Aljazeera. 2021. Floods. Accessed 3 December 2022 from https://www.aljazeera.com/tag/floods/.

Alfonso, J. 2017. ET61:Integrated Pest Management in grain value chain. Accessed 17 December 2022 from http://farmertofarmer.crs.org/wp-content/uploads/2017/05/Alfonso-Joseph-ET61-IPM-Grain-Value-Chain.pdf.

Atemanke, J. 2022. Caritas Internationalis Draws Attention to "unprecedented food crisis" in Parts of Africa. Aciafrica. Accessed 17 December 2022 from https://www.aciafrica.org/news/6502/caritas-internationalis-draws-attention-to-unprecedented-food-crisis-in-parts-of-africa.

Ballano, V.O. 2021. Why Is Catholic Social Teaching Difficult to Implement in Society? A Theological-Sociological Analysis. *The International Journal of Religion and Spirituality in Society*, 11(1), 93–106.

Bauerfeind, S.S. and Fischer, K. 2013. Testing the plant stress hypothesis: stressed plants offer better food to an insect herbivore. *Entomologia Experimentalis et Applicata*, 149(2), 148–158.

Baxter W. 2019. Responding to hunger crisis in Ethiopia. Catholic Relief Services. Accessed 18 December 2022 from https://www.crs.org/stories/responding-hunger-crisis-ethiopia.

Bordoni, L. 2019. South African Jesuits condemn xenophobia, call for political action. Vatican News. Accessed 13 December 2022 from https://www.vaticannews.va/en/church/news/2019-09/south-africa-xenophobia-jesuit-institute-catholic-church.html.

Botha, A., Kunert, K.J., Maling'a, J. and Foyer, C.H. 2020. Defining biotechnological solutions for insect control in sub-Saharan Africa. *Food and Energy Security*, 9 (1), e191.

Brockhaus, H. 2020. Vatican urges Catholics into action on Pope Francis' Laudato si'. Catholic News Agency. Accessed 16 December 2022 from https://www.catholicnewsagency.com/news/44898/vatican-urges-catholics-into-action-on-pope-francis-laudato-si.

Canon, G. 2022. Wildfires in US west fueling extreme weather in other states, study finds. *The Guardian*. Accessed 24 October 2022 from https://www.theguardian.com/us-news/2022/oct/17/wildfires-us-west-extreme-weather-central-states.

Carson, R. 1962. *Silent Spring*. Boston: Houghton Mifflin.

Carvalho, N. 2022. Indonesian church helping earthquake victims.Crux. Accessed 13 December 2022 from https://cruxnow.com/church-in-asia/2022/11/indonesian-church-helping-earthquake-victims.

Catholic Relief Services. 2009. Getting to market: From agriculture to agroenterprise. Accessed 16 December 2022 from https://www.crs.org/sites/default/files/tools-research/getting-to-market.pdf.

Chimtom, N.K. 2022b. Caritas head says food insecurity, violence, and climate change linked in Africa. Crux. Accessed 15 December 2022 from https://cruxnow.com/church-in-africa/2022/08/caritas-head-says-food-insecurity-violence-and-climate-change-linked-in-africa.

Chimtom, N.K. 2022c. Horn of Africa facing food crisis affecting 20 million people. Crux. Accessed 19 December 2022 from https://cruxnow.com/church-in-africa/2022/06/horn-of-africa-facing-food-crisis-affecting-20-million-people.

Damien, M. and Tougeron, K. 2019. Prey–predator phenological mismatch under climate change. *Current Opinion in Insect Science*, 35, 60–68.

Day, R., Abrahams, P., Bateman, M., Beale, T., Clottey, V., Witt, A. 2017. Fall Armyworm: Impacts and Implications for Africa. *Outlooks on Pests Management*, 28(5), 196–201.

Deberri, E.P., Hug, J.P., Henroit, P.J. and Schultheis, M.J. 2003. Catholic Social Teaching: Our Best-Kept Secret. Fourth Revised and Expanded ed. Maryknoll, NY: Orbis Books.

DeLucia, E.H., Casteel, C.L., Nabity, P.D. and O'Neill, B.F. 2008. Insects take a bigger bite out of plants in a warmer, higher carbon dioxide world. *PNAS*, 105 (6), 1781–1782.

Diseko, L. 2019. Pope Francis in Africa: Is the continent the Catholic Church's great hope? BBC. Accessed 29 November 2022 from https://www.bbc.com/news/world-africa-49564397.

Dunne, D. 2020. Q&A: Are the 2019–2020 locust swarms linked to climate change? Carbon Brief. Accessed 7 December 2022 from https://www.carbonbrief.org/qa-are-the-2019-20 locust-swarms-linked-to-climate-change/.

European Commission. 2022. European Forest Fire report: Three of the worst fire seasons on record took place in the last six years. Accessed 24 November 2022 from https://ec.europa.eu/commission/presscorner/detail/en/IP_22_6465.

Evans, E.W., Carlile, N.R., Innes, M.B. and Pitigala, N. 2013. Warm springs reduce parasitism of the cereal leaf beetle through phenological mismatch. *Journal of Applied Entomology*, 137(5), 383–391.

FAO. 2020. Alarm over Desert Locusts increases as new generation of the destructive pests starts breeding in Horn of Africa. Accessed 7 December 2022 from https://www.fao.org/news/story/en/item/1258877/icode/.

FAO. 2021. Fight against fall armyworm: good progress, more efforts needed. Reliefweb. Accessed 18 December 2022 from https://reliefweb.int/report/world/fight-against-fall-armyworm-good-progress-more-efforts-needed.

Food and Agricultural Organization. 2022. FAO welcomes €17 million contribution from Germany to combat the impact of the Desert Locust upsurge in East Africa. Accessed 7 December 2022 from https://www.fao.org/news/story/en/item/1264151/icode/.

Food and Agricultural Organization of the United States. 2022. Pest and Pesticide Management. Accessed 16 December 2022 from https://www.fao.org/pest-and-pesticide-management/ipm/integrated-pest-management/en/.

Food Security Information Network (FSIN). 2022. Global Report on Food Crises 2022 Mid-Year Update. ReliefWeb. Accessed 18 December 2022 from https://reliefweb.int/report/world/global-report-food-crises-2022-mid-year-update.

George, M.L. 2011. Effective Grain Storage for Better Livelihoods of African Farmers Project: Completion report June 2008 to February 2011. International

Maize and Wheat Improvement Center (CIMMYT). Accessed 16 December 2022 from https://repository.cimmyt.org/handle/10883/1323.

Goedde, L., Ooko-Ombaka, A. and Pais, G. 2019. Winning in Africa's agricultural market. McKinsey & Company. Accessed 30 November 2022 from https://www.mckinsey.com/industries/agriculture/our-insights/winning-in-africas-agricultural-market.

Grace, G. and Wodon, Q. 2021. Catholic and faith-based schools in sub-Saharan Africa: introduction to the special issue, International Studies in Catholic Education, 13(2), 154–162. https://doi.org/10.1080/19422539.2021.2010448.

Hellweg, J. 2019. Africa's Catholic churches face competition and a troubled legacy as they grow. *The Conversation*. Accessed 28 November 2022 from https://theconversation.com/africas-catholic-churches-face-competition-and-a-troubled-legacy-as-they-grow-122611#:~:text=Africa%20has%20the%20world's%20third,%E2%80%93%2019.2%25%20%E2%80%93%20is%20Catholic.

Horowitz, J. 2021. Pope Widens Church Law to Target Sexual Abuse of Adults by Priests and Laity. *The New York Times*. Accessed 13 December 2022 from https://www.nytimes.com/2021/06/01/world/europe/vatican-priests-sexual-abuse.html#:~:text=ROME%20%E2%80%94%20Pope%20Francis%20has%20broadened,laypeople%20who%20abuse%20their%20offices.

International Organization for Migration (IOM). 2022. 15 Million People Face Humanitarian Crisis Due to Drought in the Horn of Africa. Accessed 18 December 2022 from https://www.iom.int/news/15-million-people-face-humanitarian-crisis-due-drought-horn-africa.

IPCC. 2021. Climate change widespread, rapid, and intensifying. Intergovernmental Panel on Climate Change Press Release. Accessed 21 December 2022 from https://www.ipcc.ch/site/assets/uploads/2021/08/IPCC_WGI-AR6-Press-Release_en.pdf.

Isenjia, S. 2022.Catholic Bishops in Ethiopia "strongly" Condemn Killings in Gambella and Western Wollega. Aciafrica. Accessed 13 December 2022 from https://www.aciafrica.org/news/6142/catholic-bishops-in-ethiopia-strongly-condemn-killings-in-gambella-and-western-wollega#:~:text=Gambella%2C%2024%20June%2C%202022%20%2F,Gambella%20and%20Western%20Wollega%20regions.

Johnson, S.N., Waterman, J.M. and Hall, C.R. 2020. Increased insect herbivore performance under elevated CO2 is associated with lower plant defence signalling and minimal declines in nutritional quality. *Scientific Reports*, 10: 14553

Johnston, D., Egan, A. and McInerney, P. 2018. The rise, fall and re-establishment of Trinity Health Services: Oral history of a student-run clinic based at an inner-city Catholic Church. *Studia Historiae Ecclesiasticae*, 44(3), 1–21. https://doi.org/10.25159/2412-4265/4701.

Kiarie, G.K. 2020. Environmental Degradation: What is the Role of the Church in Environmental Conservation in Kenya from 1963–2019? *Studia Historiae Ecclesiasticae*, 46(2), 1–12. https://doi.org/10.25159/2412-4265/6762.

Longland, J.M. 2015. ET-58 Integrated Pest Management (IPM) in Agriculture. Accessed 16 December 2022 from https://farmertofarmer.crs.org/wp-content/uploads/2017/05/Longland-Julie-ET58-IPM.pdf.

Ma, C., Ma, G. and Pincebourde, S. 2021. Survive a Warming Climate: Insect Responses to Extreme High Temperatures. *Annu. Rev. Entomol.* 66: 163–84

Mafongoya, P., Gubba, A., Moodley, V., Chapoto, D., Kisten, L., Phophi, M. 2019. Climate Change and Rapidly Evolving Pests and Diseases in Southern Africa. In: Ayuk, E., Unuigbe, N. (eds) New Frontiers in Natural Resources Management in Africa. Natural Resource Management and Policy, vol 53. Springer, Cham. https://doi.org/10.1007/978-3-030-11857-0_4.

McKeown, J. 2022. Pope Francis: In the name of God, 'stop this massacre' in Ukraine. Catholic News Agency. Accessed 13 December 2022 from https://www.catholicnewsagency.com/news/250655/pope-francis-calls-for-an-end-to-the-massacre-in-ukraine-in-the-name-of-god.

Mwalusepo, S., Tonnang, H.E.Z., Massawe, E.S., Okuku, G.O., Khadioli, N., Johansson, T., et al. 2015. Predicting the Impact of Temperature Change on the Future Distribution of Maize Stem Borers and Their Natural Enemies along East African Mountain Gradients Using Phenology Models. PLoS ONE 10(6): e0130427. https://doi.org/10.1371/journal.pone.0130427.

Nche, G. C. 2020. Beyond spiritual focus: Climate change awareness, role perception and action among church leaders in Nigeria. *Weather Climate & Society*, 12(1), 149–169.

Nche, G.C. 2022. Five years after: How African Catholics have responded to the Laudato Si Call on Creation Care. In C. Ezra, E. Conradie, & S. Kilonzo (Eds.), *African Perspectives on Religion and Climate Change*, pp. 120–146. Routledge.

Ngumbi, E.N. 2017. Africa's most notorious insects—the bugs that hit agriculture the hardest. The Conversation. Accessed 7 December 2022 from https://theconversation.com/africas-most-notorious-insects-the-bugs-that-hit-agriculture-the-hardest-83107.

Ngumbi, E.N. 2021. Bugs and climate change: What a warmer, wetter world means for insects and the crops they eat. The Conversation. Accessed 5 December 2022 from https://www.dailymaverick.co.za/article/2021-09-08-bugs-and-climate-change-what-a-warmer-wetter-world-means-for-insects-and-the-crops-they-eat.

Nzwili, F. 2020a. Church action against hunger in Africa reemphasized as report shows under-nutrition. World Council of Churches. Accessed 18 December 2022 from https://www.oikoumene.org/news/church-action-against-hunger-in-africa-reemphasized-as-report-shows-under-nutrition.

Nzwili, F. 2020b. Church leaders in eastern Africa call for urgent action on devastating locust invasion. World Council of Churches. Accessed 20 December 2022 from https://www.oikoumene.org/news/church-leaders-in-eastern-africa-call-for-urgent-action-on-devastating-locust-invasion.

Ogunje, V. 2022. Rescue Africa from Insecurity, Catholic Bishops Beg Foreign Forces. ThisDay. Accessed 28 November 2022 from https://www.thisdaylive.com/index.php/2022/08/12/rescue-africa-from-insecurity-catholic-bishops-beg-foreign-forces/.

O'Neill, L. 2009. Seeking survivors after Asian-Pacific catastrophes. ReliefWeb. Accessed 13 December 2022 from https://reliefweb.int/report/american-samoa/seeking-survivors-after-asian-pacific-catastrophes.

Onyalla, D.B. 2020. Church in Nigeria Grants Government "full access to" All Catholic Health Facilities. *Aciafrica*. Accessed 30 November 2022 from https://www.aciafrica.org/amp/news/1117/church-in-nigeria-grants-government-full-access-to-all-catholic-health-facilities.

Patil, B.I. and Fauquet, C.M. 2009. Cassava mosaic geminiviruses: actual knowledge and perspectives. *Molecular Plant Pathology*, 10(5), 685–701.

Pope, F. 2015. Laudato Si': Encyclical Letter of the Holy Father Francis on Care for our Common Home. Paulines Publications Africa, 176 pp.

Pratt, C.F., Constantine, K.L. and Murphy, S.T. 2017. Economic impacts of invasive alien species on African smallholder livelihoods. Global Food Security, 14, 31–37.

Ringuette, D. 2017. ET99 Farmer-to-Farmer program promotes economic growth and agricultural development in East Africa. Accessed 16 December 2022 from https://farmertofarmer.crs.org/wp-content/uploads/2018/09/Ringuette-David-ET99-Final-Report-1.pdf.

Rwomushana, I., Bateman, M., Beale, T., Beseh, P., Cameron, K., Tambo, J. 2018. Fall armyworm: impacts and implications for Africa. CABI Evidence Note Update. Accessed 8 December 2022 from https://www.cabdirect.org/cabdirect/abstract/20193363039.

Skendžić, S., Zovko, M., Živković, I.P., Lešić, V. and Lemić, D. 2021. The Impact of Climate Change on Agricultural Insect Pests. *Insects*, 12(5), 440. https://doi.org/10.3390/insects12050440.

Thomson, L.J., Macfadyen, S. and Hoffmann, A.A. 2010. Predicting the effects of climate change on natural enemies of agricultural pests. *Biol. Control* 52, 296–306.

Tocco, C., Foster, J., Venter, N., Cowie, B., Marlin, D. and Byrne, M. 2021. Elevated atmospheric CO_2 adversely affects a dung beetle's development: Another potential driver of decline in insect numbers? *Global Change Biology*, 27(19), 4592–4600.

UNEP. 2022. COP27 ends with announcement of historic loss and damage fund. Accessed 30 November 2022 from https://www.unep.org/news-and-stories/story/cop27-ends-announcement-historic-loss-and-damage-fund.

Union of Catholic Asian News. 2015. Catholic Church supports striking Indian workers. Accessed 13 December 2022 https://www.ucanews.com/news/catholic-church-supports-striking-indian-workers/74195.

United Nations. 2022. WMO: Climate change in Africa can destabilize 'countries and entire regions'. Accessed 28 November 2022 from https://news.un.org/en/story/2022/09/1126221#:~:text=The%20State%20of%20the%20Climate,to%20aggravate%20conflict%20and%20displacement.

United States Conferences of Catholic Bishops (USCCB). 2005. Seven themes of Catholic Social Teaching. Accessed 12 December 2022 from https://www.usccb.org/beliefs-and-teachings/what-we-believe/catholic-social-teaching/seven-themes-of-catholic-social-teaching#:~:text=The%20Catholic%20tradition%20teaches%20that,things%20required%20for%20human%20decency.

United States Environmental Protection Agency. 2022a. Integrated Pest Management (IPM) Principles. Accessed 16 December 2022 from https://www.epa.gov/safepestcontrol/integrated-pest-management-ipm-principles.

United States Environmental Protection Agency. 2022b. Climate Change Indicators: Heavy Precipitation. Accessed 28 November 2022 from https://www.epa.gov/climate-indicators/climate-change-indicators-heavy-precipitation.

Wright K.S. 2017. The principles of Catholic social teaching: A guide for decision making from daily clinical encounters to national policy-making. *The Linacre quarterly*, 84(1), 10–22. https://doi.org/10.1080/00243639.2016.1274629.

Zalot, J.D. 2002. The Roman Catholic Church and economic development in sub-Saharan Africa: Voices yet unheard in a listening world. *Dissertations (1962–2010) Access via Proquest Digital Dissertations*. AAI3059393. https://epublications.marquette.edu/dissertations/AAI3059393.

CHAPTER 8

Farming God's Way and Climate Change in Zimbabwe

Obvious Vengeyi

Introduction

This chapter is a critical appraisal of the deliberate deployment of biblical verses by Foundations for Farming in Zimbabwe in crafting a method of farming called Farming God's Way. The centrality of the Bible in the establishment of the principles of farming for poverty alleviation is what motivated this undertaking. Ordinarily, the Bible and Farming are not mentioned together. This does not however imply that the Bible has said nothing about farming, environment or climate through related subjects like forests, droughts, floods, poverty and pestilences. Farming, with its intrinsic relationship with the land and climate was the main activity during the biblical era. There are numerous stories and parables about such subjects as seeds, crops, trees, farmers, seasons, weather, hunger, pests and diseases among others. Often times, the climate contributed significantly to peasant poverty in the biblical world. To date, peasant poverty,

O. Vengeyi (✉)
Department of Philosophy Religion and Ethics, University of Zimbabwe, Harare, Zimbabwe

L. Maseno et al. (eds.), *Religion, Climate Change, and Food Security in Africa*, Sustainable Development Goals Series,
https://doi.org/10.1007/978-3-031-50392-4_8

particularly in the Global South, is directly connected to climate change. Critical questions that inform this chapter include the following: What is the role of the Bible in the diagnosis by Foundations for Farming of the causes of peasant poverty in Zimbabwe and Africa at large? What are the positives and negatives of Foundations for Farming's diagnosis and prescription? What has been the impact of Foundations for Farming in respect of poverty alleviation? What lessons can biblical scholars and scholars of religion derive from the activities of Foundations for Farming?

Foundations for Farming and the Bible

Foundations for Farming is a Christian organisation in Zimbabwe whose major concern is poverty alleviation. It is engaged in promoting sustainable, environmental and climate friendly, low-cost farming methods and technologies. In its activities, it regards the Bible as an indispensable resource: they refer to the Bible in an endeavour to identify not only the causes of poverty but also principles of farming as poverty alleviation strategies. The Foundation lists the Bible among the tools of farming. The Bible is a reference text, a manual for farming techniques; an enterprise that is not uncommon in African societies. It is not an overstatement to claim that in Africa the Bible is read by everybody; the old and the young, the rich and the poor, believers and non-believers alike; for guidance on moral issues, (Gunda 2010) political critique (Gunda and Kuegler 2012; Gunda 2018) religious counsel, socio-cultural and economic instructions and advice. The Bible is consulted in all matters of life: marriage, wedding, birth, naming, sickness, health celebrations, farming and at funerals. Research conducted by the Bible in Africa Studies (BiAS) series at the University of Bamberg confirmed this position that Africans read and interpret the Bible to confront virtually all contemporary challenges: be it climate (Berman et al. 2021), environment (Taringa 2014), violence (Hunter and Kügler 2016), gender (Kügler et al. 2019), HIV/AIDS (Muneja 2012), children (Togarasei and Kügler 2014), church economics (Wessels 2015), among others. The Bible is indeed a manual for everyday living (Kuegler and Gunda 2009).

Various verses from the Bible are interpreted as giving instructions about conservation farming as a Godly way of ending poverty in Africa, hence, Farming God's Way.

Unlike other agricultural or farming training centres that may issue worksuits or hoes or other tools, on the first day, the Foundations for

Farming's first issue is a copy of the Bible. The Bible is consulted to motivate farmers to embrace zero-tillage methods of farming under the 'sacred principles'[1]: on time, up to high standard, with minimum wastage and with joy, as strategies for poverty alleviation.

The Foundation started as an individual journey by Brian Oldreive in 1978. Oldreive is a white commercial farmer who claims to have received a revelation from God concerning how humanity must farm, to end poverty. Literature produced by the Foundation to guide farmers acknowledge the centrality of God in all activities.[2] For him, God provided lessons in nature as recorded in Romans 1:20; hence, the Foundation was initially coined Farming God's Way. Since Oldreive is a born-again Christian, he makes God and Jesus the centre of everything, a typical characteristic of Pentecostals as observed by Masenya (2009: 152–165). The Foundation's literature claims that God later led Oldreive to 1 Corinthians 3:11 for a new prophetic name: Foundations for Farming.[3] With the new name, the Foundation has attracted a global constituency including in the Muslim world and governments.

Pfumvudza *as Farming God's Way*

Since 2020, the Government of Zimbabwe adopted the *Pfumvudza* (zero-tillage) model of maize farming that was developed by Foundations for Farming in its quest to end poverty in Zimbabwe. The organisation was given responsibility to train provincial and district leaders who later would cascade the skills to ordinary peasants across the country. The World Food Programme, Swiss Agency for Development and Cooperation and others have partnered with Foundations for Farming and a manual on *Pfumvudza* Alternative Crops (*Pfumvudza* 2020) covering a variety of food crops has since been developed to assist farmers. This saw the method of *pfumvudza* being incorporated in the Zimbabwean Government's National

[1] We call them sacred principles because Oldreive the founder of Foundations for Farming claims to have received them from God, at different instances.

[2] Cf. The Story Within a Story; Foundations for Farming Vegetable Garden Handbook for a Climate Smart Nutrition Garden; Farming and General Handout; Nutrition and Health Handouts; Pfumvudza Alternative Crops and Trainer's Manual. Unfortunately, the literature does not indicate date and place of publication and the author. However, see https://www.farming-gods-way.org/Resources/FGW_Vegetable_Guide_Large.pdf accessed 11.7.2023.

[3] The Story Within a Story, p. 30. unfortunately, no author's name and publication details like date and place are shown.

Development Strategy 1 as one important intervention in the reduction of peasant poverty.

The *pfumvudza* model was instrumental in fulfilling two National Development strategy 1 goals that are sustainable environmental protection and resilience and food security and nutrition (NDS1, 2021). For the government, the *pfumvudza* model fulfils the seventh of the eight priority areas of Vision 2030 (Vision 2030: 2018) that specifically focusses on improving the livelihoods of ordinary citizenry. Vision 2030 is Zimbabwe's transformation blueprint that was enunciated in 2018 with the sole aim to turn Zimbabwe from low-income society to a prosperous middle-income society by year 2030. Zimbabwe's seventh priority area in the Vision 2030 is a local adaptation of Sustainable Development Goal number 2 that is committed to attaining zero hunger. Because of the great potential to solve the problem of hunger, various international organisations have made an undertaking to partner Foundations for Farming in the implementation of *pfumvudza*.[4]

Ian Scoones (2021) credits the *pfumvudza* model, especially among other government interventions, for the bumper harvest that Zimbabwe recorded in the 2020/21 farming season. Zimbabwe harvested 'an estimated 2.7 million tonnes of maize; triple the amount in the previous season'. According to him, 'one response to uncertain rainfall has been the Pfumvudza/Intwasa programme—a system of zero tillage cultivation involving the construction of small pits allowing water and fertility to concentrate. ... [Y]ields on such plots increased from on average 1.2 tonnes per hectare in extensive dryland fields to 5.3 t/ha on Pfumvudza plots'. The same assessment has been made by other scholars (Mujere 2021).

The Foundation has since spread to the regional countries and other countries across the Globe including Muslim countries. And the success is attributed to the strong biblical foundation that anchors the principles. The Bible is central not only in the development of the *Pfumvudza* principles but also in every literature of Foundations for Farming. A litany of biblical verses is always cited to support various claims about the effectiveness of the proposed methods of farming. Apart from Mark 13:28–29, Isaiah 58:11 also informs the discourse about *Pfumvudza*'s model of farming.

[4] Cf. Foundations for Farming Manual, *Pfumvudza* Alternative Crops, was sponsored by World Food Programme, Swiss Agency for Development and Cooperation and Liberte Elagilite Fraternite of France (2020).

Foundations for Farming claims that the Bible carries instructions and justification about Farming God's Way, from preparation of the land, planting, caring until harvest of crops. In the Foundation's manuals for various farming activities, one or more verses are listed as the starting point. According to the founder, Oldreive, it is not huge machinery and mechanisation that assures one of a good harvest. A family with just a hoe, coupled with following the principles laid down in the Bible, is guaranteed of a bumper harvest; hence, it can defeat poverty.

Diagnosis of Africa's Poverty

According to Brian Oldreive, poverty is prevalent in Africa. His controversial claims regarding the causes of poverty are a subject of discussion. For him, Africa needs to be humble and turn to Christ. Second, Africa needs to make profit from farming. Lastly, Africa needs to be faithful in her implementation of God's principles of farming. Furthermore, Oldreive claims that God told him that Africa was 'not fulfilling its God given mandate' (Trainer's, p. 18), of feeding the world from her abundant arable lands. The main problem, according to Oldreive, is that Africa blames others and not herself for her predicament. Africa, according to him, must be bold enough to accept the truth as is commanded in John 8:31-32: 'Jesus then said to those Jews who believed in him, "If you remain in my word, you will truly be my disciples, and you will know the truth, and the truth will set you free." And this truth must be accepted although it may be, "hurtful, humiliating or disloyal to our culture and heritage"' (Trainer's Manual, p. 15).

Africa Needs Christ

For the Foundations for Farming, 'the base of our problem is pride and selfishness, which we have inherited from Adam and Eve when they followed Satan in rebellion'. Therefore, Africa will not a get cure until 'we (*Africa*) truly follow Christ in taking up our crosses in humility and unselfishness' (Trainer's, p. 15). The connection between African poverty and rebellion from God by following Satan is consistent with the perspectives of the organisation expressed in almost all their literature. Africa's poverty has been explained as caused by ancestral spirits, demons and other dark forces, all from the devil.

Africa Must Learn to Make Profit

Because Africa is groping in darkness following the devil, since the days of Adam and Eve, according to Oldreive, Africa does not know how to make profit. This is the second problem causing poverty in Africa. According to Oldreive, people he reached 'did not lack knowledge of methodology, but that their ability to actually implement that knowledge into a profitable farming enterprise was lacking' (Trainer's, p. 8). This is true in many communities in Africa. A number of factors stand against these African communities. Poverty is the chief culprit.

In spite of the facts, Oldreive claims that after gruelling sessions with God, God responded and instructed him to, 'teach Africa to make profit'. This profit has to be made from individual, community, national and continental level. Africa for Oldreive has not been making profit because Africa did not know the secret of investment. To elaborate on the principle of profit, Oldreive anchored his lessons on Proverbs 3:5–6 and 14:23. Proverbs 3:5–6 calls upon people to trust in God and never in their own strength or mind (Trainer's, pp. 8–15). It is encouragement to surrender to God for it is God who Oldreive believes is in charge of all existence. As a Christian, Oldreive considers anyone who lives their life independent of God as being rebellious. For him, Africa needs to trust in God. In this way, Africa will defeat poverty because Africa will begin to make profit.

The actual text that however mentions the issue of profit is Proverbs 14:23 which says, 'All hard work brings a profit but mere talk leads only to poverty' (Trainer's, p. 29). But this is a general perspective that is difficult to apply in contemporary affairs. The text does not talk about farming at all. It seems the motivation for Oldreive to pick it is the word 'profit'. However, the claim is not verifiable in every day realities. Those who toil do not always get the profit. Profit is made by the owners of enterprises at the expense of those who work hard. However, when taken by an ordinary farmer, who is the prime target of Oldreive's biblical hermeneutics, the text has some resonance.

The assumption of Oldreive by citing Proverbs 14:23 seems to be that Africans are not working hard, hence Africa is poor (Trainer's, p. 30). Africa's production is too meagre that there is no possibility of profit. According to Oldreive, this explains why Africa is one of the poorest continents. In trying to eradicate poverty, Africa takes a humiliating route of stretching the 'begging bowl'. Yet according to him, begging and theft are the greater causes of poverty in the wider community. The practical intervention of debt cancellation or debt relief which is always advocated for by African leaders is for Oldreive not the solution (Trainer's, p. 30). In fact, for him, it promotes begging. It in a way increases poverty in Africa rather than reversing it.

He again turned to the God whom he always calls Father, just like Jesus did, with a further question: how can we (Africa) make profit? Then God responded giving him four principles to make profit. God said, 'Everything that people do must be done: on time, at standard, without wastage and with joy' (Trainer's, p. 8). These are the principles the Foundations in Farming adopted, in order to fight poverty in Africa. They are all anchored on verses in the Bible.

Farming on Time Solves Poverty

For Foundations for Farming, whether it is planting, weeding or harvesting, all must be done on time. There is a time for everything. All these times must be planned and prepared for, on time. The guiding biblical texts are Ecclesiastes 3:1, 2; Titus 1:3; Proverbs 20:4 and Mark 13:28, 29. According to Ecclesiastes 3:1, 2, 'There is a time for everything, a season for every activity under heaven ... a time to plant and a time to uproot' (Climate Smart, p. 17). For the Foundations for Farming, this is the simple message that Africa has not grasped. Yet this is what Africa needs to obey. In order for Africa to make profit to escape excruciating poverty that she is always drowning in, there is need for planning on time. Preparation of fields, buying seed, planting, weeding and harvesting must be done on time. If these activities are not done on time, the results are obvious: loss and more loss.

Titus 1:3 was brought in to reinforce the idea that there is a time for everything. Yet the appointed season does not talk of time in a chronological way as is suggested by Ecclesiastes. The appointed season relates to the fullness of time (*kairos*); God's time which is not related to seasons as we have them. For Foundations for Farming however, the phrase 'season' is enough to put across their message. Also, unlike Ecclesiastes which implies a farming activity, Titus 1:3 has nothing to do with farming.

Proverbs 20:4 mentions lazy people who do not plough in season; as a result, they will have nothing at harvest. It says, 'sluggards do not plow in season; so at harvest time they look but find nothing' (New International Version). Since the Foundations for Farming is quite sensitive to ploughing for they promote zero-tillage, they put a footnote to explain the word plough. According to their research, a plough during the biblical times was shallow, it did not cut deep into the soil as does modern ploughs. Ploughing therefore during biblical times was like stretching the surface only to facilitate seed and soil contact. This for the Foundations for Farming is similar to zero-tillage. What is important for the Foundations for Farming is not just to plough but to do so on time: 'in season'. This resonates with Ecclesiastes 3:1, 2.

For Foundations for Farming, God instructed people to observe the signs of the seasons in nature. Mark 13:28 is very central to the Foundations for Farming projects from which they derived the *pfumvudza* concept. *Pfumvudza* is defined by Foundations for Farming as a deep Shona word which indicates the change of the season when the trees begin to bud with new leaves (*Pfumvudza*, p. 6). This is where the connection with Mark 13: 28 is made. It says, 'Now learn this lesson from the fig tree: As soon as its twigs get tender and its leaves come out, you know that summer is near' (New International Version). According to Foundations for Farming, the verse refers to being ready for the (farming) season. Realising that contexts are different. Foundations for Farming adds that 'this (verse) is relevant to our farming calendar (in Zimbabwe) and also our walk with Christ as we need to be ready for His return' (*Pfumvudza*, p. 6).

The implication is that Africans are poor and do not make profit from farming because they do not plan their farming on time. Resultantly, they are like the lazy ones who did not plough in season (Prov 20:4), did not follow the recommendations from Ecclesiastes 3:1, 2 and Mark 13:28, but still expect an abundant harvest (Proverbs 20:4).

Farming to a High Standard Ends Poverty

The second principle that Oldreive claims to have been given by God is farming up to high standard. According to Foundations for Farming, everything from preparation to ploughing in season must be done up to a high standard. Being on time but doing everything without standard results in poor yields and that would not take Africa out of poverty. Therefore, Foundations for Farming teaches all minute details regarding measurements, from in-row to inter-row spacing; depth of the hole to the direction of the soil from the hole; from where to place fertilizer to how to place it; from how many seeds per hole to how to close the hole; from when and how to weed to details about what instrument to weed; from when and how to harvest to how to take care of the Stover. The same details are given about every crop and on how to make a compost. No detail is taken as insignificant to warrant less attention.

Three Bible texts were singled out to anchor the discussion: Genesis 1:31, Ezekiel 40:3 and Colossians 3:23 (Climate Smart, p. 17). Genesis 1:31 says, 'God saw that all that he had made, and it was very good…'. The phrase 'very good' is what Foundations for Farming regards as proof that God did everything up to a high standard. As indicated already, Oldreive is a Christian white farmer whose interest is to preach through sharing farming methods with poor black peasant farmers. In Oldreive's view, God set the example that must be followed by his creation.

This dovetails with the development of the zero-tillage farming being promoted by the Foundations for Farming. After experiencing loss after loss at a farm where he was the manager, Oldreive engaged agricultural experts to inquire about the possibility of zero-tillage. But they told him that it would not work in Zimbabwe. This is despite the fact that Odrerive had researched about zero-tillage's historical origins in the United States of America around the 1930s where it had yielded positive results. Disappointed, but nevertheless determined, Oldreive claims that he turned to the word of God for the solution. In the first chapter of Romans, he was struck by how creation pointed to the truth, revealing both God's nature and His ways (Romans 1:20) (The Story, p. 9). It is claimed that immediately, Oldreive began to reflect on nature and realised that God takes care of nature with zero-tillage. God created everything with perfection, up to a high standard without deep inversions of the land.

Farming up to a high standard involves having to research, plan, measure, and implement with precision and according to recommendations by experts. This is what Foundations for Farming encourages. For them, Ezekiel 40:3 is all about high standards. The measuring rod implies the indispensability of measurements. In all the Foundations of Farming manuals, there are specific details about all the crops (Climate Smart, Farming and General, Nutrition and Health, Pfumvudza, Trainer's Manual); the quantity of everything required is calculated: amount of manure, water and seed among others. A measuring rod is always listed among tools that a farmer needs. Farming in this way, according to the Foundations for Farming, does not only glorify God, but it also guarantees good results.

In Colossians 3:23, there is nothing to do with farming *perse*. But for Oldreive, it specifically encourages farmers to work as if they were working for God: doing everything up to a high standard. Considering that farming involves hard labour, the purpose is mainly to motivate farmers. The text shares some commonality with 1 Corinthians 10:31, which motivated Oldreive to stop growing tobacco. The key phrases in both cases are 'whatever you do' and, 'for the (glory of the) Lord'. Working as if one is working for the Lord assures that one does all up to a high standard. According to Oldreive, when Africa does this, Africa will make profit and harness the ever-menacing poverty.

Farming Without Wastage

Although the principle is coined in some instances as farming without wastage in other manuals, it is farming with minimum wastage. This could have been a realisation that wastage/loss is ingrained in farming. There is always loss but also net gain as amplified in the parable of the sower (Matthew 13:1–9). Foundations for Farming therefore implores farmers to ensure minimum wastage of resources, space and time. For Oldreive, farmers must not waste the soil through bad farming practices, especially through deep inversions into the land, which accelerates soil erosion. Farmers, again, must ensure that there is mulch in order to save water. Nutrients are saved by not tilling and not burning stover. This is the principle on which Foundations for Farming was established. Oldreive claims to have received this principle from God after he encountered loss each season, for, 'costs continued to rise … (but) yields and profits continued to drop' (The Story, p. 9). Conservation of the soil and the environment became central to the operations and practice of Foundations for Farming since the day he received the secret from God. Indeed minimised loss or wastage results in profit.

In order to communicate the revelation of God about the importance of zero-tillage and mulch in farming, Oldreive appealed to the metaphor of characterising mulch as God's blanket. Oldreive claims that God revealed to him the secrets about this through Romans 1:20. Various experiments each year had the same conclusion that God's blanket protected the soil from erosion as it minimised runoff. But to motivate farmers to adopt the principle of minimum waste, John 18:9 and John 6:12 are utilised (Climate Smart, p. 17). There does not seem to be any special reason why this is the only principle anchored on verses from one biblical author. Other principles were supported by texts from not only different biblical authors but also different genres and different eras.

Ironically, both biblical texts cited from the gospel of John do not speak specifically about wastage in farming. John 18:9 is a statement by Jesus during his arrest. He is most likely speaking about his disciples that he lost none. Interestingly, Foundations for Farming established a principle of minimum wastage from this text. But John 6:12 has some connotations to the principle of minimum wastage. It is an instruction from Jesus to his disciples to gather all the leftovers after the feeding of the five thousand. The key phrase that must have attracted Foundations for Farming is, 'Let nothing be wasted' (New International Version). Yet, this text is not necessarily establishing a principle of avoiding wastage. New Testament scholars believe that it is a text promoting sharing even the little: little is more when shared with love. This is another piece of evidence that Oldreive is not a trained biblical scholar writing and communicating to equally untrained biblical readers, people that use the Bible as a manual.

Farming with Joy

This is the fourth and last principle that Oldreive claims to have received from God as a revelation to offset poverty in Africa. According to him, although farming involves hard work, it must be done with joy. All activities from preparation of the land, planting, weeding, caring and harvesting must be done ideally in groups, either with family, friends, community, church members or neighbours. Foundations for farming discourages individual work. This for him enables farmers to work with joy. To enable farming with joy, Oldreive offers some practical suggestions to always consider. Oldrive suggested the concept of 'eating an elephant' which is, 'one bite at a time'. Therefore, he encourages farmers to, 'mark a small portion at a time' ... devise ways of working with joy such as singing, or one can ,'turn on music' ... making sure that, 'everybody takes a break while they still have joy' (Climate Smart, pp. 20–21).

The principle of farming with joy is anchored on Nehemiah 8:10, Proverbs 17.22 and Philippians 4:4 (Climate Smart, p. 17). Nehemiah 8:10 is surprisingly talking about celebration and not working. The only reason why it has been brought up here is the presence of, 'joy'. This is the same reason may be why Philippians 4:4 is also cited here. It encourages believers in Christ to 'Rejoice in the Lord always'. But this is the joy of the Lord that comes with following Christ. It is encouragement to focus on Christ despite challenges and persecution. The joy of the Lord can be enjoyed by prisoners and slaves, kings and servants. This joy is not the same as happiness. For Foundations of Farming, cheerfulness is equal to joy in the Lord. This is why Proverbs 17:22 was also included to promote joy while working.

The four principles discussed so far are considered foundational for any farming undertaking. On top of them, Oldreive claims that God continued to reveal other things that are necessary to end poverty in Africa. For him, Africa can graduate out of poverty through the realisation that faith and faithfulness leads to fruitfulness and that giving (into the land) yields prosperity.

Faith and Faithfulness Lead to Fruitfulness

As already established, according to Foundations for Farming, Africa is poor because Africans does not follow God's principles clearly spelt in the Bible. Africa for Oldreive is like the servant in Matthew 25:14–30 who received one talent but did not invest it until the master returned. The master cursed him, took the talent and gave to the ones that had doubled the talents they received (Trainer's, p. 25). To put across this crucial message, Oldreive encourages instructors (and readers) of his manual to organise a skit comprising the master and the three servants. This for Oldreive is important especially in Africa where story telling is an essential medium of conveying crucial information and shaping worldviews. He is so meticulous as to provide an appendix (Appendix 1) (Trainer's, pp. 87–88), where he explained all the details. Oldreive provides all the minute details of the drama including the background of the cast, the words to be spoken by each actor, when and how. The tone of the voice and posture are also described.

Oldreive turned this first-century parable into a contemporary allegory describing the economic wellness of the world. One cannot avoid the parallelism implied in the parable as interpreted by him. Matthew represents the global North/First World, whereas Mark represents the Second World/Middle Income countries and Luke represents the Global South/Third World generally and Africa specifically. Africa is the unfaithful servant who did not invest the one talent that she was given by God. The implication is clear: First World and Second World countries are the faithful servants who invested their talents, but Third World countries are the unfaithful servant. The one talent that Africa was given will be allocated to the First World. As the First World is prospering, Africa is sinking into poverty each day. From this interpretation, Africa has been condemned by God because of her unfaithfulness.

According to Foundations for Farming, the talent that Africa has not invested to get profit is the soil, rainfall, sunlight, heat units, seed and manpower (Trainer's, p. 25). Yet for Foundations for Farming, God created the first man on the land, for man to utilise the land for sustenance. That was the first job that Adam got. Africans must therefore start from here before they are given responsibility over other bigger things, such as cities. To bolster this argument, Oldreive cites Luke 19:12–27, and not Matthew 25 this time. Luke talks about cities given as rewards for investing the talents. Matthew is silent about that. The lesson here for Oldreive is, once Africa is faithful with land, Africa will be able to develop and manage cities. Rhetorically he asks, 'Can we be honest and say that perhaps Africa has not developed as much as other continents because she has not faithfully attended to this first basic job that she has been given?' (Trainer's, p. 25).

In line with the argument that communities in the First World were successful because they were faithful with the few and basics until God added cities, Oldreive argues that God does not build from lintels. Any building is developed from the foundation. He argues, 'if you look back in history, at all successful economies of the world, you will find that they were faithful with the first things God gave them, and then God added to that good foundation and continued to do so as they were progressively faithful with the other aspects of development' (Trainer's, p. 26). A closer scrutiny of this perspective, however, shows that the opposite is true. The more the West exploited others (the Third World, Africa in particular), the more they prospered. And the more they turned their backs on faith, the more they built cities. If faith in Christ was the ingredient for prosperity, the Third World would have been the First World.

According to Foundations for Farming, however, Africa must realise that farming requires giving. Giving for Oldreive is the first step towards receiving (benefits/profit). He therefore discourages selfishness. This recurrent theme of Africa being poor because Africa is selfish is pertinent to Foundations for Farming. Its instruction to Africa is, give selflessly in order to receive. This principle has been well thought out. The biblical text chosen to motivate giving into the land is Luke 6:38. Unlike how it is ordinarily approached by evangelicals who emphasise the prosperity gospel's (Trainer's, p. 9;26) connotation of mysterious powers rewarding the giver, Oldreive interpreted giving as investment. This is done so well that it related to his interpretation of the parable of talents.

According to him, the servants got more talents through investment. The amount they invested was the amount they received in return as profit. In the same way, farming for him requires giving energy, thought, resources, effort, among others (Trainer's, p. 26). In other words, farming is giving. Harvest reflects how much was given. Those who gave little, harvest very little. Those who gave more, harvest more.

Another dimension of giving that Foundations for Farming emphasises is zero-tillage, not burning stover from the previous year and using mulch. According to Oldreive, all these are related with the principle of giving. In the principle of giving, he says, 'we realise that we cannot keep taking from the land without giving back to it. If we simply take from the land, never being willing to give an extra effort, or put money into supplying the nutrients which we have extracted back into the soil, then we cannot expect to receive. When we give extra effort ensuring that very high standards are maintained, then we are able to reap a bountiful crop. If we are lazy and unwilling to give this effort, standards will drop and so will yields' (Trainer's, p. 26). Indeed, receiving without giving back is a violation of the principle of nature. Nature eventually retaliates by withholding. Climate change is in a way a result of humanity taking from nature without giving back. For Oldreive, however, farmers must give back to the land because God set the example by giving Christ to humanity.

Critique of Foundations for Farming

Facts on the ground do not agree with the diagnosis given by Foundations for Farming as causes of poverty in Africa. The diagnosis of Foundations for Farming falls into already existing stereotypes about Africa: that Africans have a problem of not properly managing resources and being on time. Also that Africans are poor because they are ungodly, unfaithful, evil and lazy, as insinuated in the diagnosis of Foundations for Farming is quite contrary to facts on the ground.

The diagnosis by Foundations for Farming is contradictory to the mitigatory measures recommended. Conservation Agriculture through *pfumvudza* and other methods as advocated by the Foundation is a response to low yields due to climate change. The correlation between climate change and poverty cannot be denied anymore. Research has unequivocally shown that like other countries in the global south whose economies are weak, climate change is responsible for much of the poverty in Zimbabwe (Dube and Phiri 2013: 11–25; Brazier 2015; Nyahunda and Tirivangani 2019:

1–9). In poor countries, this is especially more visible in form of 'extreme temperatures, storms, droughts and floods' (Chanza and Gundu-Jakaza 2020). This has resulted in Global commitments to reduce carbon emissions by 45% by 2030 and to reach net zero by 2050 (Paris Agreement 2015).

Despite poor countries having an insignificant contribution to global warming, they are the worst affected. Africa, especially Sub-Saharan Africa, is the least contributor, but most affected. Zimbabwe, for instance, is one of the 20 most affected countries in the world between 2000 and 2019 (Newsham et al. 2021). In comparison to emissions by other countries in the developed world, Zimbabwe's peasants particularly pay the price for the damage done elsewhere with their very own livelihoods (Planet Aid 2021). Since, poor countries have little capacity to mitigate the effects of climate change, compared to developed nations, climate change directly contributes to poverty. This is particularly so when we appreciate that more than 62% of the Zimbabweans are peasants and small holder farmers whose livelihoods are directly dependent on the land (Borgen Magazine 2014).

The effects of climate change are dire for livestock and crops, thereby contributing directly to peasant poverty. Statistics show that between 2014 and 2019, some regions within Zimbabwe faced from three to six bad rainfall seasons. Experts' analyses also indicate that due to global warming, the rainy season across Zimbabwe has been shortened by 30 days and broken up by longer dry spells which last up to 20 consecutive days. Apart from droughts, frequencies of occurrences of natural disasters such as cyclones, storms and floods are on the rise. Zimbabwe was hit by Cyclone Eline (UN report 2000; Reason and Keibel 2002) and Cyclone IDAI (ReliefWeb Report 2021) leaving more than 100,000 in humanitarian disaster.

Regarding the methods of farming, especially *pfumvudza* that was presented by Foundations for Farming as revealed knowledge, a critical analysis shows that it is neither entirely new nor a revelation. For instance, the mission of the Interfaith Center for Sustainable Development in Jerusalem is similar to that of Foundations for Farming. Both organizations want to demonstrate that the Bible has directions to address poverty and combat climate change caused by environmental destruction by human activity. In Zimbabwe, Conservation Agriculture with specific focus on poverty alleviation and with the same modus operandi as Foundations for Farming was introduced by an American missionary Amory Alvord, from 1921 to 1960s. The same farming method has been in place for over two decades under the tutelage of a humanitarian NGO, CARE International—it's popularly known in rural communities in Masvingo as *dhiga udye* (dig and reap in abundance). The only difference is that Oldreive uses the Bible to teach about the same principles and methods. Could this be a mere coincidence?

Conclusion

The chapter has established that the *pfumvudza* model of climate smart farming was successfully implemented in Zimbabwe. The method is a poverty alleviation strategy with the capacity to satisfy the goals of the NDS1, Vision 2030 and SDG2. Its success is, however, dependent not on faith, spirituality or the Divine. Various coordinates need to be related, chief of which is government support through implements, fertilisers and seeds. The chapter has therefore dispelled the claims by Foundations for Farming relating to spiritual causes of poverty in Africa. Instead, the chapter premised climate change as the main culprit. Without a wholesale dismissal of the four 'biblical' principles of *pfumvudza* farming proposed, the chapter has, however, bemoaned the citing of the Bible out of context. Foundations for Farming serves as a classic example of an institution run by ordinary readers of the Bible, whose reading, although literalistic and out of context, offers practical solutions to the problems in society. This is an indictment on African biblical scholars whose knowledge is somewhat unable to offer practical solutions to societal problems.

References

Gunda, R. *The Bible and Homosexuality in Zimbabwe: a Socio-historical analysis of the political, cultural and Christian arguments in the homosexual public debate with special reference to the use of the Bible*. Bamberg: University of Bamberg Press, 2010.

Gunda, M.R. and Kuegler, J. (Eds), *The Bible and Politics in Africa*. Bamberg: University of Bamberg Press, 2012

Gunda, M.R., *At the Crossroads: A Call to Christians to Act in Faith for an Alternative Zimbabwe*. Bamberg: University of Bamberg Press, 2018.

Berman, S. K., Leshota, P.L., Dunbar, E.S., Dube, M.W., and Kgalemang, M. (Eds), *Mother Earth, Mother Africa and Biblical Studies: Interpretations in the Context of Climate Change*. Bamberg: University of Bamberg Press, 2021.

Taringa, N.T. *Towards an African-Christian Environmental Ethic*. Bamberg: University of Bamberg Press, 2014.

Hunter, J. & Kügler, J. (Eds.), *The Bible and Violence in Africa*. Bamberg: University of Bamberg Press, 2016.

Kügler, J., Gabaitse, R. and Stiebert, J. (Eds.), *The Bible and Gender Troubles in Africa*. Bamberg: University of Bamberg Press, 2019.

Muneja, S.M., *HIV/AIDS and the Bible in Tanzania: A Contextual Re-reading of 2 Samuel 13:1-14:33*. Bamberg: University of Bamberg Press, 2012.

Togarasei, J. & Kügler, J. (Eds.), *The Bible and Children in Africa*. Bamberg: University of Bamberg Press, 2014.

Wessels, J. M. *Offering the Gospel ADAPANON: An interpretation and application of 1 Corinthians 9:18*. Bamberg: University of Bamberg Press, 2015.

Kuegler and Gunda, From Text to Practice: *The Role of the Bible in Daily Living of African People Today*. Bamberg: University of Bamberg Press, 2009.

Mujere, Never, "Assessing the potential contribution of Pfumvudza towards climate smart agriculture in Zimbabwe: A review ", 2021.

Scoones, Ian (2021) Zimbabwe's Bumber Harvest: What Explains the Success? https://www.future-agricultures.org/blog/zimbabwes-bumper-harvest-what-explains-the-success/

Chanza, N. and Gundu-Jakaza, V. Deciphering the Climate Change Conundrum in Zimbabwe: An Exposition. In Tiefenbacher, J. P. (ed.), Global Warming and Climate Change. Texas State University, 2020.

Newsham, A, et al. (2021) 'Lessons from Zimbabwe's tobacco farmers for the COP26 climate change talks', The Conversation. https://theconversation.com/lessons-from-zimbabwes-tobacco-farmers-for-the-cop26-climate-change-talks-169868 (Accessed on 05 Nov 2021)

Focus on Climate Change and Poverty in Zimbabwe. 2021. *Planet Aid Blog*. https://www.planetaid.org/blog/curtailing-climate-change-and-poverty-in-zimbabwe (accessed 05 Nov 2021)

Climate Change and Poverty in Zimbabwe. *Borgen Magazine*, 28 March 2014. https://www.borgenmagazine.com/climate-change-poverty-zimbabwe/ (Accessed on 05 Nov 2021)

Reason, C.J.C. and Keibel, A. 'Tropical Cyclone Eline and Its Unusual Penetration and Impacts over the Southern African Mainland' *American Meteriological Services: Weather and Forecasting*. Vol 19. No. 5. 2002. pp. 789–805.

United Nations Country Team, Floods Disaster in Zimbabwe report, 2000. https://reliefweb.int/sites/reliefweb.int/files/resources/5DEECD113DEC5728C125689E004B7F2C-zimappeal.pdf (Accessed on 05 Nov 2021)

ReliefWeb report ,'Zimbabwe: Tropical Cyclone Idai Final Report, DREF Operationn°:MDRZW014'https://reliefweb.int/report/zimbabwe/zimbabwe-tropical-cyclone-idai-final-report-dref-operation-n-mdrzw014 (Accessed on 05 Nov 2021)

Dube, T and Phiri, K. 'Rural Livelihoods under Stress: The Impact of Climate Change on Livelihoods in South Western Zimbabwe', American International Journal of Contemporary Research. Vol. 3 No. 5 (2013), pp. 11–25.
Brazier, A. Climate Change in Zimbabwe: Facts for Planners and Decision Makers. Harare: Konrad Adenauer Stiftung, 2015.
Nyahunda, L. & Tirivangani, H.M., 'Challenges faced by rural people in mitigating the effects of climate change in the Mazungunye communal lands, Zimbabwe', *Jàmbá: Journal of Disaster Risk Studies* 11(1), 2019, a596, pp. 1–9.
Paris Agreement, United Nations, 2015. Rabbi Yonatan Neril and Rabbi Leo Dee (eds.), *Eco-Bible: A Jewish Ecological Commentary on the Hebrew Scriptures.* Vol 1 & 2. Jerusalem: Interfaith Center for Sustainable Development, 2021.
Foundations for Farming Manual, *Pfumvudza Alternative Crops*, sponsored by World Food Programme, Swiss Agency for Development and Cooperation and Liberte Elagilite Fraternite of France (2020).
Masenya, M. 'The Bible and Poverty in African Pentecostal Christianity: The BOSADI (Womanhood) Approach', in Peter Paris (Ed.), *Religion and Poverty: Pan-African Perspectives.* London: Duke University Press, 2009, pp. 152–165.
Zimbabwe National Development Strategy 1 (NDS1): January 2021–December 2025. (2021). https://www.dpcorp.co.zw/assets/national-development-strategy-1_2021%2D%2D-2025_goz.pdf
Zimbabwe's Vision 2030. Government of Zimbabwe, 2018.

CHAPTER 9

Crop Diseases and Food Insecurity in Africa: A Hindu Perspective

Elizabeth P. Motswapong, B. R. Gaie Joseph, and D. Saudu Siamisang

Introduction

Religion has generally been at the forefront of food security in everything it does. These efforts resonate with the Sustainable Development Goals' commitment to reach Zero Hunger by 2030. However, in Africa, climate change coupled with a host of crop diseases have thwarted efforts that religion has envisaged for its adherents. Hinduism has not been spared. Despite the challenges, Hinduism has continued to be an active player in the fight against food insecurity and crop disease in Africa. Furthermore, Hinduism has always been an environmentally sensitive philosophy. The religion's sacred texts, for example, the Mahabharata, The Ramayana, Vedas and others, contain the earliest messages for preservation of the environment and ecological balance. In addition, the teaching of

E. P. Motswapong (✉) • B. R. G. Joseph • D. S. Siamisang
Department of Theology and Religious Studies, University of Botswana, Gaborone, Botswana
e-mail: motswaponge@mopipi.ub.bw; GAIEJBR@ub.ac.bw

L. Maseno et al. (eds.), *Religion, Climate Change, and Food Security in Africa*, Sustainable Development Goals Series,
https://doi.org/10.1007/978-3-031-50392-4_9

non-violence or *ahimsa* has also catapulted the ethics of a symbiotic relation between human beings and the environment. The chapter will unpack the causes of food insecurity in relation to crop disease in Africa using the lenses of Hinduism. It will further delve into the extent to which Hinduism serves as a resource in addressing crop disease as one of the causes of food insecurity in Africa.

Sustainable Developmental Goals 1 and 2

Before we address food insecurity in Africa due to crop disease, it is imperative to briefly address SDGs 1 and 2, which talk of eradication of poverty and hunger by 2030. This is farfetched because by 2030, the problem of poverty and hunger will continue to ravage Africa, especially the sub-Saharan region. The region seems to be the hardest hit in terms of food production and distribution. Eradicating poverty in all its forms remains without doubt one of the greatest challenges facing humanity, simply because many are still struggling for the most basic needs. Women and children are the most affected by poverty. Furthermore, malnutrition and stunted growth in under-fives continue to be a challenge, not to mention nutritional needs for mothers and lactating ones. Therefore, there is a need to ensure sustainable food production systems implement resilient agricultural practices that increase productivity and production, that help to maintain ecosystems. The move will strengthen capacity for adaptation to climate change, extreme weather, drought, flooding and other disasters, thus progressively improving land and soil quality. These adopted measures will ensure the proper functioning of food commodity markets and their derivatives and facilitate timely access to market information. These will include among others access to food reserves, in order to help limit extreme food price volatility. Although these are very important, we argue that they are farfetched because of emerging infectious crop/plant diseases that frustrate any policies put in place to address food insecurity in Africa. As long as sub-Saharan Africa continues to be a hub for these crop diseases due to its moderate to hot temperatures, the region will continue to experience food insecurity causing unimaginable devastation.

Food Insecurity in Africa: An Overview

Food insecurity in Africa, especially in sub-Saharan Africa, has been a thorn in the flesh for a very long time, and measures in place to address the problem have proved futile because many Africans still experience malnutrition and are living under the poverty datum line. In 2022, over 20 million people and at least 10 million children faced severe food shortage in Africa due to crop failure and for four consecutive dry seasons (UNICEF 2022). East Africa alone lost close to 2 million livestock in a year due to recurrent drought and low response capacity (Bloomberg 2022). Covid-19 further escalated food insecurity and the ramifications were widely visible in sub-Saharan Africa (see Onyeaka et al. 2022). There are a number of pointers that have escalated the current predicament Africa finds herself in. One of the most significant challenges that aggravate food insecurity in Africa is crop diseases. The emergence of diseases in crops is often driven by innumerable factors, and understanding these factors can expedite preparedness, deterrence and repression of epidemics, hence eradicating food insecurity in Africa. Although scientific evidence is key to the development of management practices for Emerging Infectious Plant Diseases, EIPDs, the actual execution is strongly influenced by political, economic and societal pressures (Almeida 2018). Consequently, once outbreaks of Emerging Infectious Plant Diseases (EIPDs) start to unravel, it has often been too late to contain them. However, we believe there are important lessons to be learnt from past epidemics in order to prevent future disasters. That is why it is important to identify and prioritize EIPDs of food crops that merit immediate research, development and regulatory actions provide a synthesis of their current and future threats to food security; and identify the underlying factors favouring disease emergence in Africa (Almeida 2018).

African Perspective on Food Security and Crop Diseases

Before addressing food insecurity in religions, it is imperative that we address the African perspective on food security and crop diseases. The Food and Agriculture Organization has defined food security as "when all people, at all times, have physical, social and economic access to sufficient, safe and nutritious food that meets their dietary needs and food preferences for an active and healthy life" (FAO Policy Brief Food Security, Issue 2 2000). There are, however, many factors that render food security problematic in Africa. Mahamat (2020) outlines, among others, healthcare where there is a link between food insecurity, HIV and recently, Covid-19 pandemics. Furthermore, there are conflicts, policies, politics, leadership, strategic vision and the environment (Mother Nature). These challenges are not new to Africa because the continent has been struggling with them for a very long time now and have become perpetual. The challenges become even more hostile if the afflicted member of the family was a bread winner, especially when it reaches a state where her body becomes incapacitated, vulnerable and weak. This is when food security will then be experienced and poverty trap would have taken over, affecting everyone in the household. Therefore, achieving food security would automatically mean that there is enough food in the storage, and readily available, or having individuals who have adequate incomes or other resources or barter for food. Furthermore, food has to be properly processed and stored, as well as sound knowledge of nutrition and child care to enable them to access adequate health and sanitary services (The Four Dimensions of Food Security in ©FAO. EC-FAO Food Security Programme 2008).

There is undoubtedly a prevalence of malnutrition in the continent. A number of pointers to this challenge would, among others, include the never-ending political instability and crises, short or long protracted civil conflicts and wars, corruption, misdirected pecuniary policies and maladministration, but most importantly, the sheer neglect towards farmers and the lack of clear financial and economic investment into the agricultural sector (Mahamat 2020). Similarly, Africa's food insecurity has been exacerbated by other factors like low productivity, shortage of technology inputs, poor quality seeds and missing viable value chains, lack of access to micro financing, and large-scale farming by African farmers (Viswanathan & Mishra 2020).

Crop Diseases as Threat to Food Security in Africa

As much as there are a number of threats to food security in Africa, there is one in particular that has been overlooked despite it being catastrophic. It is that of crop disease. This has compelled scholars to engage in dialogues to establish which one is deadlier, human or crop epidemic. Hence, we have to bear in mind the fact that the global food system is highly interconnected (Puma 2019), and if a major epidemic occurs in one part of the world, its effects can rapidly spread throughout the food system (Godfray and Mason-D'Croz 2016). For example, Godfray and Mason-D'Croz (2016) used a partial equilibrium economic model to study a severe but short-lived epidemic that reduces rice yields by 80% in Asia. The results demonstrate some of the worst effects will be experienced by rice-importing countries in sub-Saharan Africa (SSA) due to increases in rice prices in the global market (Godfray and Mason-D'Croz 2016). Therefore, the emergence and re-emergence of plant diseases is increasingly threatening global food security in its entirety, as many researchers have maintained. These include Almeida (2018), Anderson et al. (2004), McDonald and Stukenbrock (2016), Savary et al. (2019), and Vurro et al. (2010). It is, therefore, imperative that emerging infectious plant diseases (EIPDs) should be defined as new diseases with increased virulence, geographical spread and/or host range, resulting in epidemics or pandemics. Once this model is adopted, then we can address and apply the SDGs 1 and 2 on poverty and hunger eradication because only then would these two goals be implemented and sustainable in the long run.

Notwithstanding, the emerging infectious plant diseases (EIPDs) are defined as new diseases with increased virulence, geographical spread and/or host range resulting in epidemics or pandemics (see Anderson et al. 2004; Bertin et al. 2020; Fletcher et al. 2010; Savary et al. 2017; Vurro et al. 2010). These pose a threat and a challenge at the same time to food security in Africa, hence the prevalence of poverty and hunger. When looking at EIPDs, these are mainly characterized by epidemics that seem to deepen over sequential emergent periods. Unlike endemic diseases, these EIPDs are caused by impositions of unidentified hostile pathogens and/or emergence of hyper-contagious pathotypes. Hypervirulent (also called supervirulent) pathotypes are unusually virulent strains of a pathogen capable of breaking down existing varietal resistance genes (Hébrard 2018). EIPDs add to the endemic disease and pest burden directly affecting all components of food security (see Godfray and Mason-D'Croz 2016; Savary and Willocquet 2020; Savary et al. 2017, 2019; Sileshi et al. 2019). Savary et al. (2017) identified six components of food security that can be impacted by plant diseases, namely production, stockpiles, physical access, economic access, stability and nutritive value. Emerging infectious plant diseases (EIPDs) thus constitutes an immediate threat to food security, from local to global scales, because of preharvest and postharvest losses, and also because the losses can affect both economic and physical access to food and the quality of food (Savary et al. 2017, 2019; Willocquet et al. 2020).

Religion and Food

Although the focus of the chapter is on Hinduism, it is helpful to have a more general understanding of "religion and food." The connections between religion and food vary extensively and are often quite complex and multifaceted. There are, however, several common connections between religion and food that scholars have begun to document in recent decades. In the functionalist view, for instance, religion provides meaning, identity, and structure within what (Geertz 1973, 87) has called "cultural systems." Therefore, religion reflects the human desire for order, but it provides order because people believe it has its origins in the divine. Food often figures prominently in functional interpretations of religion. Lévi-Strauss (1966) described food as a type of language that helps human beings express their basic perceptions of reality. He observed that rules about eating cooked and raw foods in some cultures are dictated by sacred stories (myths) and prohibitions (taboos). These rules reflect underlying notions about differences between nature and culture.

Mary Douglas (1972) has shown how food communicates ideas about holiness that provide identity and order. Functionalist deliberations are predominantly accommodating for exploring the connections between religion and food. They also illumine the influences between religion, food, and other culturally constructed systems such as gender norms. Furthermore, women may be religious experts through their control of food in societies where previous scholarship has focused on the male exclusivity of sacred knowledge. Among the Thai Buddhists that Van Esterik (2000) studied, it is the women whose feeding of monks and deities primarily determines attainment of merit and thus shapes the eternal destiny of their people both dead and living.

Subsequently, insecurities and religion are highly interconnected. There is a strong correlation between insecurity and strength of faith; often those less off rely more on faith. This dependency is bolstered by one study where religious coping was found to play a part in the trauma recovery from drought (Zeligman et al. 2020). However, food supply chains being disrupted causing resource stress have also been shown to be more positively correlated to labour and food sharing than belief in higher gods. It is clear that religion and food justice are intertwined. Hence food plays a large part on religious identity and contributes to the communal aspect of religion. The community draws religious people to serve others in pursuit of food justice, and those who face insecurities turn to religion in the face of food injustice (Boyd, Forsthoff, Hentz and Seranno 2023). Hence food shortage will be viewed as a very severe predicament to humankind. In the following section, the chapter focuses on food and its importance in Hinduism.

Food and Its Importance in Hinduism

The relevance of Hinduism to Africa has been overlooked over a long time; however, it is worth noting that the religion has firmly established itself in the continent. Despite its diversity and a large following, there is very limited literature on Hinduism in diaspora, especially in Africa. However, the sources have reached a consensus that the religion infiltrated Africa around the nineteenth century with the colonization of African countries by the British (Pratap 2000). Hinduism was introduced by Indians who migrated as casual labourers from British India, some of those who chose to remain in Africa. The religion has been embraced in several countries in Africa, but the majority of the followers are of Indian origin with very few African "converts." South Africa is one of the few nations in Africa with a large population of Hindus (Pratap 2000; Motswapong 2022). In West Africa, Hinduism is relatively a minority religion, but it is seen growing fast in Ghana due to a recent Indian migration to this part of Africa (Atiemo 2020). It is worth noting that when the Indians settled in Africa, they did not only bring their labour, but also brought their culture, especially their religion, dress and most importantly food.

Food in Hinduism is considered a very significant part of life. Actually, it is viewed as a God. It follows that food in Hinduism is treated with a lot of respect because of the role it plays, namely nourishment of the physical, mental and emotional portions of the body. Largely influenced by the doctrine of *ahimsa*, many Hindus are lacto-vegetarian, shunning all meat, fish and eggs, but those less strict will certainly avoid beef. During temple worship, foodstuffs are regularly offered to the Deity (sacred image), and such sanctified food is called *prasad* (Gita 3.13) "the mercy of the Lord." This *prasad* is distributed to visitors who receive a little respectfully in the palms of their right hand. Some Hindus will refuse to eat anything but *Prasad* and hence in their household all meals are first offered at the shrine. Likewise, in some African Traditional Religions, before eating, the adherent will throw a morsel first to the floor for the ancestors. In the Bhagavad Gita (17.8–10), a Hindu Classical Text, food is categorized according to the three *gunas* (material qualities), as follows: Foods in the mode of goodness (*sattva*) increase the duration of life, purify one's existence and give strength, health, happiness and satisfaction. Such foods are juicy, fatty, wholesome, and pleasing to the heart. Foods that are too bitter, too sour, hot, pungent, dry and burning are dear to those in the mode of passion *(rajas)*. Such foods cause distress, misery and disease. Foods prepared more than three hours before being eaten, food that is tasteless, decomposed and putrid, and food consisting of remnants and untouchable things is dear to those in the mode of darkness (*tamas*). It is commonly known that Hinduism does not allow eating of beef and pork; a cow is a sacred animal and is actually considered a mother. "Indians have reasoned with me that killing a sick cow is like killing your own mother and that is unthinkable." Similarly, there are many rituals which are associated with food eating in Hinduism. The first solid food that is given to a baby and when a person passes away, he/she will be offered food during the funeral rites to enable the departed soul to complete the journey to the ancestors. These are part of the *samsakaras* or rites of passages for any individual in Hinduism.

Thus the Hindu teaching that all beings that exist on earth are being born of food. Thereafter, they live by food, and ultimately, they go back to it and merge to become food. So hypothetically, food is the eldest of all creatures. On that basis, food is called the medicine (*aushdham ucchyate sarvam*) of all. Those who intercede upon Brahman as food will obtain all food. From food are born all being and after being born they grow by food; food is eaten by all beings and in the beings all beings are eaten by food. Therefore food is called *annam* (Traitreya Upanishad, Ch.2. Anuvagha 2. Shloka 1).

Hindu Sacred Texts and Annam (Food)

Crop diseases need to be targeted head on to guarantee continuous surplus of security in Africa. Even the Hindu sacred texts acknowledge the importance of food. Food is considered dear to those in the mode of goodness, because it increases the duration of life, purify one's existence and give strength, health, contentment and fulfilment. That is why it is imperative that food was eaten while being seated (*Rigveda Saṃhitā* 4.30.3) and was eaten only twice a day (*Taittirīya Brāhmaṇa* 1.4.9). During the meal, talking was kept to a minimum while eating (*Baudhayana Dharmasūtra* 2.7.2); this is consistent with most African traditions where the expectation is to keep the talking to a minimum during eating.

However, in times of emergencies, there were undeniably no restrictions on food (*Brahma Sūtra* 3.4.29-31). In addition, the people are asked to greet the food, honour it, rejoice upon seeing it, and pray that they may always obtain it *(Manusmṛti 2.54-55).*

Likewise, the Vedic texts have verses that have been translated by scholars to mean either acceptance or resistance to food dependent on meat. Early Vedic texts, such as the Rig Veda, denounce all killings of animals, cattle and horses, and appeal to the god Agni to punish those who attack. The Upanishads give a lot of importance to food. Without food, the world cannot exist. We can discuss *atma*, meaning, talk of high intellectual and spiritual matters, but at the most practical level, "*anna hi satya hai*"—food is truth. The Shatapatha Brahmana (3.1.2.2.21) denounces as a sin the eating of beef from cows and oxen. Accordingly, vegetarianism has been a well-accepted modern Hindu practice since ancient times.

Furthermore, in matters of Ayurvedic Food, the Brihadaranyaka Upanishads (Ch. 5) and Sutra texts (Maharatnakuta Ch. 5) evoke the notion of virtuous self-restraint, whereas the Sushruta Samhita (Ch 20) addresses when and where such foods are necessary. A few Hindu texts incorporate them, such as Hathayoga Pradipika (1.57–1.63), which recommends diet moderation *(mitahara)*, where one must "eat only when one feels hungry" and "neither overeat nor eat to completely fill the capacity of one's stomach; rather leave a quarter portion empty and fill three quarters with quality food and fresh water."

Furthermore, The Gītā is recited by all Hindus before partaking of food as an offering to God, a thank you. In all religions, food is treated as a gift from God or Mother Nature.

> *Brahmārpaṇam brahma haviḥ, brahmāgnau brahmaṇā hutam; Brahmaiva tena gantavyam, brahma karma samādhinā* (Gītā 4:24)
>
> "The offering process is Brahman, the offered oblation is Brahman, offered by Brahman, in the fire of Brahman; by that, Brahman alone is to be reached by one who is in the *Samadhi* of Brahma-action." This means everything is Brahman or God. God is all pervasive.

This is also stated in Chandogya Upanishad (3.14.1): *"Sarvaṁ khalu idaṁ brahma."* All this is verily Brahman. *Sāttvik*, "Pure food that promotes health, vitality, mental calmness, and appetite, which is wholesome and agreeable" (Gītā 17:8). It may include fresh fruits, vegetables, nuts, and legumes. But it does not include meat, fish or alcohol.

Rājasik, “Food that is bitter, sour, saline, excessively hot, pungent, dry, and burning; productive of pain, grief and disease” (Gītā 17:9). It may consist of meat, fish, egg, onion, garlic and hot chilies. It contributes to the intensification of hostility and desire.

Tāmsik, “Food that is stale, tasteless, stinking, cooked overnight, refuse, and impure” (Gītā 17:10). It may include preprocessed, musty food, meat, tobacco and alcohol that may lead to idleness, bad-mannered and temperamental conduct. Furthermore, Manusmriti discusses diet in Chap. 5, where he strongly discourages meat eating. In 5.48–5.52, Manusmiriti says:

> One can never obtain meat without causing injury to living beings … he should, therefore, abstain from meat. The man who authorizes, the man who butchers, the man who slaughters, the man who buys or sells, the man who cooks, the man who serves, and the man who eats—these are all killers.

In contrast, *Manusmriti* (5.33) states that a human may eat meat in a time of adversity, verse 5.27 recommends that eating meat is okay if not eating meat may place a person’s health and life at risk, while various verses such as 5.31 and 5.39 recommend that the meat be produced as a sacrifice *(jhatka method)*.

Ahimsa Versus Humans and Environment—Symbiotic Relation

The understanding of nature in Indian culture can be perceived in classical religious documents in ancient legal documents and in literary narratives. It can also be in the role of veneration for the various features of the landscape that is not always preserved in documentary form. In various parts of India today, groves are dedicated to a deity that is understood to reside in it (Rinehart 2004). Likewise, vegetarianism in India, like *ahimsa*, has as much, if not more, to do with concerns about reincarnation, one’s personal degree of purity, and place in society than with concern for animals. The Hindu and Jain set taboo against killing animals “has more to do with personal purity and caste than with the principles of *ahimsa* and *aghnya* (non-killing). Spiritual merit and advancement, dissociation from being involved in the slaughter of cattle and other animals for consumption leads to vegetarianism.”

Animal killing offsets the natural equilibrium and brings about more distress on earth and delays further evolution of the earth. Hindus prefer eating with family members and helping others not well off (Visram, J. 2011). *Ahimsa* is an important idea in several prominent religions originating from India including Hinduism, Jainism, and Buddhism. *Ahimsa* is also a term used to describe the traditional (non-violent) vegetarian diet of ancient India that was taught by Lord Buddha and many other religious teachers and scriptures. In the Bhagavad Gita (16.1-3), "*Ahimsa* is a quality belonging to godly men endowed with divine nature." More recently, Mahatma Gandhi was a great exponent of *ahimsa*, saying, "The way to truth lies through ahimsa" (Gandhi 1927). To treat the land with respect, to farm sustainably and organically; to produce foods that are beneficial for people and do not harm them; to be kind to animals and all living beings, to not eat them: this is *ahimsa*.

Likewise, one does not have to be a follower of any of these religions to believe in *ahimsa* or to practice it. Anyone who is a vegetarian is actually practicing *ahimsa*, because *ahimsa* means kindness and non-violence towards all living things. It has been shown that food is a very significant aspect of the Hindu religions. Food among the Hindus is treasured and taken with a lot of care. To the Hindus, food is considered a god because it gives life. It is emphasized in the texts to remind us about the meaning of existence. *Annam is* called Brahmin, the way *bhasha* is called Brahmin. God resides in food, the one who eats is god, what you are eating is also god. The concept that "god is life and life is food" is repeated continuously in *mantras* and other writings so that Hindus do not forget the significance of food. When taking food, a number of rituals have to be observed which ensure that positive eating habits are maintained. One's food should not be a source of pain to another living organism and that is why Hindus are largely vegetarians.

In addressing crop disease as one of the causes of food insecurity in Africa, how can Hinduism address the predicament? We wish to introduce Vrikshayurveda as an alternative to fighting crop disease and food insecurity in Africa.

Vrikshayurveda: The Science of Plant Life

As a way of curbing crop diseases, Hinduism from ancient times had a long-standing remedy called *vrikshayurveda*. The Hindu model can be used as a remedy that would guarantee crop diseases eradication resulting in food security not only in the African continent but globally. Before, outlining how this model of agriculture benefits African food production, we will briefly explain what it entails. *Vrikshayurveda* is the ancient science of plant life. Its name basically means "Ayurveda fortrees" (Devesh et al. 2015). The model deals with all species of trees and their healthy development and efficiency. It further, boosts the production of plants and allows the control of pests and diseases without using any chemicals. This is what Africa needs considering the costs involved in the purchase of chemicals that help fight crop diseases in Africa, a continent already struggling with capita. Vrikshayurveda is the original science of what is today called permaculture or agroforestry.

It will not be farfetched to maintain that the Hindu ancient sages had a deep understanding of the relationship of plants with other plants and animals, soil moisture and other agricultural phenomenon; hence, *Vrikshayurveda* becomes substantial in giving insights into what an ideal, all-encompassing agricultural and horticultural system would look like in that case scenario. Two main scriptures are specifically dedicated to *Vrikshayurveda: the first was* written by Salihotra (around 400 B.C). The writer goes into lengths in describing techniques which can assist plants and trees to bear fruits throughout the year irrespective of the season, climatic conditions and other underlying factors. Something we feel will be very beneficial to agriculture in Africa in the drive towards self-sustenance. The second *Vrikshayurveda* was written by Surapala (1000. A.D), where in detail it outlines countless skills about how to fertilize the soil or produce larger fruits. It is interesting to note that the scripture makes reference to the cultivation of about 170 plant species, it gives a detailed description and guidance on various issues, including water management, soil preservation, manures, the various diseases affecting the plants and their treatment.

In addition, the document deliberates on various topics connected with the science of plant life such as obtaining, conserving, and handling of seeds before planting, preparing plots for planting samplings, choice of soil, and technique of watering, nourishment and fertilizers. Most importantly, plant diseases and plant protection from internal and external diseases is outlined meticulously. It describes in detail the cultivation and preservation of the plant from its seed till its maturity. Due to the degradation of soils worldwide, especially in Africa, *Vrikshayurveda* can be the answer, hence addressing the problem of food insecurity in Africa which is now becoming rampant and uncontrollable. The necessity of a controlled, quality cultivation becomes significant in the sector of medicinal plants. To overcome this, the protocols and methods of *Vrikshayurveda* are promising.

Conclusion

In conclusion, *Vrikshayurveda* can be an alternative to Africa, especially when it comes to the problem of food insecurity. These texts should not be ignored because they provide other means to counteract deterioration of African agricultural practices. We propose that Ancient sciences should make a comeback as there is a huge demand of the conservation and ecological utilization of forest products as well as the medicinal plant sciences. Worldwide, we identify an increasing need for recognition of the need for incorporating the traditional systems to meet the limitations of both modern medicine and agriculture. Notwithstanding, the accessibility of quality medicinal herbs with the desired pharmacological and biological markers has become a challenge. However, it is worth noting that agroforestry can occur at a variety of spatial scales (e.g., field or woodlot, farm, watershed) in different ecosystems and cultures, thus becoming a welcome development in Africa. When properly applied, we are of the view that *Vrikshayurveda* can improve livelihoods through upgraded well-being and sustenance, enriched profitable improvement, and strengthened environmentally friendly flexibility as well as environs sustainability. In turn, such developments can underwrite to augment communal sustainability in which human essentials are satisfied in a way that fosters ecofriendly compliance. Furthermore, farm modification is a promising method for profit-making efficiency, especially all the way through the developed temperate zone that we come across in Africa, especially sub-Saharan Africa. Hence, *Vrikshayurveda* offers boundless potential for the ecological production of specially nut and fruit crops, high-value medicinal, dairy and beef cattle, sheep, goats, and biomass for biofuel. These are very important in the African context and can be the answer to the sustenance of the economy; hence, food insecurity can be a thing of the past. Therefore, we wish to maintain that the African continent is not a lost case because this model can benefit her in more ways than one has already reiterated in the discussion above.

References

Primary Sources

SS Sushruta Samhita
BD Baudhayana Dharmasūtra
BS Brahma Sutra
BU Brihadaranyaka Upanishads
CU Chandogya Upanishad
HP Hathoyoga Pradipika
M Maharatnakuta
MS Manusmriti
R.S Rigveda Saṃhitā
SB Shatapatha Brahmana
TB Taittirīya Brāhmaṇa
TU Traitreya Upanishad
V Vrikshayurveda

Almeida. 2018. *Emerging plant disease epidemics: biological research is key but not enough*. PLoS Biol 16(8): e2007020. https://doi.org/10.1371/journal.pbio.2007020.

Atiemo, A. O. 2020. "Religion's contribution to development: The case of African Hindus in Ghana." In E. Chitando, M.R. Gunda, and L. Togarasei (eds.), Religion and Development in Africa. *Bamberg: University of Bamberg Press*, 165–184.

Bertin, S., Mangalli, A., McLeish, M., & Tomasali, L., 2020. "Genetic variability of watermelon mosaic virus isolates infecting cucurbit crops in Italy." *Archives of Virology* 165, 937–946.

Boyd, Forsthoff, Hentz and Seranno. 2023. *Religion and Food Justice*. https://ecommons.udayton.edu/cgi/viewcontent.cgi?article=3268&context=stander_posters.

2000. *FAO Policy Brief Food Security, Issue 2*. Available from: www.fao.org/fileadmin/template/faoitaly.

Devesh, N.S., Mangal, A.K., 2015. "The Science of Plant (Vrkshayayurveda in archaic literature: An insight of Botanical, Agricultural and Horticultural aspects of Ancient India." *World Journal of Pharmacy and Pharmaceutical Science* 4(6): 388–404.

Douglas, M. 1972. *Deciphering a Meal*. Daedalus, 101(1), 68–81.

Fletcher J., Luster D., Bostock R., Burans, J., Cardwell, K., Gottwald, T., Mc Daniel l., Royer, M., Smith, k., 2010. "Emerging infectious plant diseases," *Trends in Ecology and Evolution* 19(10): 535–544. https://doi.org/10.1128/9781555816803.ch18.

Geertz, C. 1973. *The interpretation of Cultures: selected Essays*. New York Basic, INC, Publishers.

Hébrard, E.,. 2018. "Identification of a hypervirulent pathotype of rice yellow mottle virus: a threat to genetic resistance deployment in west-central Africa." (Phytopathology).

Lévi-Strauss, C. 1966. *The Culinary Triangle*. New Society, 22, 937–940.

Mahamat, K.B. 2020. "Understanding Africa's Food Security Challenges." In *Food Security in Africa*,)ed) Mahmoud B. https://doi.org/10.5772/intechopen.91773.

McDonald, Bruce A. and Stukenbrock, Eva H. 2016. Rapid emergence of pathogens in agro-ecosystems: global threats to agricultural sustainability and food security Phil. *Trans. R. Soc.* B3712016002620160026 http://doi.org/10.1098/rstb.2016.0026

Motswapong, E.P. 2022. "Hinduism and Climate Change in Africa": In *African Perspectives on Climate Change*. https://doi.org/10.4324/9781003147909.

Onyeaka, H., Tamasiga, P., Guta, N.H., Teshome, A., 2022. Food insecurity and outcomes during Covid -19 pandemic in sub Saharan Africa (SSA) Agriculture and Food Security. 11 Article 56.

Pratap, P.K. 2000. *Hindus in South Africa: Their Traditions and Beliefs*. Durban: The University of Durban Westville.

Puma, Micheal. 2019. "Resilience of the global food system," *Nature Sustainability, Nature*, vol. 2(4), pages 260–261, April.

Rinehart. 2004. *Contemporary Hinduism : Ritual, Culture and Practice*. Santa Barbara, Carlifonia: ABC-CLIO, Inc.

Savary, S, and Willocquet, L, 2020. "Modeling the impact of crop diseases on global food security," Annual review of phytopathology 58: 313–341.

2008. *The Four Dimensions of Food Security in ©FAO. EC-FAO Food Security Programme*. Available from: www.foodsec.org.

Van Esterik, P. 2000. *Materializing Thailand*. Oxford: Berg Books.

Viswanathan & Mishra. 2020. *"India-Africa Partnership for Food Security: Beyond Strategic Concerns"*. ORF Occasional Paper No. 242, April, Observer Research Foundation.

Vurro, M., Bonciani, B., Vannacci, G., 2010. "Emerging infectious diseases of crop plants in developing countries: impact on agriculture and socio-economic consequences" *Food Sec.* 2, 113–132, https://doi.org/10.1007/s12571-010-0062-7.

Zeligman, M., Majuta, A.R., Shannonhouse, L.R., 2020. "Posttraumatic growth in prolonged drought survivors in Botswana: The role of social support and religious coping." *Traumatology* 26(3). https://doi.org/10.1037/trm0000237.

Online References

Anderson PK. Cunningham A.A., Patel N.G., Morales, F.J., Epstein P.R., & Daszak, P., 2004. "Emerging infectious of plants: pathogen pollution, climate and agro technology drivers," *Trends Ecol Evol.* 19(10):535–44. At: https://

www.sciencedirect.com/science/article/abs/pii/S0169534704002186 Date Accessed 03.05.2023.

Bloomberg. 2022. "'On Brink of Catastrophe: Horn of Africa Drought kills over 1.5 million Livestock.". At: https://www.bloomberg.com/news/articles/2022-02-15/un-horn-of-africa-drought-kills-over-1-5-million-livestock. Date Accessed 29. 07.2023

Godfray, H.C.J & Mason-D'Croz, D., 2016. Food system consequences of a fungal disease epidemic in a major crop,. *Philos Trans R Soc Lond B Biol Sci.* 5; 371(1709):20150467. https://doi.org/10.1-98/rstb.2015.0467, Date Accessed 04.06.2023.

Savary, S. Bregaglio, S., Willocquet, L., & Gustafson, D., 2017. "Crop health and its global impacts on the components of food security.*Food* Security volume 9, pages 311–327 At: https://link.springer.com/article/10.1007/s12571-017-0659-1, Date Accessed 03.05.2023.

Savary, S., Willocquet, L., Pethybridge, S.J., 2019. "The global burden of pathogens and pests on major food crops," *Nature Ecology & Evolution*, volume 3, pages 430–439, At: https://www.nature.com/articles/s41559-018-0793-y#citeas Date Accessed, 01.06.2023.

Sileshi, G.W., Gebeyehu, S., Mafongoya, P.L., 2019. The threat of alien invasive insect and mite species to food security in Africa and the need for a continent-wide response: https://doi.org/10.10007/S41559-018-0793-Y. Date Accessed 01.06.2023.

The Four Dimensions of Food Security in ©FAO. EC-FAO Food Security Programme. At: https://fscluster.org/sites/default/files/documents/What%20is%20Food%20Security.pdf. Date Accessed: 12.05.2023.

UNICEF. 2022. "At least 10 million children face severe drought in the Horn of Africa." United Nations Children's Fund. At: https://www.unicef.org/press-releases/least-10-million-children-face-severe-drought-horn-africa-unicef. Date Accessed, 04.05. 2023.

Visram, J. 2011. *The concept of Hinduism annam, Food.* At: https://www.google.com/search?q=Visram%2C+J.+2011.+The+concept+of+Hinduism+annam%2C+Food&tbm=isch&ved=2ahUKEwizsrD21LuAAxW7nycCHfIAD94Q2-cCegQIABAA&oq=Visram%2C+J.+2011.+The+concept+of+Hinduism+annam%2C+Food&gs_lcp=CgNpbWcQAzIECCMQJ1CpHlipHmDYNmgAcAB4AIAB5wKIAY8FkgEFMi0xLjGYAQCgAQGqAQtnd3Mtd2l6LWltZ8ABAQ&sclient=img&ei=qRXJZPO5KLu_nsEP8oG88A0&bih=898&biw=803. Date Accessed, 03.07.2023.

CHAPTER 10

Crop Diseases, Pests and Food Insecurity in Africa: An Islamic Perspective

Newton Kahumbi Maina

Introduction

Climate change is a major global challenge, hence the need to use a multi-pronged approach to address this challenge. Studies on culture, values and worldviews on climate change shift the focus to religion. In this connection, there is need to understand the fundamental role of religion in climate change and food security. In interrogating the role of religion in climate change, this chapter has picked on Islam and the regions of Africa that are predominantly Muslim. Muslims constitutes approximately 50% of the population of Africa (*Muslim population in Africa* 2023) and form the second largest faith group in Africa after Christians. Muslims are concentrated in regions that are particularly affected by global warming and its effects such as rising heat temperatures and dwindling fresh water supplies. This chapter examines "crop diseases, pests and food insecurity in Africa: An Islamic perspective" by probing the role of Islam on food

N. K. Maina (✉)
Kenyatta University, Nairobi, Kenya
e-mail: kahumbi.newton@ku.ac.ke

L. Maseno et al. (eds.), *Religion, Climate Change, and Food Security in Africa*, Sustainable Development Goals Series,
https://doi.org/10.1007/978-3-031-50392-4_10

security and climate change in regions of Africa dominated by Islam. These regions include North Africa, the Horn of Africa and Sahel region. The chapter attempts to answer the following broad questions: first, how can Islam combat climate change and improve food security in line with the United Nations' Sustainable Development Goal (SDG) 13?; second, in what ways could Islam be an avenue to end hunger, achieve food security, improve nutrition and promote sustainable agriculture in Africa as per SDG 2? Through the use of secondary sources and desk review, the chapter hypothesizes that as an ideology and a way of life, Islam offers mechanisms that can mitigate effects of climate change, ensure economic equity and food security. The chapter draws heavily from the two main scriptural sources of Islam (Muslims)—Qur'an and Sunna (teachings of Prophet Muhammad)—and interpretations of scholars on the two main sources, in order to explore the interplay among Islam, food security and climate change in Africa. The chapter demonstrates how these two sources provide principles that could mitigate climate change and food insecurity. The chapter concludes that for Muslim regions of Africa to align and achieve SDG 13 and SDG 2, there is need to operationalize the principles of Islam in order to mitigate the effects of climate change and improve food security in Africa.

Crop Diseases, Pests and Food Insecurity in Africa

According to FAO (2021: 2), climate change is the increase in combined surface-air and sea-surface temperatures. Climate change may also refer "to any change in climate over time, whether due to variability or as a result of human activity" (Islam and Won 2017: 1). Climate change has resulted in increasing global-average mean annual air temperatures and also increased the amount of atmospheric greenhouse gases (GHGs) (Islam and Won 2017: 1). The increased concentration of GHGs in the atmosphere is expected to trap more heat on earth, which in turn results in an increase in global mean temperatures (Islam and Won 2017: 3). Africa is considered a minor player in total global greenhouse gas emissions which are known to contribute to global warming. Africa generates less than 4% of the world's total; however, the continent is most predisposed to the consequences of climate change (Pickson and Boateng 2021). Studies show that countries in Africa, especially those in Sub-Saharan Africa (SSA), will experience an increase in temperature of 2–4°C by 2100, which is about 1.5 times higher than the expected mean increase in global temperature (Pickson and Boateng 2021).

Climate change represents an unprecedented challenge to the world's biosphere and to the global community. Climate change causes adverse impacts on both ecosystems and human societies. Increasing temperatures and sea levels gives propensity to occurrence of floods; changing precipitation patterns cause droughts and other climate-related disasters that have the potential to affect human health, farmlands and livestock (Islam and Won 2017: 3). Climate change has contributed to pest infestation and altering the distribution of many pest species. The term pest refers to organisms, often insects, any species, strain or biotype of plant, animal or pathogenic agent that can cause damage to plants or plant products (FAO 2021: 2; Delaney 2021). Crops are at risk of being affected by diseases or pests throughout their lifecycle. Plant diseases and pests impact on availability and safety of plants for human and animal consumption, affect quality and reduce crop yield. The situation is worse in developing countries where in addition to post-harvest losses, mycotoxins in food threatens the health of humans and livestock as farmers are ill-prepared and ill-equipped to contain new pests. Indeed, crop pests, pathogens and diseases are causes of food insecurity, resulting to average yield losses of over 20% for many crops such as wheat, rice and maize (Delaney 2021).

Food security is defined as "a situation that exists when all people, at all times, have physical, social and economic access to sufficient, safe and nutritious food that meets their dietary needs and food preferences for an active and healthy life" (Islam and Won 2017: 2). Climate change affects food security through increasing temperatures and changing precipitation.

Food security is threatened by the emergence and spread of crop pests and pathogens. The UN Food and Agriculture Organization (FAO 2021: 28) estimates the crop production that is lost to pests globally at between 10% and 28%; this is a major contributor to food insecurity in the world. It is in this connection that the SDG 2 focuses on ending hunger through food security and enhanced nutrition, as well as sustainable agriculture. Recent estimates show that 704 million people (9.2%) of the world's population were subject to severe food insecurity in 2018 (Pickson and Boateng 2021).

Climate change has contributed to food insecurity in Africa, thus becoming a critical concern. According to Pickson and Boateng (2021), climate change and variability could adversely affect Africa's food security in two ways. First, climate extremes may influence the supply of food in Africa by reducing the area of viable arable land suitable for crop production, leading to a surge in the prevalence and number of undernourished people. Secondly, a rise in temperature by 4°C or more will have disastrous effects on farmers' livelihoods in Africa. Agriculture being the backbone of Africa's economy and accounting for the majority of livelihoods across the continent, climate variability and change in rainfall patterns make Africa vulnerable and more susceptible to food insecurity, because of heavy reliance on rain-fed agriculture.

Food insecurity has been on the increase in virtually every sub-region of Africa. Its prevalence rose from 210.7 million to 277 million between 2014 and 2018. This affected 21.5% of the population in 2018 (Pickson and Boateng 2021). In 2018, Southern Africa had the highest prevalence of severe food insecurity (30.6%), followed by Eastern Africa (25.9%), Western Africa (17.6%), and Northern Africa (8%) (Pickson and Boateng 2021).

Food insecurity is escalating across Sub-Saharan African (SSA) countries. In 2022, 123 million people or about 12% of population of SSA were projected to be acutely food insecure, i.e., they suffered from high malnutrition and were unable to meet minimum food consumption needs. Climate change is set to further intensify food insecurity and potentially jeopardize hard-earned development gains. Increased food insecurity could weigh on child nutrition and educational attainment, unraveling decades of improvements in SSA health and education outcomes (Baptista et al. 2022: 1).

The Horn of Africa is currently experiencing the longest and most severe drought, which is in its sixth consecutive year of below-average rainfall in the past 40 years. This has worsened the food crisis pushing 26 million people across Ethiopia, Kenya and Somalia into crisis or worse levels of food insecurity, severe hunger and water shortages (Horn of Africa Food Crisis Explained 2023). A majority of the population depend on pastoralism and livestock production as their main sources of livelihood. This means that climate change and severe droughts have impacted negatively on millions of people in the Horn of Africa. Widespread livestock deaths, severe shortages of food and water and increasing prices of goods are forcing millions of people to flee their homes, including vulnerable groups like refugees and internally displaced people. The long-term impacts of COVID-19, climate change, protracted conflict, the war in

Ukraine and reduced funding globally have all contributed to food insecurity in the Horn of Africa (Horn of Africa Food Crisis Explained 2023).

Religion and Climate Change

Religion is part and parcel of human life and shapes the views of believers regarding their perceptions of different aspects of life. Religion shapes the perspectives of their followers on how they approach nature. Hence religions have the capacities to address environmental effects of climate change, especially those that are human induced. Religious organizations, leaders and networks form an important category of shaping public opinion regarding myriad of decisions on climate change. These decisions include and are not limited to environmental activities, such as recycling or tree-planting projects, and environmental education to sensitize adherents on the importance of environmental conservation (Koehrsen 2021: 2).

The potential of religions in making the world more environmentally sustainable by mitigating climate change is recognized (Bergmann 2009; Gottlieb 2008; Holmes 2006). The 2015 "Interfaith Climate Statement" to the Paris meeting affirmed that all faiths recognize a moral obligation to avoid harm, respect fairness, and care for the vulnerable. That follows closely the 2015 exhortation to climate action from the Parliament of World Religions, "Embracing Our Common Future," which also mentions duties to future generations and respect for all life (Jenkins et al. 2018: 91). This calls for the need to understand the role of religion in climate change and its effects.

Notwithstanding the role of religion in climate change, much of the scholarship in the area of religion and climate change mitigation focuses on Christianity in the Global North (Koehrsen 2021: 2). There is an apparent dearth of information on the relationship between Islam and climate change. This is despite Islam being the world's second largest religion and with many Muslim regions being highly vulnerable to climate change. This chapter is a humble attempt to fill the lacuna by examining the interplay between Islam, climate change and food security.

Conceptual Perspectives on Islamic Principles on Environment and Climate Change

Islam is the second largest religion in the world after Christianity, and Muslims constitute the second largest religious group after Christians, thus forming an important segment of the world population (World Population Review 2023). Latest estimates show the population of Muslims to be about 2 billion with a projection of outnumbering Christians by the year 2050 given the high birth rates of Muslims (World Population Review 2023). The global statistics of the Muslim population is replicated in Africa where Islam is major religion in many African countries. Many countries in North Africa such as Algeria, Egypt, Libya, Tunisia and Morocco have more than 90% Muslim population. Sub-Saharan Africa countries such as Mali, Mauritania, Senegal, Nigeria, Chad, Gambia and Niger have substantial Muslim populations (Janson 2018: 951–52). In the Horn of Africa countries, except for Ethiopia, Sudan, Somalia and Djibouti have majority Muslim populations.

Many of the Muslim predominant regions of Africa are highly vulnerable to climate change. These regions include North Africa; parts of Sub-Saharan Africa, such as Northeast Nigeria and Mali; and the Horn of Africa. These areas experience arid or semi-arid climate which makes them vulnerable to effects of climate change such as rising heat stress and shrinking fresh water availability (Koehrsen 2021: 5).

In view of the role of religion in shaping believers' perspectives about nature, undoubtedly the teachings of Islam should influence Muslims in these countries regarding their understanding on matters of environment and climate change. This is in view of Islam as not only a religion but also an ideology, and a way of life. Fundamentally, the life of Muslims is governed by the Qur'an, the Holy Scripture revealed to Prophet Muhammad; then *Sunna* (*Hadith*) which are the practices, conduct and way of life of Prophet Muhammad. The other sources of Muslim conduct are *al-i'jma* (consensus of the learned Muslim jurists) and final source is *al-qiyas*—the analogical deductions. The Qur'an and Sunna and interpretations of scholars on the two main sources based on *al-i'jma* and *al-qiyas* provide a framework of Islamic principles that could mitigate effects of climate change, ensures economic equity and food security.

Within the framework of Islamic principles from the Qur'an, *Sunna*, *al-i'jma* and *al-Qiyas*, I will explore the interplay between Islam and climate change. The Qur'an which Muslims consider sublime is the first source of reference and source of Muslim conduct.

The term "climate change" is not explicit in the Qur'an, but the 759 verses of the Quran that reflect on the importance of ecology and nature (Abdallah 2010: 5) buttress the importance that Islam attaches to issues of environment. In a nutshell, the Qur'an and Sunna emphasize the importance of preserving the Earth and its resources and encourage sustainable practices, responsible stewardship of the environment, and awareness of the consequences of human actions on the natural world. These teachings provide principles which coupled with interpretations of Islam by scholars provide a basis of Islamic environmental ethics. The main principles could be grouped under *Tawhid, Mizan, Khalifa* and *Maslahah* (Hayat et al. 2023: 398–9; Koehrsen 2021: 3–4). These principles support the responsible and sustainable use of natural resources and the preservation of biodiversity and the environment.

Tawhid refers to the unity or oneness of God in Islam. It is the absoluteness, transcendence and nature of God which is expressed in the first words of the *Shahada* (confession of faith)—the first pillar of Islam—"there is no God but Allah," attesting there is no being worthy of worship except Allah (Qur'an 2:163; 21:22; 112:1-4). *Tawhid* is the bedrock of the Islamic faith and occupies the highest place in Islam. *Tawhid* gives answers to all metaphysical questions regarding the universe. It points to the supremacy of the law governing the cosmos, the all-pervading unity behind the manifest diversity (Mawdudi 1932: 43–50; Philips 2005: 16–25).

Tawhid shows that everything, including nature, comes from Allah as the source of life, and all are connected to the source (Abdallah 2010: 6; Hayat et al. 2023: 398). *Tawhid* is the foundation for Islamic environmental ethics and emphasizes that everything in creation is interdependent and interconnected and everything in the universe works together harmoniously (Hayat et al. 2023: 398). *Tawhid* emphasizes the interconnectivity of the cosmos, the intrinsic value of every part of creation, and the significance of maintaining equilibrium and harmony in everyday life. Muslim environmentalists interpret this principle to clarify the unity of all creation to which humans belong (Koehrsen 2021: 3; Mawdudi 1932: 43–50). As the "bedrock of the holistic order" approach in Islam, *Tawhid* affirms the interconnectedness of the natural order. The Qur'an (33:72) emphasizes the value of nature as a trust and sign from Allah while the hadith underscores the significance of conserving the environment through planting trees and taking care of them (Al-Bukhari, Sahih, Vol. 3, Book 39, Hadith 513).

Muslim environmentalists frequently connect the principle of *Tawhid* to that of *Mizan*, which means balance, and describes the universe as one harmonious entity (Abdallah 2010: 6; Koehrsen 2021: 3). Qur'an verses (54:59; 55:5-9;15: 19–20.) illustrate how Allah created everything with due connection and concern for balance (Abdallah 2010: 6). The harmony and balance of the natural world are considered to be fundamental aspects of the cosmos as a whole. The importance of the natural world in Islamic cosmology serves as the essential foundation for Islamic environmental ethics implying that the well-being of people and the environment is interdependent (Hayat et al. 2023: 398). Every detail of creation has been created to stand in perfect relationship to the other parts of creation. This principle indicates the interconnectedness of ecosystems. The idea that the planet was created as a place of harmony and balance for all living things underscores the importance of safeguarding and preserving the natural environment from the Islamic perspective (Hayat et al. 2023: 399). And when applied to the context of climate change, *Mizan* will explain the effects of global warming as greenhouse gases affect the balance of the creation that is interconnected by Allah through the natural order of creation (Koehrsen 2021: 4).

With regard to Islam and Ecology, quoting Muslim scholars, Koehrsen (2021: 4) avers that the principle of *Khalifa* is the single most important principle. *Khalifa* refers to the role of humans as vicegerents of God on earth. This principle is based on Qur'an (2:30–31) which shows how human beings are exalted to the noble and dignified position of being God's deputies on earth. Their responsibility is to fill the will of God on earth. Scholars and Muslim environmentalists interpret this role of *Khalifa* as being stewards of God's creation (Abdallah 2010: 5; Koehrsen 2021: 4). God afforded humans with trusteeship over creation. It is their responsibility to take care of nature. Therefore, caring for nature means serving God. Being stewards of God's creation, humans have to carefully manage natural resources and protect the environment and its resources for future generations. For example, Qur'an (7:31) warns Muslims against wastage while a hadith lays emphasis on the need to conserve water (Daily Hadith Online 2013). From this viewpoint, climate change appears to be a failure of the assigned trusteeship and an apparent consequence of humanity's impact on the climatic system (Koehrsen 2021: 7). Koehrsen provides a summary showing how Muslims as individuals and collectively should execute the responsibility of *Khalifa*:

> Protection, conservation and development of the environment and natural resources is a mandatory religious duty to which every Muslim should be committed. This commitment emanates from the individual's responsibility before God to protect himself and his community. It is also a common social

> duty which rulers, administrative and municipal (...) agencies and organizations undertake in accordance with the responsibilities assigned to them. The primary duty of the ruler and his assistants, whether they are administrative, municipal or judicial authorities, is to do their best to realize the interests of individuals for the betterment of life and society as a whole. This also includes protection, conservation and development of the environment and natural of resources. (Koehrsen 2021: 7)

The fourth principle is *Maslahah* which is used in Islamic law and translates to "public interest," "public good" or general welfare of the Muslim community. This principle is an illustration of the application of Islamic jurisprudence (*Fiqh*) to environmental ethics (Hayat et al. 2023: 399). *Maslahah* pursues a "sustainable achievement of good, welfare, advantages, and benefits of creatures." *Maslahah* prioritizes the public welfare over individual and private interests (Koehrsen 2021: 4). *Maslahah* can be regarded as a means of incorporating the public interest into considerations of Islamic law and is therefore an important interpretive tool in analyzing potential issues of climate justice (Koehrsen 2021: 4). Through the application of *Fiqh*, environmental issues such as pollution, deforestation, and climate change can be addressed (Hayat et al. 2023: 399). In regard to the threat of climate change, Muslim environmentalists could apply the principle of *Maslahah* to think, argue and act globally with the obligation to preserve resources and care for future generations (Koehrsen 2021: 4).

The common position of Muslim countries on climate change reflect the principles of Islam and their interpretations by Muslim scholars. Due to the vulnerabilities of Muslim regions to climate change and its effects on food security, the 57-member Organization of Islamic Cooperation (OIC) drafted "The Islamic Declaration on Climate Change" (2015). The declaration specifically calls on oil-producing states to phase out greenhouse gas emissions and calls on all nations to commit to a zero emissions strategy (Jenkins et al. 2018: 91). The Declaration drew from the principles of *Khalifa* and *Mizan* combining them with scientific insights on global warming to create greater awareness for climate change, calling on decision-makers in the Muslim world and beyond to take sound actions (Koehrsen 2021: 4). The declaration holds that although humans are *Khalifa* of God on earth, their caretaking role has been compromised and their corrupt performance has instead wrought damage and destabilization in the management of nature and natural resources.

The Islamic Declaration on Climate Change (2015) is an appropriate illustration of how Islamic environmental principles could provide interpretations that apply to climate change. Scholars have argued that the message of the Islamic Declaration could have an impact on Muslim mitigation efforts "if the imams in mosques are committed to reminding the rapidly increasing number of Muslims of their moral responsibilities as taught in the Qur'an and by their revered prophet" (Koehrsen 2021: 8).

Islamic Teachings on Food Security

This chapter has alluded to food insecurity in many regions of Africa which includes areas that are predominantly Muslim. The COVID 19 pandemic and Russian-Ukraine war escalated the state of food insecurity in Africa. This section of the chapter examines how Islamic teachings can mitigate food insecurity. Muslims should seek food security as part of faith and worship. Several verses of the Qur'an and hadith support this argument. Qur'an (2:57, 172) commands Muslims to select and eat appropriate good food, while 17:27 and 55:8 warns of prodigality, wasting food and excessive eating. In addition, all works related to agriculture and food production are considered a continuous charity in Islam. This means, Muslim nations should ensure that their citizens engage in agriculture to produce sufficient food supply to mitigate food insecurity (Haddad 2020).

Islam lays strategies—through the teachings of Quran and Hadith—of ameliorating poverty and ensuring equitable and fair distribution of wealth. Islam underscore enhancement of all socio-economic welfare and food security measures through hard work, self-reliance, establishment of *Waqf* (endowment) and *Zakat* institutions which cater for the welfare of Muslims (Kihara 2013: 5). Islam gives guidelines on the various ways that can help to mitigate the challenges of hunger and food insecurity in general. These include following Islamic teachings and taking cognizance that all wealth belongs to Allah and human beings are just trustees (Kihara 2013: 43). According to the Quran (2:284), Allah has absolute ownership of wealth and people are *Khalifa* who should hold the wealth in trust for Allah.

Islam offers guidelines in alleviating poverty through dispensation of *Zakat* (obligatory alms) and *Sadaqa* (voluntary alms). This calls for moderation on spending and using the wealth and assisting others through

Zakat in order to earn the blessings of Allah (Quran, 57: 7). Poverty has a bearing on food security since the poor may not have the wherewithal to purchase food. One of the roles of *Zakat* in Muslim society is to mitigate to a minimum the sufferings of the needy and the poor members of the society.

Zakat establishes a self-supporting society through fulfillment of the basic needs of the poor by affording them an opportunity to lead a respectable life, one of which is being food secure. Some Muslims states, for example, Libya, Pakistan, Sudan, Saudi Arabia, Kuwait, Jordan, Bahrain and Malaysia, have institutionalized *Zakat* as a source of revenue for social welfare sectors like education and health to meet financial obligations of the poor and needy citizens. Islam underscores proper and timely storage of food in times of abundant harvest (Qur'an 12:54–56). The Hadith reports how the Prophet used to store food for his family that could cater for their needs for a whole year or until another harvest season (Kihara 2013: 59; Al-Bukhari, vol. 7: 270). Muslims could use this age-old practices to mitigate food insecurity during drought and other periods of poor harvest occasioned by effects of climate change such as irregular weather patterns and rainfall.

Conclusion

Islamic environmental principles of *Tawhid, Mizan, Khalifa, and Maslahah* emphasize the harmony between humans and nature. These principles offer important perspectives on the place of religion in climate change and food security. These principles have the potential to contribute to global efforts of promoting environmental sustainability, mitigate climate change and promote food security. Islamic principles thus could supplement the global efforts in attaining the SDG 13 and SDG 2. For Muslim regions of Africa to align and achieve SDG 13 and SDG 2, there is need to operationalize the principles of Islam. This could help to mitigate the effects of climate change and improve food security in Africa. However, inasmuch as Islam through these principles have the potential in mitigating the effects of climate change and food insecurity, the principles cannot operate in isolation. They should be integrated with "knowledge and scientific insights on global warming and climate change" (Koehrsen 2021: 4).

References

Abdallah, D. 2010. Religion′s role in development, ecology & climate Change: An Islamic perspective. Paper presented at the Conference on the role of Religion in Development-focus on Ecology and Climate Change. 2010, April 22; Finnish Evangelical Lutheran Mission. Helsinki: Finland. Available from http://hdl.handle.net/10138/24691.

Al-Bukhari, M.I. Sahih Bukhari. The book of sharecropping, the chapter on the virtue of planting and planting if one eats from it. USC-MSA web (English) reference. Vol. 3, Book 39.

Baptista, D.M.S., Farid, M., Fayad, D., Kemoe, L., Lanci, L.S., Mitra, P., et al. [Muehlschlegel, T.S., Okou, C., Spray, J.A., Tuitoek, K. & Unsal, F.D] 2022. Climate Change and Chronic Food Insecurity in Sub-Saharan Africa. Available from International Monetary Fund: https://www.elibrary.imf.org › journals › article-A001-en. 20 July 2023.

Bergmann, S. 2009. Climate change changes religion. *Studia Theologica*—Nordic Journal of Theology, 63(2), 98–118. Available from: https://doi.org/10.1080/00393380903345057.

Daily Hadith Online, 19th August 2013. The Teachings of Prophet Muhammad. Hadith on Water: Conserve water even by a running stream.

Delaney, O. 2021. Food security: Pests and diseases Report - EA Forum. Available from: https://forum.effectivealtruism.org›posts›food-security.

FAO [Food and Agriculture Organization of the United Nations]. 2021. Scientific review of the impact of climate change on plant pests: A global challenge to prevent and mitigate plant pest risks in agriculture, forestry and ecosystems. FAO: Rome.

Gottlieb, R. 2008. You gonna be here long? Religion and sustainability. Worldviews: Global Religions, Culture, and Ecology, 12(2),163–178. Available from: https://doi.org/10.1163/156853508X359967.

Haddad, M. 2020. Food production and food security management in Muslim civilization. Muslim Heritage. Available from: https://muslimheritage.com ›...›Sustainability.

Hayat, Malik, Ali, Husnain, Sharif, Haleem 2023. The Role of Islamic Environmental Ethics In the alleviation of Climate Challenges and the Preservation of Ecosystem. *Russian Law Journal* Vol. 11 No. 11s.

Holmes, R. 2006. Caring for nature: What science and economics can't teach us but religion can. Environmental Values, 15(3), 307–313.

Horn of Africa Food Crisis Explained (April 06, 2023) [Internet]; available from: USA for UNHCR. The Un Refugee Agency Horn of Africa Food Crisis Explained - USA for UNH Available from https://www.unrefugees.org › news › horn-of-africa-foo...

Islam, M. S & Won, A.T. 2017. Climate change and food in/security: A critical nexus. MDPI. Available from: https://www.mdpi.com.

Islamic Declaration on Climate Change. 2015. United Nations Framework Convention on Climate Change. Available from: https://unfccc.int › news › islamic-declaration-on-climate change.

Janson, M. 2018. Islam in Sub-Saharan Africa. In Shanguhyia, Martin S. and, Toyin, F. (eds.) Palgrave Handbook of African Colonial and Postcolonial History. New York: Palgrave Macmillan, pp. 951–977.

Jenkins, W. Berry E. & Kreider, L.K. 2018. Religion and climate change. Annual Review of Environment and Resources; pp. 85–108, environ.annualreviews.org. Available from: https://doi.org/10.1146/annurev-environ-102017-025855.

Kihara, R. Challenges and prospects of Muslim organizations and institutions in mitigating food insecurity in Isiolo County, Kenya [MA dissertation]. Kenyatta University; 2013.

Koehrsen, J. 2021. Muslims and climate change: How Islam, Muslim organizations, and religious leaders influence climate change perceptions and mitigation activities. Wiley Interdisciplinary Reviews. Available from https://wires.onlinelibrary.wiley.com›doi›full›wcc.

Muslim population in Africa, by country. 2023. [Internet). [updated 28 Apr 2023; cited 2023, July 19]. Available from: https://www.statista.com › Society › Religion Muslim population in Africa 2023.

Mawdudi, S.A. 1932. Towards understanding Islam. UKIM Dawal Centre.

Philips, A.A.B. 2005. The fundamentals of tawheed. International Islamic publishing house.

Pickson, R.B. & Boateng, E. 2021. Climate change: a friend or foe to food security in Africa? https://doi.org/10.1007/s10668-021-01621-8. National Library of Medicine.

Qur'an, The Holy. 1989. Text, translation & commentary by Abdullah Yusuf Ali. Maryland (USA): Amana Corporation.

World Population Review 2023. Muslim Population by Country 2023. Retrieved from https://worldpopulationreview.com › country-rankings.

PART III

Religion, Climate Change, Food Security and Human Flourishing

CHAPTER 11

Indigenous Knowledge, Food (in) Security and Climate Change: A Case Study of *Bota Reshupa* (Herbal Porridge) Among the Ndau of South-Eastern Zimbabwe

Tenson Muyambo

Introduction

Africa is the largest continent with over 20 per cent of its population facing hunger (Delgado et al. 2023). It is also the hardest hit continent by climate change; it carries the heaviest burden from the impact of climate change, yet it is not responsible for causing climate change. According to Food Agriculture Organisation of the United Nations (FAO) in 2021, a total of 20.2 per cent of the African population was facing hunger, compared to 9.1 per cent in Asia, 8.6 per cent in Latin America and the Caribbean, 5.8 per cent in Oceania and less than 2.5 per cent in North America and Europe (FAO et al. 2022). Current projections indicate that

T. Muyambo (✉)
Great Zimbabwe University & Research Institute for Theology and Religion, College of Human Sciences, University of South Africa, Masvingo, Zimbabwe

L. Maseno et al. (eds.), *Religion, Climate Change, and Food Security in Africa*, Sustainable Development Goals Series,
https://doi.org/10.1007/978-3-031-50392-4_11

the situation is going to be worse in the coming years. Among the key drivers of this increasing levels of food insecurity is climate change coupled with violent conflicts, the COVID-19 pandemic and the cost-of-living crisis (Delgado et al. 2023). The argument is that Africa is host to a large proportion of the world's armed conflicts, in Mozambique, The Democratic Republic of Congo, Cameroon and recently, Sudan. It is also exceptionally exposed to climate change as compared to other regions. In addition, the World Bank (n.d) indicates that since 2020, the COVID-19 pandemic has pushed around 40 million people in Africa into extreme poverty. The pandemic in cohort with food and commodity price inflation triggered by Russia's invasion of Ukraine in February 2022 has generated the worst cost-of-living in a generation, hitting people living in or near poverty the hardest (Delgado et al. 2023).

At a World Food Summit, Rome, in November 1996, the Summit's Plan of Action stated that food security exists when an entire population at all times has physical and economic access to sufficient, safe and nutritious food that meets its dietary needs and food preferences for an active and healthy life (World Food Summit, Rome 1996). Food security is a critical component of Sustainable Development Goals, and most governments are bent on ensuring that people have sufficient, safe and nutritious food. The Zimbabwean's president, Emmerson Mnangagwa, like his predecessor, is on record saying that no one will starve in Zimbabwe. Nonetheless, the reality is that food insecurity is a major concern in most of the African countries. World Food Programme has estimated that in Zimbabwe more than half the population need food aid. Arguments have been that sub-Saharan Africa, to which Zimbabwe is a part, had been food secure prior colonialism (Bjornlund et al. 2020a, 2020b). Historical events such as the slave trade are testimonies that Africans were well-nourished, strong and of regular stature (Diffie et al. 1977; Hawthorne 2003). Whenever scarcity occurred, it was due to drought, flooding, pests or conflicts (Bjornlund et al. 2022). Food security was enabled by endogenous agriculture (Amin 1973). It is against this backdrop that this chapter assesses how indigenous knowledge systems derived from African religio-cultural heritage on food can be a source of food security and human flourishing on a continent that is food insecure owing to several drivers of food insecurity. This chapter (as a part of my doctoral thesis) does this by assessing how *bota reshupa* among the Ndau people of south-eastern Zimbabwe could be a source of food security in the midst of climate change. The following section presents the theoretical framework, the prism through which issues raised here are gleaned, followed by the method, and presentation of findings and discussion. The chapter ends with a conclusion and some recommendations for policy formulation and implementation.

Theoretical Framework

This study was informed by the postcolonial theory. The focus of postcolonial studies is to examine the long-standing impact of cultures in contact with each other and with a particular emphasis on Western colonial interactions with non-western countries (Said 1978). Quayson (2002: 2) conceptualises postcolonial theory as a theory that "involves a studied engagement with the experience of colonialism and its past and present effects, both at the local level of ex-colonial societies as well as at the level of more general global developments thought to be the after effects of empire." It is a theory that offers a critical perspective to the study of social phenomena, particularly issues surrounding power relations between various groups in society (Velautham 2015). For Quayson (2002), the effects of colonial power did not end with independence. Instead, as Velautham (2015) argues, current social practices which have been established as a result of the colonial encounter such as inherited governments, law, media, journalism, business and a range of other interactions are fraught with issues that stem from unequal power relations. The writings by Said (1978, 1985, 1993), Spivak (1988, 1991, 1993) and Bhabha (1984, 1990, 1992, 1994) as the leading postcolonial theorists provided the lenses through which the study analysed and interpreted *bota reshupa*. Given this, data collected was gleaned through the theory with an intent to assess how *bota reshupa* as Ndau herbal porridge can be part of a raft of coping strategies being employed to deal with climate change–induced food insecurity. How can this indigenous knowledge be tapped into for addressing food insecurity? This calls for reverting back to indigenous knowledge which was and still continue to be annihilated and vilified during colonialism and neocolonialism as no knowledge or quasi knowledge in the very least. This knowledge has been pushed to the margins as 'complementary or alternative' knowledge (Kilonzo 2022; Siwila 2022), descriptors meant to '*epistemicide*' (killing and displacing existing knowledges), '*linguicide*' (killing and displacing the languages of a people and imposing one's) and '*culturecide*' (killing and displacing the culture of a people) (Ndlovu-Gatsheni 2019) the African ways of knowing. More important to note is that if

Africa removes colonialism physically without removing it epistemically, it will not disappear (Ndlovu-Gatsheni at the sidelines of an African Studies Association (ASA) annual meeting held in Boston (November 21–23, 2019)). Using the postcolonial lens that involves decoloniality, I am making the thesis that indigenous knowledge needs to be valorised if Africa needs to squarely deal with climate change–induced food insecurity.

Methodology

This chapter is part of a qualitative research design that was carried out among the Ndau people of south-eastern Zimbabwe for my doctoral studies from 2014 to 2018 with the University of KwaZulu Natal. I adopted a poly-methodic paradigm, comprising the historical, sociological and the phenomenological approaches. Although the qualitative research design, where interviews are a predominant tool for gathering data, was adopted for this thesis, I was alive to its limitation that as a researcher I had no control of the outcome of the conversations with interviewees. It was the interviewees who determined the information communicated which guided my thoughts. The poly-methodic stance taken was informed by the observation that there is no one method or discipline that can lead to an exhaustive and all-encompassing understanding of man's religious aspects (Mapuranga 2010). Since the study was about the Ndau people and their indigenous practice of *bota reshupa*, a historical approach to the study could not be avoided. At the same time, the Ndau are a social ethnic group, and an attempt to study them could not avoid the sociological approach. In addition, the Ndau people are believers of their own practices; hence, the phenomenological approach as a method that represents a people's beliefs, practices and values was indispensable. Hence, a combination of the historical, sociological and phenomenological approaches.

The rationale for adopting the historical approach was premised on the understanding that every aspect of human society has a past and is worthy to be studied (Mapuranga 2010). The historical approach is "an examination of religious people, their ideas, beliefs and practices within concrete historical epochs" (Streng 1985: 220). Nevertheless, the historical approach could not work in isolation from the sociological one given that the Ndau people are a social ethnic group. The sociological approach emanates from the sociology of religion. Sibanda (2014) understands the sociology of religion as the study of beliefs, practices and organisational forms of religion by employing the tools and methods of the discipline of sociology. At the centre of the sociology of religion is society. Smart (1996) defines the sociology of religion as the study of human relationships and interaction. This points to the inseparability of religion and society. The sociological approach emphasises the extent to which society influences individual and group status and behaviour in a particular context. To ensure that methodological glaring loopholes were minimised, that is the shortcomings of the historical and sociological approaches, the descriptive phenomenological approach was also employed where the emphasis was on *epoche*, eidetic intuition and empathetic interpolation. The argument is simply that since the Ndau people constitutes a single religious grouping in south-eastern Zimbabwe, studying them would require 'bracketing out' pre-conceived notions about them, and allow them to speak for themselves as well as attaining an intuitive insight into their essential forms, structures and patterns of religion (Sibanda 2014). This enabled the carrying out of the study with the objective to understand the meaning the Ndau people attach to the practice of *bota reshupa*. The phenomenological principle of empathy interpolation was also practised during the study. This is understood as entering into the experiences of the adherents under investigation. Cox (1992) further argues that empathy involves the cultivation of a 'feeling for,' whereby the researcher identifies with the attitudes, thoughts and activities of the adherent without necessarily being converted and looking into and through the mind of the believer. For Chitando (2010), this is cultivating empathy. Chitando summarises cultivating empathy thus:

> The major argument for promoting an empathetic understanding of religion is that the focus of religious studies is fellow human beings. Unlike other disciplines that tend to concentrate on inanimate objects, the study of religion dwells on human beings, who have emotions and feelings of their own. The duty of the scholar of religion is to respect these feelings and emotions. (2010: 269)

This principle encourages researchers to 'feel for and feel with' followers of the religion under study. The phenomenology of religion recommends 'methodological conversion.' This 'conversion' is a matter of methodology for the access to data. It was temporary and once the research was done, I distanced myself from the immersion. The 'immersion' was only intended to overcome the distance between me and the focus of the study. Though being embedded in empathy is the danger of being immersed totally, practising this principle of empathetic interpolation enabled me to approach the Ndau with their practice of *bota reshupa* with compassion. I predominately used the face-to-face in-depth interviews with key informant interviewees as well as focus group discussions with men and women who had deep knowledge of *bota reshupa*. A total of 34 participants took part in the study. Purposive and snowballing were key in identifying the participants. All research ethical issues were considered during the study and the consent of the participants was sought before discussions took place.

Findings

It must be stated on the outset that this chapter, being a part of a bigger research work, focuses only on what came out of the study regarding *bota reshupa* which can be tapped into for food security purposes in a society not only ravaged by climate change but food insecurity, thereby threatening, in a big way, the achievement of sustainable development goals 2 and 3.

The Nature of Bota reshupa

What constitutes *bota reshupa*, according to respondents MS 5, MS10, MS23, MS32, MS34, is a porridge that is prepared using a mixture of herbs, rapoko, millet flour or maize meal, depending on the region and water. The mixture is stored in a calabash. The calabash, according to the respondents, can be made of either clay, *chidhuwa* (clay pot), or *danga* (gourd). The respondents indicated that when the calabash has been prepared and thoroughly cleaned, traditional herbs are cut into small pieces and placed in it. Water and some millet flour or maize meal is also added. The mixture is left overnight to brew. When it has fermented the watery mixture, which is brownish in colour depending on the ingredients mixed, it is drawn into a pot and put on the fire to prepare the porridge. This porridge is referred to as *bota reshupa*. Muyambo (2015: 65) concurs by stating that *shupa* is "porridge prepared using a mixture of traditional herbs soaked in water." The use of the herb *mungurahwe*, a sweety plant which flavours the porridge, serves as the modern sugar. One respondent MS4 from Musani village put it thus:

> *Kare shupa yaibarwa ngemutanga. Yapera kunasirwa, yaiiswa mvura popenywa mitombo yakasiyana-siyana yoiswa mushupa muya. Mitombo iya yaisan-*

> *ganiswa nemvura yemushupa pamwe neuswa hwemungoza kana kuti mafunde kana uswa hwemabonore. Zvaisiiwa kwenguwa yakati kuti zvimboita nguwa yekusangana imwomwo. Kana zvaakuti zhazhirei, paimiminwa mvura iya kubva mudumbu yobikiswa bota. Iri ndiro bota reshupa. Mungurahwe waiswa kutapirisa bota uye kupa kunhuhwira kwakanaka Kubota. Ndizvo zvatakakura teiona anamai edu naanambuya veiita. Taipuwa bota riya torya. Rainaka seraiswa chokera.* [When the calabash from a gourd plant was ready, traditional herbs of different kinds were put in the calabash. Water and millet meal was added and left to brew for some time, especially overnight. The next morning, the mixture which would have turned a bit sour was used to prepare porridge. Among the herbs added was *mungurahwe*, a sweety and flavoury plant that acted as modern sugar. This is what we grew up seeing our mothers and grandmothers doing. They gave us this porridge to eat. It was very sweet as if sugar had been added].

The respondents MS 4, MS 8, MS 11 and MS 14 agreed that the term *shupa* is derived from a gourd prepared from a *mutanga* plant. The respondents indicated that when the gourd is transformed into a container, the container [*dumbu*] is referred to as *shupa*. *Shupa* is the container wherein a mixture of water, traditional herbs and *hlama* [boiled watery meal] is kept. Once the soaked herbs in the *shupa* are ready for use, the solution can either be used to prepare herbal porridge, which they refer to as *bota reshupa* or can be taken directly as a beverage (MS 14, MS 22, MS 28).

Indigenous Knowledge

For purposes of avoiding ambiguity, an attempt at defining or rather conceptualising indigenous knowledge and its efficacy is briefly given here. Indigenous knowledge is institutionalised local knowledge that the natives build on (Senanayake 2006; Nyong, Adesina & Osman 2007). This resonates with what Tume, Kimengsi and Fogwe (2019) submit. They conceptualise indigenous knowledge as a cumulative and complex body of knowledge, practices and representations maintained and developed by peoples through extended histories interactions with the natural environment. It is shown through a people's language and other aspects of human beings and is imparted to the young generation by elders (Senanayake 2006).

The efficacy of indigenous knowledge on some of the world challenges such as climate change, pandemics, disasters, environmental degradation and food insecurity have been researched from many perspectives. More specifically no such studies have been made on the Ndau people of south-eastern Zimbabwe, particularly on *bota reshupa*. Literature emphasises, for instance, the importance of indigenous knowledge on preservation of the environment. There are sacred places that people cannot tamper with as they wish (Shoko 2022; Nyawo 2022; Maseno and Mamati 2022; Siwila 2022; Kilonzo 2022). For these scholars, the environment is the provider of life and needs to be guarded against degradation. Siwila (2022) takes up the argument by questioning the privatisation of Mother Earth, an indigenous resource for all indigenes in different settings. It is against this that this chapter re-centres Ndau indigenous knowledge of *bota reshupa* as a food security safe valve in the midst of the devastating effects of climate change. The next section is an attempt to unravel *bota reshupa* from the Ndau perspective.

Bota reshupa*: Ndau Perspectives*

The participants in the interviews and the FGDs indicated that the most important ingredient of *shupa* was traditional herbs from tree roots, bark and leaves. While the herbs were similar in some communities, they differed in others. Chavunduka (1994) confirms that most traditional herbs are obtained from either tree bark, roots or leaves. Other scholars who hold this view include Shoko (2007) and Chinsembu (2015). For these scholars, most indigenous trees in African communities are an agency for primary health care needs for the communities, owing to the vast use of the trees' roots, bark and leaves. Okello et al. (2010) confirm this when they state that the Sabaots in the Kopsiro division of Mt. Elgon district in Kenya use medicinal plants to treat colds by chewing *Tagetes minuta* plant leaves that belong to the Asteraceae family; malaria is treated by boiling and drinking *Artemisia afra* leaves and ulcers are treated by boiling and drinking the root of *Heteromorpha*. The participants, especially from the FGD, further illustrated that small grain crops such as millet, rapoko, sorghum as well as maize were also ingredients for *shupa*. The use of small grain crops as well as maize differed from community to community. The variations were due to the climatic region in which the research site fell. For example, in the dry areas, small grain crops such as rapoko, sorghum and millet were mentioned as the meal used in the mixture for *bota reshupa*. These areas include Mutema and the western parts of Musikavanhu. The areas receive very low rainfall, and drought resistant small grain crops such as rapoko, sorghum and millet are grown instead of maize. The highland areas of Musikavanhu and parts of Garahwa receive average to above average rainfall. For this reason, maize is their main crop. Maize meal was often cited as an additive in *bota reshupa*.

Challenges That Affect the Efficacy of Bota reshupa

Amongst a plethora of challenges that threaten the efficacy of *bota reshupa* is climate change. When asked about the status of *shupa* in the research sites, one female respondent MS08 clearly stated that the tree roots, leaves and bark were no longer accessible. She complained about the drastic changes in rain patterns which have resulted in the trees drying up. She said:

> *Mitombo yekuisa mushupa yonesa kuona. Anditi unoona nyika yaagwenga kudai. Makomo eshe ambova makwasha hino mati ruru kutama mimbiti kudaro. Rangova dutu rega-rega mugowa muno mvura aichanayi.* (It is now difficult to get herbs for *shupa*. [The country side is now a valley. The mountains no longer have trees. There used to be forests in those mountains. The area is now windy and there are no rains].

In an interview with grandmother Mufunda concurred by stating that *mutanga* from which *dumbu reshupa* is made was no longer available. She stated, "*mvura yekunaya yaashoma muno mwaMutema*" [we receive very little rainfall here in Mutema]. For that reason, *mutanga*, which requires average to above average rainfall to germinate and mature into *dumbu*, cannot grow in the dry lands of the Mutema chieftaincy. For that reason, practitioners of *shupa* have turned to other containers for *shupa*. For instance, one respondent showed a metal container for *shupa* and in another instance, a Delta Beverage container popularly referred to as *scud* (because it looks like a scud missile) were used as alternative containers. When asked for the reason for such modifications and their implications, the respondents stated that the traditional *dumbu* was no longer available due to erratic rainfall patterns. The reason why the *dumbu* container was no longer in use was because of the scarcity of the *mutanga* plant. The improvisation where the metal containers and the *scud* were used demonstrates the ingenuity of the local people. It also demonstrates their adaptability as coping strategies to the changing times contrary to colonial thinking that traditional people are conservative and refuse to change. But it can be argued that the improvisations point to the unfortunate reality that things are changing and have impacts on the practice of IKS.

Globally, Africa is one of the most affected continents by climate change. According to Christian (2014), the reason is because of its low adaptive capacity. Climate change discourses pervade the United Nations summits, conferences and symposiums. It has become a highly topical issue because it threatens human existence. Given its devasting effects on the environment, climate change has not spared the practice of people's beliefs, values, customs and norms in Africa. Christian (2014) succinctly summarises the effects of climate change below:

> With the indiscriminate melting of the glaciers, the devastating and chronic deluge in several countries, the rise in sea level, the gradual annihilation of some plants and animal species and the evidence of extreme weather conditions around the globe, the reality of climate change has become undisputable. (2014: 1)

For Christian (2014), climate change is the significant increase in the earth's temperature over a long period of time. This temperature increase affects biodiversity, which also affects the plant species upon which *shupa* thrives. Due to erratic rainfall patterns, people have to walk long distances to obtain the tree roots, leaves and bark necessary for the herbal porridge. This has not only made *shupa* difficult to practise but has resulted in acute shortage of sources of Ndau indigenous food. The respondents indicated that the plants that were useful in the herbal porridge preparation could be found across the Mozambican border with Zimbabwe. Apart from the long distance, the political situation in Mozambique, where incidents of sporadic fighting between the Mozambican soldiers and opposition fighters are rife, is volatile.

To illustrate the effects of climate change, Boko, Niang, Vogel, Githeko and Medany state that the impact:

> range from decreases from grain yields, changes in runoff and water availability in the Mediterranean and southern countries of Africa, to increased stresses resulting from increased droughts and floods, and significant plant and animal species extinctions and associated livelihoods impacts. (Boko et al. 2007: 433)

Furthermore, climate change impacts on the socio-economic lifestyles of people. This includes plant and animal life. It has also impacted on the traditional religious practices of the African people. Christian (2014: 1) points out that "anything that affects nature affects religion, and anything that distorts the natural course of events, affects the traditional religious practices of the Africans." He further argues that since rain patterns are no longer predictable because of changes brought about by climate change, the traditional religious practices such as rainmaking ceremonies, rituals and other agriculture-related rituals are no longer being practised. This has either resulted in the rituals being discontinued or practised at a lower rate. Shifts in seasons due to climate change has also resulted in shifts in seasonal ritual ceremonies. This affects the timing of the rituals as some ritual are period-specific. If they are meant for September, they should be done in September. To emphasise this, Christian (2014) argues that climate change has a serious impact on how festivals and ceremonies are performed in communities. Festivals associated with ecological cycles are the most affected in that if planting rituals, for example, are supposed to be done in October and because of changes in weather patterns, they are delayed to December, the quality of the plants is adversely affected. Because the rituals are not done when due, the religious life of the community is disturbed. In many cases, the rituals would never be performed at all. Under these circumstances, it becomes difficult to believe that these religious festivals and ceremonies have not been affected in one way or the other.

To show the desperate situation, I was reminded of the fruits that my friends and I grew up eating in some of the areas mentioned by the respondents. Some plant species have gone into extinction. Grandmother Mabhuya, one of the elderly interviewees, lamented the lack of nutritious fruit that used to be plentiful in Manzvire village. She stated:

> *Mukoma iwewe, maongororo, nthengeni, matokorotsviyo neinthani muchero yawakakura weirya anditi akuchina. Vana vako ava avaiziye michero yandinoronza iyi. Akuna. Anditi matuntu awo aamapuwe ega amuchina michero yaidai kuwanda muchiri zvikomana. Hino unoti mitombo yeshupa ndiyo ingasara. Anditi kana woda mungurahwe wotoenda Mozambique kootsvaka iyoyo.* [The fruits you grew up eating are now extinct. Your children will never know that there were such fruits, yet they used to be plenty. How then can herbs for *shupa* be spared? Nowadays people have to travel to Mozambique for herbs like *mungurahwe*.]

In the FGD, similar observations were expressed. The FGD stated that it was difficult to find most of the tree species they once could. For the group, the scarcity was due to the extinction of the special tree species. The extinction was either because people wantonly cut down the tree species without replacing them or it was due to lack of respect for sacred places where these tree species used to grow. The sacred places had been cleared to build homes or turned into land for farming. Grandmother Simango in the FGD stated:

> *Apa peshe apa pamunoona paiya nemumbiti yataishandisa mubota iri asi iye zvino apachina. Porimwa mimbuti yeshe yakapishwa. Imweni yakaoma yega uye kutemwa kuita huni ngevantu. Taakuenda kure kundotsvaka mitombo yekuisa mubota iri. Mudamhuru amuchina muno zvekuti kwaakutopera kwebota iri.* [We used to have the special tree species all over here. But now as you can see, there is nothing now. The trees were cleared, people clear land for farming. In some instances, the trees dry on their own while others were cut for firewood. We now travel long distances to find the tree species. Some like *mudamhuru* are no longer available. Preparation of the porridge is slowly diminishing.]

The above views indicate that although the respondents may not know that climatic changes are responsible for this, they were aware that things were not good. Their environment was under siege, and life was threatened. Mother Nature could no longer provide as it used to do.

Critical Reflections on Bota reshupa *as a Food Security Safety Net*

Looking at the nature and ingredients of *bota reshupa* as indicated by the participants, one is convinced that *bota reshupa* is food in its own right. Participants revealed that it can be taken as either porridge or a beverage, *maheu*, in local language. The medicinal properties that participants alluded to are meant not only to enhance good health but to also ensure that children to whom it is commonly administered develop strong immunity against a number of ailments. Adults are reported to take it as a beverage after some strenuous work in the fields. In this regard, *bota reshupa* becomes a supplement of the main meal, *sadza*, staple food in Zimbabwean communities. Similarly, Somarathna et al.'s (2023) study in Sri Lanka revealed that *kola kenda*, a mixture of curry leaves, drumstick leaves, mung bean flour and unripe banana flour is an indigenous food that reduces high levels of malnutrition among Sri Lanka children.

During the study, there were 'claims' that *bota reshupa* ensures that children grow up healthy. Although the herbs have not been 'scientifically' proven to be the panacea for a number of ailments, the respondents were convinced about their efficacy. It was argued by most respondents that *shupa* boosts the immune system of the people who ate or drank it. This resonates with the findings of Mushishi (2010) on the Budya people of Zimbabwe, where *mono* [a traditional herb] mixed with the fragments of a snake tail tip was used to protect children from evil forces such as witchcraft or alien spirits. The herbal medicine is also used "to boost the immune systems of children against diseases which are either natural or caused by evil spirits" (Mushishi 2010: 165). These traditional foods that are medicinal provide nutrition and ensure food security among the rural population.

To indicate the centrality of the *bota reshupa* among the Ndau of southeastern Zimbabwe as indigenous nutritional food and health asset, the participants bemoaned the loss of the herbs that are used as ingredients due to climate change. For instance, one major ingredient is a plant known in local language as *mungurahwe* which provides the sweet flavour to the food. There is no need to use sugar in the porridge or *maheu*. *Mungurahwe* sweetens the concoction meaning that once one gets it, one is energised. *Bota reshupa* energises especially after some strenuous work in the field. Daughters-in-law, as revealed by the study, would provide fathers-in-law the concoction as drink to quench their thirsty after having worked in the fields (Muyambo 2019). In this case, *bota reshupa* as a drink quenches thirsty and at the same time acts as food for the hungry worker. Claims were that once children were fed by the concoction, they would spend the day without complaining of hunger. What this means is that *bota reshupa* is a supplement for the main food. The ingredients are largely sources of carbohydrates like millet, rapoko, sorghum and maize depending on the region as earlier on alluded to.

To ensure that the trees from which the *bota reshupa* ingredients are taken such as roots, tree bark and leaves are protected, the Ndau enforce that the ingredients are not unwantonly exploited. There are instances where a cut root is replaced after use to ensure that the tree does not wilt and die. The Ndau also participate in the Ndau Festival of the Arts, a Ndau culture and language promoter that holds its celebration annually. The Ndau Festival of the Arts (NdaFA) has taken serious measures to protect the tree species that are threatened by extinction. The cultural board is looking after a forest in the Chikore area, in Bangira community. The board is looking after the indigenous trees in the forest as well as planting those that are under the threat of extinction because of their medicinal value like *mungurahwe*. The cultural board has gone further to educate the local community in the need to protect medicinal plant species (Muyambo 2019: 222). These are innovative ways of ensuring that *bota reshupa,* which is heavily dependent on the plant species, does not become extinct amidst the challenges posed by climate change.

Conclusion

The chapter has argued that *bota reshupa* is food in its own right, given the ingredients that make it. It is medicinal, and people used and still use it as food supplement administered to adults and children alike. The concoction reduces high level of malnutrition and serves as food security in most of the Ndau communities. This knowledge to identify the nutritional and medicinal ingredients of *bota reshupa* is passed from elders to the youth through apprenticeship, the young observing and being sent to collect from the forest by the elders. The chapter has also demonstrated that climate change has brought changes to the indigenous practices of the porridge like the loss of the *mutanga* plant, thereby forcing practitioners to improvise. The improvisation, as indicated by use of modern items, takes away the religiosity thereof. This has also a ripple effect of disturbing the food storage in the natural container, meaning changing dynamics of the *bota reshupa* practice. Climate change impact on *bota reshupa* makes the achievement and realisation of Sustainable Development Goals 2 and 3 very difficult. The study, therefore recommends protection of forests that harbour essential trees for *shupa* through use of taboos and *machinda amadzimabo* as IKS. Since modernity is in our midst and wishing it away is foolhardy, I recommend creation of herbal gardens that will result in the beneficiation and value addition to the traditional food for it to be used beyond its confines, and this can be done by preparing instant porridge for the Zimbabwean market.

References

Amin, S. 1973. *Le development inégal: Essai sur les formations sociales du capitalisme periphérique*. Les Collections de Minuit.

Bhabha, H.K. 1984. 'Of mimicry and man: The ambivalence of colonial discourse.' *Discipleship: A special issue of Psychoanalysis, Spring,* 125–133.

Bhabha, H.K. 1990. 'The third space: interview with Homi Bhabha'. In Rutherford, J. (ed). *Identity, community and difference.* London: Lawrence and Wishart, 207–221.

Bhabha, H.K. 1994. *The location of culture*. New York: Routledge.

Bhabha, H.K. 1992. 'Postcolonial criticism'. In Greenblatt, S. and Gunn, G.B. (eds). *Redrawing the boundaries: the transformation of English and American literary studies*. New York: Modern Language Association of America.

Bjornlund, V., Bjornlund, H., & van Rooyen, A. F. 2020a. Why agricultural production in Sub-Saharan Africa remains low compared to the rest of the world: A historic perspective. *International Journal of Water Resources Development, 36*(1), S20–S53.

Bjornlund, V., Bjornlund, H., & van Rooyen, A. F. 2020b. Exploring the factors causing the poor performance of irrigation schemes in post-independence sub-Saharan Africa. *International Journal of Water resources Development, 36*(1), S54–S101.

Bjornlund, V., Bjornlund, H., & van Rooyen, A. F. 2022. Why food insecurity persists in sub- Saharan Africa: A review of existing evidence. *Food Security, 14:* 845–864.

Boko M.I, Niang A.N, Vogel C, Githeko A, Medany M. 2007. 'Climate Change 2007; Impacts, Adaptation and Vulnerability. Contribution of Working Group II to the Fourth Assessment Report of the Intergovernmental Panel on Climate Change', Cambridge: Cambridge University Press, 433–467.

Chavunduka, G.L. 1994. *Traditional Medicine in Modern Zimbabwe*, Harare: University of Zimbabwe Publications.

Chinsembu, K.C. 2015. 'Bioprospecting for 'green diamonds': Medicinal plants used in the management of HIV/AIDS related conditions'. In Chinsembu, K.C., Cheikhyoussef, A., Mumbengegwi, D., Kandawa-Schulz, M., Kasanda, Choshi, D. and Kazembe, L (eds.) *Indigenous Knowledge of Namibia*, Windhoek: UNAM Press, 9–40.

Chitando, E. 2010. 'Phenomenology of religion' in Amanze (ed.) *Biblical Studies, Theology, Religion and Philosophy: An introduction for African Universities*, Eldoret: Zapf Chancery, 263–281.

Christian, N.G. 2014. 'The impact of climate change on African traditional religious practices'. *Journal of Earth Science and Climate Change*. Vol. 5, No. 7, 1–5.

Cox, J.L. 1992. *Expressing the Sacred: An introduction to phenomenology*, Harare: University Zimbabwe Publication.

Delgado, C., Tschunkert, K and Smith, D. 2023. Food insecurity in Africa: Drivers and Solutions. Sipri Research Policy Paper, 1–24. https://www.sipri.org/sites/default/files/202301/2301_sipri_rpp_food_insecurity_in_africa_1.pdf.

Diffie, B. W., Shafer, B. C., & Winius, G. D. 1977. *Foundations of the Portuguese empire, 1415–1580*. University of Minnesota Press. https://doi.org/10.5749/j.ctttsrg9.

FAO et al. 2022. The State of Food Security and Nutrition in the World 2022: Repurposing Food and Agricultural Policies to Make Healthy Diets More Affordable. Rome.

Hawthorne, W. 2003. *Planting rice and harvesting slaves: Trans- formations along the Guinea-Bissau Coast, 1400–1900.* Heinemann.

Kilonzo, S.M. 2022. Women, Indigenous Knowledge, and Climate Change in Kenya. In Ezra Chitando, Enerst M. Conradie & Susan, M. Kilonzo (eds.). *African Perspectives on Religion and Climate Change.* London & New York: Routledge, 79–90.

Mapuranga, T.P. 2010. 'A phenomenological investigation into the effects of traditional beliefs and practices on women and HIV & AIDS, with special reference to Chipinge district, Zimbabwe'. Unpublished PhD thesis. Harare: University of Zimbabwe.

Maseno, L. and Mamati, K. 2022. The nexus between indigenous beliefs on environment and climate change adaptation amongst the Sengwer in Embobut Forest, Kenya. In Ezra Chitando, Enerst M. Conradie & Susan, M. Kilonzo (eds.). *African Perspectives on Religion and Climate Change.* London & New York: Routledge, 50–64.

Mushishi, C. 2010. 'Aspects of Budya Traditional Religion which promote Human Rights', unpublished PhD thesis, University of Cape Town, South Africa.

Muyambo, T. 2015. 'Bondswomen of Culture': A gender critique of *Bota reshupa* and *Kuhaza* among the Ndau people of Chipinge, Zimbabwe'. *Greener Journal of Social Sciences,* Vol. 5 (3), 65–71.

Muyambo, T. 2019. Indigenous knowledge systems of the Ndau people of Manicaland province in Zimbabwe: A case study of *bota reshupa.* PhD thesis. Pietermaritzburg: University of KwaZulu Natal.

Ndlovu-Gatsheni, S.J. 2019. Conversation with Duncan Omanga at the sidelines of an African Studies Association (ASA) annual meeting held in Boston (November 21–23).

Nyawo, S. 2022. The climate crisis: Mitigation and control through Emaswati indigenous knowledge. In Ezra Chitando, Enerst M. Conradie & Susan, M. Kilonzo (eds.). *African Perspectives on Religion and Climate Change.* London & New York: Routledge, 34–49.

Nyong, A., Adesina, F. and Osman, E. 2007. The value of indigenous knowledge in climate change mitigation and adaptation strategies in Sfricsn Sahel. *Mitig Adapt Strat Glob Change* 12, 787–797.

Okello, S.V., Nyunja, R.O., Netondo, G.W. and Onyango, J.C. (2010). 'Ethnobotanical Study of Medicinal plants used by Sabaots of Mt. Elgon, Kenya, *African Journal of Traditional, Contemporary and Alternative Medicines,* 7(1), 1–10. http://journals.sfu.ca/africanem/index.php/ajtcam/index. [Date accessed 3 August 2017].

Quayson, A. 2002. *Postcolonialism: Theory, practice or process?* Cambridge: Polity Press.

Said, E 1985. 'Orientalism reconsidered' in *Cultural Critique* 1, 89–107.

Said, E, 1993. *Culture and Imperialism.* London: Chatto and Windus.

Said, E. 1978. *Orientalism*. New York: Vintage Books.

Senanayake, N. 2006. Indigenous knowledge as a key to sustainable development. *Journal of Agricultural Sciences* 2(2), 87–98.

Shoko, T. 2007. *Karanga indigenous religion in Zimbabwe: Health and Well-Being*. Aldershot: Ashgate Publishing Company.

Shoko, T. 2022. African Traditional Religion and climate change: Perspectives from Zimbabwe. In Ezra Chitando, Enerst M. Conradie & Susan, M. Kilonzo (eds.). *African Perspectives on Religion and Climate Change*. London & New York: Routledge, 22–33.

Sibanda, F. 2014. 'Quest for identity: Rastafari cultural identities as expressions of liberation in postcolonial Zimbabwe'. Unpublished PhD thesis, Harare: University of Zimbabwe.

Siwila, C.L. 2022. An African ecofeminist appraisal of the value of indigenous knowledge systems in responding to environmental degradation and climate change. In Ezra Chitando, Enerst M. Conradie & Susan, M. Kilonzo (eds.). *African Perspectives on Religion and Climate Change*. London & New York: Routledge, 65–78.

Smart, N. 1996. *Dimensions of the Sacred, An Anatomy of the World's Religions,* London: Harper and Collins.

Somarathna,E.G.P.S.,Lakmali,K.M.,Bandara,E.M.S.M&Samarakoon,E.R.J.2023. Development of *kola kenda* mixture utilizing herbal plant leaves, mung bean flour, and unripe banana flour. 3rd International Symposium on Agriculture, 9 March, Sri Lanka.

Spivak, G.C. 1988. 'Can the subaltern speak?' in Nelson, C. and Grossberg, L. (eds). *Marxism and the interpretation of culture*. Urbana: University of Illinois Press, 271–313.

Spivak, G.C. 1991. 'How to teach a culturally different book' in Donna Laundry and Gerald MacLean. (eds). *The Spivak Reader*. New York: Routledge.

Spivak, G.C. 1993. *Outside the teaching machine*. New York: Routledge.

Streng, F.J. 1985. *Understanding Religious Life*. California: Wadsworth Publishing Company, Belmony.

The World Bank, (n.d.). COVID-19 (coronavirus) response in Africa.

Tume, S., Kimengsi, J. and Fogwe, Z. 2019. Indigenous knowledge and farmer perceptions of climate and ecological changes in the Bamenda Highlands of Cameroon: Insights from the Bui Plateau. *Climate* 7, 138. https://doi.org/10.3390/cli7120138.

Velautham, L. 2015. 'Voice and Representation. A postcolonial approach to higher education promotional media and the international postgraduate student experience', unpublished PhD thesis, The University of Adelaide, Adelaide.

World Food Summit, Rome Declaration on World Food Security and World Food Summit Plan of Action, Rome, 13–17 Nov. 1996.

CHAPTER 12

Indigenous Knowledge Systems, Climate Change and Food Security: Perspectives from Bungoma County, Kenya

King'asia Mamati and Simon Gisege Omare

INTRODUCTION: FOOD CRISIS IN KENYA

It is estimated that about 278 million people in Africa, representing 20% of the continent's population, suffer from chronic hunger (World Economic Forum 2022). This dire condition on the continent has been exacerbated by climate change and civil wars. Several decades after independence, food security remains to be an elusive and serious challenge. In Kenya, over 10 million people suffer from chronic food insecurity and malnutrition, and between 2 and 4 million survive on emergency food assistance from the government and non-governmental organizations

K. Mamati (✉)
University of Cologne, Cologne, Germany
e-mail: kingasiamamati@gmail.com

S. G. Omare
Department of Philosophy, Religion and Theology, Moi University, Eldoret, Kenya
e-mail: omarexg@mu.ac.ke

L. Maseno et al. (eds.), *Religion, Climate Change, and Food Security in Africa*, Sustainable Development Goals Series,
https://doi.org/10.1007/978-3-031-50392-4_12

(Republic of Kenya, 2008). The 2022 Global Hunger Index ranked Kenya at position *94 out of the 121 countries*, at dire levels of hunger (*Hunger Hotspots* 2021). The World Food Program notes that access to substantial quantities of nutritious food remains a concern for many Kenyans, particularly those living in arid and semi-arid regions, which make up 80% of the country's land area. Only 20% of Kenya's land is suitable for agriculture, even though the majority of it is underutilized. Many people are economically dependent on the agriculture sector for their livelihoods, with about 75% of Kenya deriving part or all of their income from the agricultural sector, which accounts for 33% of Kenya's gross domestic product (GOK 2021). Rising temperatures and declining rainfall attributed to climate change have led to large parts of sub-Saharan Africa becoming unsuitable for multiple staple crops. As a part of its 2030 development agenda, on September 23, 2015, the General Assembly of the United Nations adopted a set of 17 goals to end poverty, protect the planet and ensure prosperity for all. The *second Sustainable Development Goal* (SDG2) is to "[e]nd hunger, achieve food security and improved nutrition, and promote sustainable agriculture." The chapter seek to contribute towards the SDG2 by gaining insights from indigenous people knowledge systems on food.

The ethnographic focus of this chapter is Bungoma County, which has a population size of 1,670,570 based on 2019 census report (KNBS, 2019). In Bungoma county, Food insecure households rate as of 2017 stood at 70% concurrently, poverty rate as of 2013 looms at 52.9% while the stunting rate stands at 30.1% (2018) (Wabwoba 2021; Chenge 2015). Main food crops include maize, beans, sorghum and millets. Cash crops include sugarcane, coffee, tea and sunflower. Maize however is the dominant crop and makes up to 80% of crops being produced in the County followed by beans 9% and other foods crops 5% (Songa et al. 2021). Over the past decades, farmers in Bungoma County have transitioned from Maize mono-cultural crop to Sugar cane mono-cultural crop. Nangendo (1992, 29) noted that "the mono-culture of maize in Bungoma has failed to ensure that there is sufficient food supply for the family as well as adequate surplus that could be used to bring in some cash income. Arguably, the mono-culture of sugarcane could prove to be even far more catastrophic than the mono-culture of maize farming. This is because while maize can be consumed as a food when it is green or dry, sugarcane can only act as a thirst-quencher".

A Note on Methodology

For several decades, the traditional concept of a field site as a bounded space containing the whole culture has been put into question with the emergence of multisite ethnography (Marcus 1995). This has opened possible modes of doing research without being spatially bounded. The internet with its emerging digital spaces has opened up new sites for social activities as well as modes of knowledge transmission (Burrell 2009). This chapter employs virtual ethnography as a primary data collection method by relying on ethnographical data collected by Aleya and posted on Twitter. Aleya is a writer from Kenya based in Nairobi who had visited Bungoma County. Even though Twitter is a public space, permission was sought from her to use the knowledge she had collected. It relies on secondary sources such as newspaper articles, published articles and books.

Food Sovereignty and Food Security

The United Nations Food and Agriculture Organization defines food security as "a situation that exists when all people, at all times, have physical, social, and economic access to sufficient, safe and nutritious food that meets their dietary needs and food preferences for active and healthy life" (FAO 2008). It further outlines four dimensions of food security to be considered, namely: food availability, accessibility, vulnerability and sustainability. However, this definition has been challenged by scholars who advocate for food sovereignty instead of food security (Patel 2009; Grey and Patel 2015). Food security has been criticized for focusing on the adequacy of food supply without specifying the means of food acquisition (Hoover 2017, 33).

Food sovereignty was first defined during a meeting that was convened in 1996 by La Via Campesina, the world's largest movement for peasant food producers, founded in 1993. The movement aimed to defend these communities against neoliberalism and the dominant market economy and champion the rights to land and seeds (Pierrick 2003).

The definition of food sovereignty was further developed and refined during Nyéléni International Forum for Food Sovereignty held in Sélingué, Mali, in 2007 (which brought together 500 delegates from various organizations in 80 countries). The outcome of the forum was the adoption of the Declaration of Nyéléni, which defines Food sovereignty under the six pillars as the right of peoples to healthy and culturally appropriate food produced through ecologically sound and sustainable methods, and their right to define their food and agricultural systems which could be useful in the achievement of the SDG2 (Nyéléni 2007). It puts the aspirations and needs of producers, distributors and consumers of food at the heart of food systems and policies rather than the demands of markets and corporations. It highlights the need to include the next generation in food production, as well as empowering food producers and artisans; (3) the importance of environmental, social and economic sustainability; and (4) the need for transparent trade, as well as equality between genders, racial groups and social classes. (5) Everyone in the food chain is positioned as a potentially powerful actor (Medina 2017). The concept of food sovereignty has been framed within a larger rights discourse based on the ability of all people freely and independently produce their foods in a political framework that recognized territorial autonomy (Edelman et al. 2014).

During the Rio 20, the *food sovereignty* movement advocated for positioning indigenous people, peasants' farmers as the penultimate solution to the food crisis, it calls for the exchange of the industrial agro-export food system for a system based on food sovereignty, that returns the land to its social function as the producer of food and sustainer of life, that puts local production of food at the centre, as well as the local markets and local processing. Food sovereignty allows us to put an end to mono-cultures and agribusiness, to foster systems of peasant production that are characterized by greater intensity and productivity, that provide jobs, care for the soil and produce in a way that is healing and diversified (Pierrick 06/06/2012). Peasant and indigenous agriculture also has the ability to cool the planet, with the capacity to absorb or prevent almost two-third of the greenhouses gases that are emitted every year (Ndabezinhle 2023). The basis of indigenous food systems is their kincentric and ecofriendly worldview.

African Religious Worldview on Food Systems

Indigenous food system also referred as traditional food systems are all foods within the particular culture that are available from local natural resources and culturally accepted, including the sociocultural meanings, acquisition and processing techniques, use, biological composition and nutritional consequences for the people using the food (Kuhnlein and Chotiboriboon 2022, 2). Indigenous people have knowledge about their origin or creation rooted within various aspects of their landscape and environment (Mbiti 1997). Food systems among diverse African cultures were dependent on the worldviews ingrained in their interconnection with other species. The religious worldview guided resource use, food utilization and conservation of various aspects of the environment. Spiritual and kinship relationships with the land, as well as other natural environmental resources such as forests, hills, rivers and animals among African communities, provided sustenance for communal well-being. Human beings were part of the universe, with a mandated responsibility of ensuring the equilibrium of life on earth (Gumo et al. 2012; Mamati 2018).

The clash of cosmologies and worldviews on the human-environmental relationship is at the root of the current food crisis. With the disruption caused by colonialism and modernity, habitats and ecological systems are degrading at an alarming rate. Of course, this is directly linkable to the colonial gaze that has characterized African ways of knowing and knowledge production. African endogenous knowledge has been "caricatured or misrepresented in the western categories of 'magic', 'witchcraft', 'sorcery', 'superstition', 'primitivism', 'savagery' and 'animism'" (Nyamnjoh 2017, 255). The result of this, partly, is that the indigenous people's land has been changing drastically as their spiritual inter-species relations become disconnected. This ritual disconnection has facilitated a departure from the traditional food systems and the adoption of commodified crops and other foreign dietary habits.

However, these new approaches, considered "western gifts" to "inferior" Africa, failed to consider African's "popular traditions of meaning making [and denied them] the right to think and represent their realities in accordance with the civilisations and universes they know best" (Nyamnjoh 2017, 255). This chapter argues, on the one hand, that because of the imposition of the so-called superior hegemonic ways of western knowing, most of the diverse indigenous cultural foodways in Kenya are now near annihilation. On the other hand, it holds that reinvigorating the hitherto dismissed indigenous knowledge systems on food production can help in achieving Kenya's elusive food security.

Indigenous Knowledge of Food Systems

Pre-colonial African societies met almost all their food and nutritional requirement through local productions. Communities depended on nutritional foods such as fish, wild game, fruits, berries, animal milk and meat, wild insects (grasshoppers, ants, caterpillars, termites) birds and vegetables. Several nutrient-dense food crops were consumed by indigenous communities in Africa in the past, including sorghum, finger millets, hyacinth (lablab) beans, bambara nuts, bottle gourds, cowpeas, yams, cowpea, lablab, pea, three African eggplants, okra, jute, kale, nightshade, Cleome, amaranths, Crotalaria (Maundu et al. 1999; Coppens d'Eeckenbrugge et al. 2019). The diversity in food production is reflected in Aleya's Twitter handle posted on September 20th 2020. Reporting, ethnographically, her encounter with a knowledgeable woman from Bungoma,[1] Aleya writes in pensive melancholia:

> It's been over a decade, but this story I haven't been able to forget. An elderly woman was telling us about how they used to farm & cook when she was much younger. Bungoma is lush with a treasure trove of delicious and nutritious indigenous vegetables. She lists them as follows:
>
> "Litoto (Amaranth), Namasaka (Black Nightshade), Sikhubi (Cowpeas), Kimiroo (Crotolania), Malenge (Pumpkin Leaves), Tsisaka (Spider Plant), Murenda (Jute Mallow) & many more."

These foods, Aleya reports, indicate the in-depth knowledge on indigenous food systems held by her informant. Her informant further states that these diverse foods not only were delicious, but also "grew easily, were drought resistant, highly nutritious, and often had medicinal properties". Alenya's informant goes on to say that these foods were also easily accessible and easy to prepare: "the informant noted how you'd just pick what was growing around and then put in a gourd over a fire and let it slowly cook as you went about your business." She goes on: "you cooked them with *munyu mesherekha* and added soured milk—the slight fermentation made them even more delicious & kept them preserved for days & days & days." Aleya notes that munyu mesherekha is high in potassium, it tenderizes tough leaves & adds a smoky flavour to the dish:

> To make the *munyu mesherekha*, you gather bean pods, groundnut kernels, maize cobbs…you dry them in the hot sun until they are paper dry, then you burn it up till it turns to ash, adding little sprinkles of water as it cools down.

The knowledge system that Aleya's informant demonstrates is no way questionable. She paints a picture of an African society that was sufficiently

[1] Bungoma is a county in western part of Kenya bordering Uganda to the west.

informed in regard to different types of food, their nutritional contents, to ways of cooking and preserving them for future use. The confidence she displays reflects Ndlovu-Gatsheni's (2018) epistemic freedom. According to Ndlovu-Gatsheni, epistemic freedom allows people the freedom to think and present their world views without being mindful of Eurocentric standards that are alien to the worlds that they know. If the society presented in Aleya's story is to go by, then such a society had in-depth knowledge on diverse food crops and food systems. It was already self-sufficient and secure. This security went as far as diversifying methods of food production. Techniques such as slash-and-burn agriculture, shift cultivation, rotation bush fallow, rotational planted fallow and mixed farming was used sustainably based on traditional knowledge (Darkoh 1989, 60). The mixed economies based on multiple sources of food, such as hunting and gathering, pastoral economy, fishing and mixed crop farming further shielded these communities from the whims of nature (Ochieng' 1988, 24).

In the fifteenth and sixteenth centuries, the discovery of sea routes to India and South-East Asia resulted in food and nutritional transitions in Africa. Through these routes, new crops such as plantain, banana, cocoyam, coconut and sugar cane were introduced into local African diets, especially among communities living along the regional trade routes (Raschke and Cheema 2008, 4). Other New World crops that became widespread, some becoming staple foods over extensive areas, were mainly introduced to Africa by the Portuguese with a profound impact on African agriculture. Some of these crops include maize, *Zea mays*, common bean, *Phaseolus vulgaris*, groundnut, *Arachis hypogaea*; sweet potato, *Ipomoea batatas*; cassava, *Manihot esculenta*; pumpkins, *Cucurbita* spp.; chillies, *Capsicum annuum* and *C. frutescens;* and tobacco, *Nicotiana tabacum*, English potato, kidney bean, pumpkin, cassava, European cabbage, sweet *potato* (Coppens d'Eeckenbrugge et al. 2019; J. W.Purseglove 2011; J. W. Purseglove 2011, 296).

Until the coming in of Europeans into Africa, not even the influence of food diversity brought by the sea routes destabilized African food production and sustenance systems. When food harvest was inadequate, African societies came up with diverse social mechanism that ensured food security to cope with famine and drought. Through reciprocal relationships among communities and between households, societies were able to withstand a crisis of food shortage occasioned by poor harvest, famine, drought or vagaries of weather. Therefore, communal capital played a significant role in shielding communities against food insecurity. For instance, Omwoyo (2015) highlights that Gusii community had strong patterns of redistributive and reciprocal gifts between households that reinforced the society's ability to withstand food shortages. In similar vein, pastoral communities like the Pokot, had elaborate coping mechanisms to shield the community against, drought famine and loss of livestock. Every member was taken care of during episodes of scarcity through the Pokot food sharing networks, mobility and hospitality (Bollig 2010, 175–182). Additionally, communities living in fragile environments developed sophisticated techniques to ensure food production and cope with the cyclical vagaries of climate. The enhancement, improvement and replication of this technology in other drought and famine prone region could be very useful in fight against climate change and hunger. For example, the Marakwet and Pokot furrow system, is an indigenous water technique ensured food security for the residents living in the Kerio valley complex. The hill furrow irrigation system with its ingenious construction dates back over 300 years ago. Today, the region has over 49 furrows totalling 265 kilometres constructed on diverse rivers flowing from the Cherangany water catchment. The pre-colonial Marakwet agriculture economy was based on subsistence food production through reliance on furrow-irrigated and rain-fed cultivation. The main food crops grown in this region were finger millet and brown sorghum with the surplus grain exchanged with their neighbouring communities such as the Tugen, Sengwer, Pokot, for meat, milk, poison for arrows and tobacco (Watson et al. 1998, 77–80).

Also, indigenous people depended on forest resources for all or part of their sustenance and livelihoods. Forests harboured spiritual and cultural significance for indigenous communities such as the Sengwer (Mamati 2018). The colonial government gave concessions for the harvesting of trees. Forests were cleared to establish mono-cultural cash crop farming (Anderson 1988). The destruction of the forest led to the decimation of indigenous trees, medicine, vegetables and wild fruits. This destruction affected local people's diets as well as their sources of medicine. Indigenous food systems from diverse cultures are associated with medical and health benefits. Iwu (2016) documents a broad range of nutritional, medicinal and aromatic plants, and plant-based food products used with an overview of recent scientific, technical and economic developments across Africa. Some vegetables such as spider plant (*Cleome gynandra*) and African nightshade (*Solanum scabrum*) are valued for their supply of iron into the body among Abagusii of western Kenya. Women who had given birth were given plenty of these vegetables to help in regaining the lost blood. African indigenous vegetables have higher quantities of nutrients associated with the prevention of cancer and diabetes than exotic vegetables such as cabbage or lettuce. These vegetables contain proteins, calories, essential vitamins, particularly A, B and C, minerals (such as calcium and iron) and other beneficial phytochemicals (Irakoze et al. 2021). The foregoing contents in these vegetables can eliminate deficiencies amongst the vulnerable population.

There is no doubt that indigenous knowledge of food systems and production within the African societies were well thought out and self-sustaining. However, the western system that was grounded on conquest of the African knowledge, what Ndlovu (2020) describes as the "epistemological inversion of its target" (1), through caricatures, introduced guilt and self-doubt in the African systems of knowing. This doubt in the African ways of knowing was further delivered, second-hand, by Africa's elites, most of whom were western trained and by acts of omission or commission, adorned Eurocentric lenses on indigenous foods. Nyamnjoh (2017) argues that "educated and steeped in the dualisms of colonial ways of knowing and producing knowledge, this elite tends to despise endogenous African ways of knowing and knowledge production" (255). These elites are the same ones who are in influential government positions and in charge of policy. Consequently, indigenous knowledge of food systems have continued to face assault from dominant forces in society, ranging from the government, intellectuals, urbanization, education, media and the church. Sustenance of Eurocentrism in food production has led to the neglect of many indigenous cultures and knowledge of food. This neglect has jeopardized indigenous knowledge of human-environmental relationships, which heavily influenced food sovereignty and ensured food-secured communities. Aleya, for example, points out how her informant was able to point out the inadequacy of the supposedly superior ways of food production:

> So the elder was telling us about how things were before … then one year, she remembers that the colonialists brought with them kale, and "convinced" them to grow these in the farms. Strangely there was a bumper harvest that year, as the kale DID SO WELL!
>
> Now there was an abundance of this one crop-but it just was not very tasty. Even when they added their *Munyu musherekha*, the kale was just well

bland. However, they discovered that kale didn't take long to cook which meant they used up less firewood.

They figured out if they added tomatoes, it made it more palatable, and the colonials told them if they cooked it with fat instead of *munyu mesherekha*, it would be tasty it's true, with fat & tomatoes it was tasty Just that many people didn't grow tomatoes or make their fat.

But she said after that, she remembers their diet began becoming less varied.

Many decades later, she told us she wished kale had never been introduced because she's seen how the community's health has been affected by this over-dependence on one type of green.

This memory that she has of that specific time is one that was repeated by other elders within her age-set. Felt this moment was such a pivotal moment, and the appearance of Kale had such long-lasting effects on their lives.

I've never been able to forget that story (and I can still taste the smokiness in the *mrenda* she served us) (I know this, and yet) It is so striking to see the tangible ways in which decisions made by others up there have such real-life impact on people.

A detailed explanation of this transition is explained in the succeeding section.

Impact of colonialism on Indigenous Food systems

The root of the present food crisis in Kenya has its origin in colonialism. Colonialism imposed a social, political, cultural and economic order alien to many of the indigenous communities in Africa. Through the principle of *Terra nullis* (empty land), African lands were divided and demarcated by European powers. Indigenous peoples' rights were dismissed and relegated (Nyangaga 2022, 126). The upshot was that Africans lost control over their land. They also lost their cultural identity and food sovereignty. Through the crown lands ordinance of 1902 and 1915, Africans became tenants of the crown and were confined to the native reserves as the best fertile and well-watered lands became white highlands for the white settlers (Okoth-Ogendo 1991).

The colonial government in Kenya aimed to advance an agricultural system which predominantly served the interest of European settlers and their metropolis industries. Africans were forced to enter the world of money as wage labourers rather than independent producers (Brett 1973, 167). The imposition of colonial rule in Kenya thus had a debilitating effect on African communities. During this period, indigenous subsistence production systems were suppressed, and drought-tolerant and resistant crops depleted setting in motion a process of lessened food production. Colonialist at that time regarded African foods as inferior and encouraged the adoption of other crops such as wheat, maize. Ironically, research conducted during the colonial period proved that most of the African indigenous foods had high nutrition value. The well tried and food preparation methods such as fermentation were nutritionally valuable but discarded (Culwick 1944; Mackenzie 2000).

The current economic order has edifices that favour and serve hegemonic structures represented by corporations. Neo-liberal economic policies have impacted agriculture by usurping decision-making authority on food production and distribution (Bjornlund et al. 2022). It has been noted that most countries in Africa with acute food shortages have well-developed agricultural blueprints whose main objective is to pacify internal political dissension and serve as a compliance instrument to access international funding (Mwichabe 2013, 415).

Over the decades after independence, Kenya has undergone a massive food and nutritional transition with an increased reliance on imported foodstuff and processed food such as carbonated soft drinks, wheat, vegetable oils, rice and noodles. According to the Kenya National Bureau of Statistics Economic Survey (2020), the total import dependency rate for food products in Kenya decreased to 12.7%, against 14.5% registered in the previous year 2020 (KNBS 2022, 170). The food import dependency rate indicates how much of the available food for human consumption has been imported. Kenya has been plagued with a dire shortage of essential food for many years, paradoxically she continues to grow horticulture and other cash crops under conducive conditions for export to the industrialized nations of the world. Kenya is a leading exporter of the following main cash crops, including tea, coffee, sugarcane, cotton, sunflower, pyrethrum, barley, tobacco, sisal, coconut and bixa that contribute about 69.7% of agricultural exports (GOK 2021). Many communities have also witnessed reduced diversity of local food crop production as small-scale farmers turn to cash crops. Indigenous people as well as local communities have lost food sovereignty and security as the focus shifted to producing for the market at the expense of their consumption.

The impact of colonialism also worked on the mental sphere of people by implanting cultural inferiority. Africans started despising their foods and acknowledging and assimilating foreign foods. Imperial culinary and cultural hierarchies of foods were introduced (Kobuthi 2019). People took over foreign foods and placed cultural attachment to them despite some having insignificant nutritional value compared to traditional foods. The consumption habits of local people have been heavily influenced by the marketing strategies adopted to promote foreign foods and drinks leading to a repugnance of indigenous food and deference to foreign corporate foods. Today, majority of the people depend on highly processed and industrialized foods. The effect of these dietary changes has seen arise in non-communicable disease and diet-related complications, contributing to over 50% of inpatient admissions and 40% of hospital deaths (Onyango and Onyango 2018). The change of dietary preferences to modern processed foods that are deficient in essential nutrients is characterized with high levels of fat and sugar, with relatively lower content of fibre and complex carbohydrates. It is argued that such diets are associated with increased susceptibility to oral cavities, hypertension, diabetes, cardiovascular complications and obesity. Staple food crops such as maize and cassava have low essential amino acids and is associated with diarrhoea, dementia dermatitis and cognitive deficiencies (Bjornlund et al. 2022, 851; Goredema-Matongera et al. 2021).

Iwu (2016) makes an interesting observation about the middle-class, wealthy and urban dwellers in most African countries. He notes that due to their consumption of cheap processed foods (or so-called fast foods), they exhibit disease patterns similar to those of the urban poor in the West, while the health status of the nutrition-cautious elites in developed nations approximates those of African rural dwellers.

There has been a long-held debate till now on the adoption of genetically modified foods as a panacea for the current food crisis facing Kenya. The debate is premised on those supporting the adoption of technology-based innovation, including gene modification that would lead to a dramatic increase in agricultural production hence providing the ultimate solution to food insecurity (Makokha and Kyalo 2016). It is important to note that under the use of technology-based agriculture, there is heavy usage of fertilizers and pest control chemicals as shall be discussed in succeeding sections. The opponents of this approach argue that technology-based agriculture not only is inappropriate for Africa, but also has the potential to cause more harm than benefit. They support traditional farming systems and agroecology that are predominant in many African societies. They advocate the use of organic-based solutions that would protect biodiversity.

The current commercialized agriculture undermines the food sovereignty of indigenous communities because it advocates for mono-culture farming to meet the agricultural market demands. The focus on cash crops for the market deprives the local people of autonomy. It maintains a dependent economy that serves foreign interests as local populations continue to produce what they do not consume and don't have control over the market prices.

Not only does the present food system fail to feed vast portions of the population, but it also damages ecosystems, soil fertility and biodiversity. There is evidence that the usage of certain pesticides has long-term, detrimental impacts on both human health and the ecosystem. Chemical pesticide exposure is associated with hazardous effects on human health such as cancer, Alzheimer's, Parkinson's, disturbance of hormones and sterility developmental abnormalities (Sabarwal et al. 2018; Mostafalou and Abdollahi 2013). In addition, they are numerous repercussions associated with the use of artificial inputs, including the loss of biodiversity, the loss of aquatic life and the emergency of pesticide-resistant insects, diminishing beneficial insect populations that serve as natural adversaries of pests and lowering the nutritional value of food (Brühl and Zaller 2019).

Moreover, the indigenous farming system is usually carbon neutral as it emitts a limited amount of carbon into the atmosphere. The adoption of environmentally friendly indigenous farming systems could be more useful in the fight against contemporary challenge of climate change. Fertile organic lands and topsoil are threatened faster than it can be replenished by the overreliance on petroleum-based fertilizers. In addition, hype exploitation of natural resources and desecration of the lands has led to a dramatic increase in CO2 emissions. Thus, modern food systems rely heavily on fossil fuels accelerating greenhouse gas emissions. Ntinyari and Gweyi-Onyango (2021) note that substantial GHG emissions come from agriculture through the use of fertilizers, farm equipment production, transportation of farm produce, forestry and land use changes.

Conclusion

Different communities in Kenya have extensive indigenous food knowledge that needs to be revitalized and encouraged before they are extinct. Mono-culture farming of nutrient deficiency crops such as maize, wheat and rice have led to a systematic loss of biodiversity and knowledge of indigenous agricultural practices. In addition to food insecurity, the modern farming system has exacerbated environmental degradation, health issues and cultural erosion. In light of this, subsistence farming plays an important role in household food security. Consequently, the diversification of local food production among households and small-scale farmers can also act as a buffer against food insecurity. It is incumbent upon stakeholders to direct attention toward small-scale farmers to produce a variety of organic foods that are rich in nutrients. The government need to encourage research, formulate and implement policies that safeguard and revitalises indigenous food systems. Indigenous knowledge on food systems should be repacked and made appealing to the younger generation for posterity and a sustainable, healthy, nutrients dense food systems that fosters a thriving and secured society.

References

Hunger Hotspots. (2021). WFP and FAO.

Anderson, David. 1988. Managing the forest: the conservation history of Lembus, Kenya, 1904–63. In: Conservation in Africa: Peoples, Policies and Practice. Cambridge University Press, 249–268.

Bjornlund, Vibeke/Bjornlund, Henning/van Rooyen, André (2022). Why food insecurity persists in sub-Saharan Africa: A review of existing evidence. Food Security 14 (4), 845–864. https://doi.org/10.1007/s12571-022-01256-1.

Bollig, Michael (2010). Risk management in a hazardous environment: a comparative study of two pastoral societies. A Comparative Study of two Pastoral Societies. Dordrecht, Scholars Portal.

Brett, E. A. (1973). Colonialism and underdevelopment in east Africa. The politics of economic change, 1919–1939. London, Heinemann.

Brühl, Carsten A./Zaller, Johann G. (2019). Biodiversity Decline as a Consequence of an Inappropriate Environmental Risk Assessment of Pesticides. Frontiers in Environmental Science 7, 177. https://doi.org/10.3389/fenvs.2019.00177.

Burrell, Jenna (2009). The Field Site as a Network: A Strategy for Locating Ethnographic Research. Field Methods 21 (2), 181–199. https://doi.org/10.1177/1525822X08329699.

Chenge, Ndalila Michael (2015). Factors influencing food security in Kenya; a case of Bungoma south sub-county, Bungoma County. University of Nairobi. Available online at http://erepository.uonbi.ac.ke/handle/11295/90527.

Coppens d'Eeckenbrugge, Geo/Schiavo, Michele/Caron, Elisa/Ongwen, Dismas/Ireri Kamau, Joseph/Rono, Bernard/Leclerc, Christian (2019). Worldwide interconnections of Africa using crops as historical and cultural markers. Les Cahiers d'Afrique de lEst (52), 7–41. https://doi.org/10.4000/eastafrica.456.

Culwick, G. M. (1944). Nutrition in East Africa. Africa: Journal of the International African Institute 14 (7), 401–410. https://doi.org/10.2307/1157007.

Darkoh, M.B.K. (1989). THE UNDERLYING CAUSES OF THE FOOD CRISIS IN AFRICA. Transafrican Journal of History 18, 54–79. Available online at http://www.jstor.org/stable/24328704.

Edelman, Marc/Weis, Tony/Baviskar, Amita/Borras, Saturnino M./Holt-Giménez, Eric/Kandiyoti, Deniz/Wolford, Wendy (2014). Introduction: critical perspectives on food sovereignty. The Journal of Peasant Studies 41 (6), 911–931. https://doi.org/10.1080/03066150.2014.963568.

FAO (2008). An Introduction to the Basic Concepts of Food Security. Available online at https://www.fao.org/3/al936e/al936e00.pdf (accessed 4/11/2022).

GOK (2021). Agricultural-Policy-2021. "FOOD: OUR HEALTH, WEALTH AND SECURITY. Available online at https://kilimo.go.ke/wp-content/uploads/2022/05/Agricultural-Policy-2021.pdf (accessed 10/31/2022).

Goredema-Matongera, Nakai/Ndhlela, Thokozile/Magorokosho, Cosmos/Kamutando, Casper N./van Biljon, Angeline/Labuschagne, Maryke (2021). Multinutrient Biofortification of Maize (Zea mays L.) in Africa: Current Status, Opportunities and Limitations. Nutrients 13 (3). https://doi.org/10.3390/nu13031039.

Grey, Sam/Patel, Raj (2015). Food sovereignty as decolonization: some contributions from Indigenous movements to food system and development politics. Agriculture and Human Values 32 (3), 431–444. https://doi.org/10.1007/s10460-014-9548-9.

Gumo, Sussy/Gisege, Simon O./Raballah, Evans/Ouma, Collins (2012). Communicating African Spirituality through Ecology: Challenges and Prospects for the 21st Century. Religions 3 (2), 523–543. https://doi.org/10.3390/rel3020523.

Hoover, Elizabeth (2017). "You Can't Say You're Sovereign if You Can't Feed Yourself": Defining and Enacting Food Sovereignty in American Indian Community Gardening. American Indian Culture and Research Journal 41 (3), 31–70. https://doi.org/10.17953/aicrj.41.3.hoover.

Irakoze, Marie Lys/Wafula, Eliud N./Owaga, Eddy (2021). Potential Role of African Fermented Indigenous Vegetables in Maternal and Child Nutrition in Sub-Saharan Africa. International Journal of Food Science 2021, 3400329. https://doi.org/10.1155/2021/3400329.

Iwu, Maurice M. (2016). Food as medicine. Functional food plants of Africa. Boca Raton, CRC Press.

J. W. Purseglove (2011). The Origins and Migrations of Crops in Tropical Africa. Origins of African Plant Domestication, 291–310. Available online at https://www.degruyter.com/document/doi/10.1515/9783110806373.291/html.

Kenya National Bureau of Statistics (KNBS). 2019. Government Printers Kenya.

KNBS (2022). 2022 Economic Survey by Kenya National Bureau of Statistics. Available online at https://www.knbs.or.ke/wp-content/uploads/2022/05/2022-Economic-Survey1.pdf (accessed 8/12/2022).

Kobuthi, Joe (2019). Culinary Imperialism and the Hierarchies of Food. Available online at https://www.theelephant.info/culture/2019/09/27/culinary-imperialism-and-the-hierarchies-of-food/ (accessed 12/20/2022).

Kuhnlein, Harriet V./Chotiboriboon, Sinee (2022). Why and How to Strengthen Indigenous Peoples' Food Systems With Examples From Two Unique Indigenous Communities. Frontiers in Sustainable Food Systems 6, 808670. https://doi.org/10.3389/fsufs.2022.808670.

Mackenzie, A. Fiona D. (2000). Contested Ground: Colonial Narratives and the Kenyan Environment, 1920–1945. Journal of Southern African Studies 26 (4), 697–718. Available online at http://www.jstor.org/stable/2637567.

Makokha, K./Kyalo, W. (2016). Ethical Objections to Commercial Farming and Consumption of Genetically Modified Foods in Kenya. Thought and Practice 7 (1), 51–76. https://doi.org/10.4314/tp.v7i1.4.

Mamati, Kingasia (2018). An African Religious Worldview and the Conservation of Natural Environmental Resources: A Case Study of the Sengwer in Embobut Forest in Kenya.

Marcus, George E. (1995). Ethnography in/of the World System: The Emergence of Multi-Sited Ethnography. Annual Review of Anthropology 24, 95–117. Available online at http://www.jstor.org/stable/2155931.

Maundu, Patrick M./Ngugi, Grace W./Kabuye, Christine H. S. (1999). Traditional food plants of Kenya. Nairobi, Kenya Resource Centre for Indigenous Knowledge, National Museums of Kenya.

Mbiti, John S. (1997). African religions & philosophy. 2nd ed. Oxford, Heinemann.

medina, mabel (2017). Nyeleni commemorates tenth anniversary. La Via Campesina of 10/12/2017. Available online at https://viacampesina.org/en/nyeleni-commemorates-tenth-anniversary/ (accessed 6/15/2023).

Mostafalou, Sara/Abdollahi, Mohammad (2013). Pesticides and human chronic diseases: evidences, mechanisms, and perspectives. Toxicology and applied pharmacology 268 (2), 157–177. https://doi.org/10.1016/j.taap.2013.01.025.

Mwichabe, Situma (2013). The African Agrarian Ideology and the Food Security Challenge in Sub-Saharan Africa. Development 56 (3), 412–420. https://doi.org/10.1057/dev.2014.14.

Nangendo, Stevie Moses (1992). The Web of Poverty: Women and sugarcane farming in Bokoli Location, Bungoma district, Kenya. Organization for Social Science Research in Eastern and Southern Africa (OSSREA).

Ndabezinhle (2023). La Via Campesina @ the UN Human Rights Council: "Address the food crisis, move towards food sovereignty, implement UNDROP!". La Via Campesina of 3/17/2023. Available online at https://viacampesina.org/en/address-the-current-food-crisis-move-towards-food-sovereignty-and-also-implement-undrop-say-zainal-at-the-side-event-52nd-session-of-the-un-human-rights-council/ (accessed 6/14/2023).

Ndlovu-Gatsheni, S. (2018). Epistemic Freedom in Africa: Deprovincialization and Decolonization (1st ed.). Routledge. https://doi.org/10.4324/9780429492204.

Ndlovu-Gatsheni, S. J. (2020). Decolonization, Development and Knowledge in Africa: Turning Over a New Leaf (1st ed.). Routledge. https://doi.org/10.4324/9781003030423.

Ntinyari, Winnie/Gweyi-Onyango, Joseph P. (2021). Greenhouse Gases Emissions in Agricultural Systems and Climate Change Effects in Sub- Saharan Africa. In: African Handbook of Climate Change Adaptation. Springer, Cham, 1081–1105.

Nyamnjoh, F. B. (2017). Incompleteness: Frontier Africa and the Currency of Conviviality. Journal of Asian and African Studies 52 (3), 253–270. https://doi.org/10.1177/0021909615580867.

Nyangaga, Joseph Ooko (2022). The Doctrine of Occupation through "Terra Nullius" as a Right of Self-Determination of Peoples and the Legal Status of "Liberland" Territory under International Law. Available online at https://www.scirp.org/pdf/blr_2022031416032556.pdf (accessed 12/20/2022).

Nyéléni (2007). The Declaration of Nyéléni. Available online at https://nyeleni.org/DOWNLOADS/Nyelni_EN.pdf (accessed 12/20/2022).

Ochieng', William R. (1988). Colonial Famines In Luoland, Kenya, 1905–1945. Transafrican Journal of History 17, 21–33. Available online at http://www.jstor.org/stable/24328689.

Okoth-Ogendo, H. W. O. (1991). Tenants of the crown. Evolution of agrarian law and institutions in Kenya. Nairobi, ACTS Press, African Centre for Technology Studies.

Omwoyo, Samson Moenga (2015). Colonialism-and-the-Severity-of-Famines-and-Food-Shortages-in-Kenya-The-Case-Kenya-of-the-Abagusii-of-Western-Kenya-1890-1919-S.M.-Omwoyo. Available online at https://serek.or.ke/wp-content/uploads/2018/02/Colonialism-and-the-Severity-of-Famines-

and-Food-Shortages-in-Kenya-The-Case-Kenya-of-the-Abagusii-of-Western-Kenya-1890-1919-S.M.-Omwoyo.pdf (accessed 12/20/2022).

Onyango, Edward Michieka/Onyango, Benjamin Moranga (2018). The Rise of Noncommunicable Diseases in Kenya: An Examination of the Time Trends and Contribution of the Changes in Diet and Physical Inactivity. Journal of Epidemiology and Global Health 8 (1–2), 1–7. https://doi.org/10.2991/j.jegh.2017.11.004.

Patel, Raj (2009). Food sovereignty. The Journal of Peasant Studies 36 (3), 663–706. https://doi.org/10.1080/03066150903143079.

Pierrick (2003). Food sovereignty. La Via Campesina of 1/15/2003. Available online at https://viacampesina.org/en/food-sovereignty/ (accessed 12/20/2022).

Pierrick (2012). The people of the world confront the advance of capitalism: Rio +20 and beyond. La Via Campesina of 6/6/2012. Available online at https://viacampesina.org/en/the-people-of-the-world-confront-the-advance-of-capitalism-rio-20-and-beyond/ (accessed 6/15/2023).

Raschke, Verena/Cheema, Bobby (2008). Colonisation, the New World Order, and the eradication of traditional food habits in East Africa: historical perspective on the nutrition transition. Public Health Nutrition 11 (7), 662–674. https://doi.org/10.1017/S1368980007001140.

Sabarwal, Akash/Kumar, Kunal/Singh, Rana P. (2018). Hazardous effects of chemical pesticides on human health-Cancer and other associated disorders. Environmental toxicology and pharmacology 63, 103–114. https://doi.org/10.1016/j.etap.2018.08.018.

Songa, W/Lukalo, S/Nasambu Barasa K, Barasa K (2021). Bungoma City Context Analysis. Nutrition in City Ecosystems Project (NICE). Available online at https://nice.ethz.ch/wp-content/uploads/2021/10/NICE_City Overview_Bungoma_Kenya.pdf (accessed 5/29/2023).

Wabwoba, Mary Stella (2021). food-and-nutrition-security-status-of-households-in-bungoma-county-kenya. Available online at https://www.heraldopenaccess.us/article_pdf/33/food-and-nutrition-security-status-of-households-in-bungoma-county-kenya.pdf (accessed 6/15/2023).

Watson, Elizabeth E./Adams, William M./Mutiso, Samuel K. (1998). Indigenous Irrigation, Agriculture and Development, Marakwet, Kenya. The Geographical Journal 164 (1), 67. https://doi.org/10.2307/3060546.

World Economic Forum (2022). A fifth of people in Africa are suffering from chronic hunger. This map shows where the situation is most severe. Available online at https://www.weforum.org/agenda/2022/08/hunger-in-africa-serious-millions/ (accessed 11/4/2022).

CHAPTER 13

African Women, Religion, Climate Change and Food Security in the Context of Pandemics

Ezra Chitando

Introduction

HIV had, and continues to have, the face of a young African woman (Cullinun 2012). Food insecurity also has the face of an African woman (Van Dijk and Nkwana 2021). COVID-19, an existential threat at the time of writing this chapter, saw many African women (like other women globally) being subjected to high levels of sexual and gender-based violence (African Union Commission (AUC), UN Women, OHCHR and UNFPA 2020). Climate change is having a serious impact on African women (Steady 2014). While more examples could be cited, the foregoing ones confirm the pressing challenges that African women face.

However, focusing only on the threats to African women's flourishing retains the danger of a single story (Adichie 2009). This single story

E. Chitando (✉)
Department of Philosophy, Religion and Ethics, University of Zimbabwe, Harare, Zimbabwe

L. Maseno et al. (eds.), *Religion, Climate Change, and Food Security in Africa*, Sustainable Development Goals Series,
https://doi.org/10.1007/978-3-031-50392-4_13

reduces the experiences of African women to pain, suffering and death. This chapter counters this single story by engaging in the feminist hermeneutics of remembrance (Schüssler-Fiorenza 1983), that is, recovering and drawing attention to the achievements of religious African women in promoting food security, negotiating climate change and responding to pandemics. It also applies the hermeneutics of creative imagination utilised by African women theologians (see, e.g., Dube 2016) to highlight the areas of African women's leadership in the face of current and possible future pandemics.

Where earlier the SDGs and food security adopt entirely secular approaches (see, e.g., Valentini et al. 2019), this chapter draws attention to the primacy of religion in the African context. Where some scholars insist on limiting focus to specific African contexts, suggesting that the category "African women" is too vast and that it masks massive diversity, in this chapter I adopt a Pan-African perspective (Geiss 1967; Malisa and Nhengeze 2018). This is informed by an ideological commitment to the unity of Africa and the conviction that only united and cohesive action by African women, acting in solidarity with liberated men, can deliver sustainable development to Africa. African feminists (see, e.g., Adeleye-Fayemi and Horn 2009) and womanists (Kobo 2021) rightly contend that women's capabilities must be harnessed in order for the continent to achieve its development objectives.

Africa needs to prioritise food security and to provide effective responses to climate change and pandemics, as these have a direct bearing on the continent's efforts to achieve the United Nations Sustainable Development Goals (SDGs) and Agenda 2063 (the African Union's development blueprint). Africa bears the greatest burden of the food insecurity challenge, which has been exacerbated by climate change. In turn, food systems are estimated to contribute to more than a third of the global greenhouse gas (GHG) emissions responsible for climate change (International Food Policy Research Institute 2022: 7). Given the impact of climate change on food security globally, it is important to invest heavily in mobilising effective responses. This has seen different actors, including those from within the faith sector, prioritising climate change as an area of focus. Thus:

> Climate change is a growing threat to our food systems, with grim implications for food and nutrition security, livelihoods, and overall well-being, especially for poor and vulnerable people around the world. The imperative for urgent action on climate change—both to achieve the major emissions

> reductions needed to limit global warming and to increase adaptive capacity and resilience of food systems—is drawing global attention. (International Food Policy Research Institute 2022: 7)

This chapter is being finalised at a time when multiple forces are coalescing and threatening Africa's quest for sustainable development. The more recent and ongoing COVID-19 pandemic, the ongoing impact of HIV and AIDS, the climate emergency (see, e.g., Chitando et al. 2022), the food shortages and increasing food prices, security challenges (Chitando 2020; Kezi-Nwoha 2020) and other challenges are having a negative impact on African women in particular. Whereas these threats are real, African women have refused to adopt a fatalistic posture. Instead, they have deployed religion as social capital to negotiate and mitigate the impact of these challenges. Surprisingly, the dominant narrative remains that of African women as perpetual victims. Without denying the vulnerability of African women, it is possible, and even necessary, to identify African women's consistent resolve to act positively in the quest to achieve sustainable development.

This chapter seeks to contribute to this discourse by highlighting the contribution of religious African women to the overall response to climate change, food insecurity and pandemics. In the first section, the chapter provides reflections on African women, climate change, food security and pandemics. The responses by religious women in Africa to the challenges posed by climate change, food insecurity and pandemics constitute the focus of the second section. The third section makes some recommendations on how to enhance the effectiveness of religious African women's leadership in the face of climate change, food insecurity and pandemics. In conclusion, the chapter calls for greater acknowledgement of, and investment in, religious African women's leadership in the quest to meet the SDGs and Agenda 2063.

African Women, Food Insecurity, Climate Change and Pandemics

Climate change and food security have emerged as serious global concerns in the contemporary period. Across different parts of the world, millions of citizens are struggling to access food. Whereas food security is based on four pillars (availability, accessibility, utilisation and stability), all these pillars are shaky for different categories of people in diverse parts of the world, thereby creating food insecurity (Ahmad et al. 2021). The

COVID-19 pandemic, the war in Ukraine (at the time of writing) and climate change emerged as three key factors responsible for rising food prices, leading to challenges with availability and accessibility for many citizens in the world. Thus:

> Despite hopes that the world would emerge more quickly from the crisis and food security would begin to recover from the pandemic in 2021, world hunger rose further in 2021, following a sharp upturn in 2020 in the midst of the COVID-19 pandemic. Disparities in the impact of the pandemic and the recovery, together with the limited coverage and duration of the social protection measures, led to widening inequalities that have contributed to further setbacks in 2021 towards achievement of the Zero Hunger target by 2030. (FAO, IFAD, UNICEF, WFP and WHO 2022: xvi)

Crucially, it is helpful to recognise that food insecurity does not emerge because some people or communities are not hardworking or lack proper planning. Food insecurity is an outworking of the current skewed global food systems. Further, while for some parts of the world the challenge is obesity, for many Africans the main challenge remains one of food insecurity (Ericksen 2008: 236). The following citation is helpful in terms of both clarifying the meaning of food security, as well as accounting for food insecurity; thus:

> A person is food secure when he or she has access at all times to enough food for an active, healthy life. Accordingly, people are food secure when their consumption of food is sufficient, secure (not vulnerable to consumption shortfalls), and sustainable .... The list of causes of food insecurity is long and multifaceted: they range from political instability, war and civil strife, macroeconomic imbalances and trade dislocations to environmental degradation, poverty, population growth, gender inequality, inadequate education, and poor health. All, however, can be related in some fashion to two basic causes: insufficient national food availability and insufficient access to food by households and individuals. (Smith et al. 2000: 200)

It is crucial to retain a global outlook when examining African women and food insecurity. This is key, as often the situation of African women needs to be appreciated against the backdrop of global patterns and trends. The reality is that the ideology of patriarchy, that is, an ideology and practice that privileges men and oppresses women (Rakoczy 2004: 29–31), prevents women in most parts of the world from enjoying life in abundance. This leads to women's food insecurity in different parts of the world. Thus:

> Historical evidence reveals that women have been systematically oppressed through various laws and norms in their respective societies. Globally, women are still vulnerable to food insecurity due to lack of access and control over their household's assets such as land. While women produce 70% of food grown for consumption, the asymmetries in ownership of, access to and control of livelihood assets negatively affect their role in food production. (Masuku and Garutsa 2021: 141)

Consistent with the global trends, African women are disproportionately affected by food insecurity, climate change and pandemics. Since these themes require more space to unravel them than is available, in this chapter I shall only draw attention to the key issues at hand. There are many factors responsible for African women's greater vulnerability to food insecurity. Traditional norms prevent many women from owning land; they have less access to education and training; they also have less access to credit, and do not have access to the latest technology. Further, women face time constraints, as well as mobility challenges, alongside the impact of HIV and AIDS that increases their caretaker roles (USAID n.d.). From a social justice perspective, it is equally disturbing to realise that, "Too often, women eat last and least" (CARE 2020: 2). The feminisation of poverty is an affront to the religious tenets of justice, equality and dignity of all. It threatens the achievement of the SDGs, including SDGs 2, 3 and 13, the focus of this chapter.

African women's food security challenges have been worsened due to the impact of climate change. Climate change's impact is higher on African countries, as they rely heavily on agriculture, which is climate dependent (Maino and Emrullahu 2022: 4). Agriculture, food security and water resources are affected by temperature and precipitation, resulting in diminishing returns for farmers and other members of society. Further, the increases in temperatures, as well as changes in rainfall patterns, have an impact on health across the continent (World Meteorological Organization 2020: 5). African women, however, are more adversely affected by these developments. The following citation helps to clarify this:

> It is now widely acknowledged that climate change impacts will not be gender neutral. This is evident from current experiences of extreme climatic

> events such as droughts and floods. Men and women have different coping and adaptive capacities that translate to gender-differentiated vulnerabilities to the impacts of a changing climate. Gender-based inequalities in access to assets and gendered social roles are mainly responsible for this difference in adaptive capacities to respond to the effects of climate change. Legal and sociocultural barriers also inhibit women from effectively responding to climatic risk. (UNDP 2012: 2)

As if the above-named challenges were not enough, African women also have to contend with pandemics. They are confronted by pandemics such as HIV and AIDS (see, e.g., Masenya 2006) and COVID-19 (see, e.g., Labeodan et al. 2021). Alongside multiple factors that increase their vulnerability, their social construction as caregivers means they carry heavy loads. As Ravayi Marindo (2017) has underscored using feminist analysis, power differentials are key to understanding the gendered nature of pandemics in Africa. She submits that, "…no understanding of African epidemics can be achieved without understanding the systems of power, how these systems define the relations of power between individuals, and how this ultimately contributes to the vulnerability of women" (Marindo 2017: 202). These systems of power should not be reduced to internal contests between men and women within local African homes and communities. No. They have historical precedence, as well as global dimensions. This is clarified by Tinyiko Chauke and Puleng Segalo (2021) when reflecting on gendered pandemics in a South African context. They highlight how

> the intersections of race, gender, health emergencies, tradition, culture, sexuality, class, colonialism, imperialism and apartheid continue to shape and colour women's experiences of being "women" in South African contexts. Women continue to face social, historical, political and economic challenges within their societies. During pandemics (such as COVID-19) societal inequalities become visibilised and women bear the brunt. (Chauke and Segalo 2021: 158)

Indeed, there are notable overlaps in African women's vulnerability to food insecurity, climate change and pandemics. First is the systematic marginalisation of Africa in the global economic system (Ndikumana 2015). Africa's story of being exploited and discarded by actors in the global North is well documented (see, e.g., Rodney 1972). The same story is replicated in the global food systems where Africa's indigenous interpretation of food as a right, as opposed to capitalist notions of food as a commodity, is continuously undermined. Africa's food sovereignty is compromised by global food systems that place emphasis on profits rather than ending hunger for all (see, e.g., Alliance for Food Sovereignty in Africa 2019). Second is the thread of patriarchy that runs through African women's struggle against food insecurity, climate change and pandemics. As argued above, women's less powerful position in African societies renders them more vulnerable to multiple challenges, including those under focus in this chapter. Female headed households in rural areas tend to suffer more from all the different challenges (see, e.g., Van Dijk and Nkwana 2021: 108), although the struggles of women in African urban centres needs to be acknowledged and addressed (see, e.g., the chapter by Naicker in this volume).

Third, climate change and pandemics are themselves contributing factors to African women's food insecurity. For example, during pandemics, women's food insecurity worsens. Fourth, and quite significant for this chapter, is the extent to which religion is implicated in African women's marginalisation. Religion introduces, justifies and reinforces patriarchy, thereby increasing African women's susceptibility to the challenges of food insecurity, climate and pandemics. Religion (often in concert with particular interpretations of culture) provides the ideological justification of patriarchy, lubricates it and facilitates its mobility and apparent viability. This is clearly articulated in the following passage:

> Patriarchal beliefs are in most cases transmitted and sustained through diverse forms of cultural and religious fundamentalism, which rigidly define what is acceptable and what is offensive to a society in relation to gender beliefs and roles. By this means, violence against women is indirect and structural in nature because it is not perceptible. Religion is often used as a social instrument through which patriarchy and its cultural beliefs about men's supremacy over the women are transmitted to the public domain.... In a bid to provide justification for the sustenance of patriarchy, religion could be politicized through various forms of religious interpretations. Religious texts are often (re)interpreted to canvass for total submission of women to the authority of men. (Onwutuebe 2019: 2)

The net effect of the challenges that I have outlined in this section is that, among others, many women are forced to adopt survival strategies that compromise their health and wellbeing, including having sexual relationships with much older men (West and Haddad 2016). Food insecurity also forces some mothers to engage in sex work for the sake of their children. Overall, the inequality that is deepened by climate change and pandemics conscripts many women into basic struggles for survival, preventing them from taking up formal leadership positions in society. It is in this environment that African women within the religions have sought to make a difference. The chapter turns to this theme in the next section.

Religion as a Resource: African Women Negotiating Multiple Challenges

Having outlined the multiple challenges that African women are facing in the previous section, this section proceeds to summarise how African women have utilised religion to respond to the challenges. However, it is necessary to begin this section by indicating that the status of religion in Africa's development remains heavily contested (see, among others, Mtata 2013; Chitando et al. 2020). Whereas some scholars regard religion as a negative force responsible for destructive tendencies (gender inequality, deepening competitive ethnicity, violent extremism, etc.), others see religion as promoting social cohesion, promoting peace and inspiring development (for an excellent review, see, e.g., Joint Learning Initiative on Faith and Local Communities 2022). This chapter contends that the available evidence shows that African women have deployed religion strategically to negotiate the challenges of food insecurity, climate change and pandemics.

While Africa is characterised by radical religious pluralism (Platvoet 1996), in this chapter I focus specifically on the activities of women affiliated to African Traditional (Indigenous) Religions [ATRs], Christianity and Islam. This is not to suggest that women who are followers of the religions that command fewer adherents (e.g., the Baha'i Faith, Buddhism, Hinduism, Judaism, Rastafari, etc.) have not been actively involved in responding to food insecurity, climate change and pandemics. Indeed, it is strategic for future studies to explore how women who belong to the less subscribed religions in Africa are utilising their membership to mitigate and strive to overcome the effects of various challenges. What is clear, however, is that although religion sponsors patriarchy, women in Africa have appropriated the very same religion as a resource for their flourishing. As scholars such as Oduyoye (2019) and Vuola (2017) have asserted, although women in Africa do experience religion as burdensome in many respects, they also mine the same religion for their advancement and liberation.

Space and thematic considerations mean that it is not possible for this chapter to present an exhaustive account of the interventions by African women in religion in response to food insecurity, climate change and pandemics. Consequently, the chapter concentrates on the key themes, in anticipation of future reflections on specific women's religious groups in particular African contexts. First, African women in religion have sought to take the struggle to the higher plane, namely, challenging and exposing the oppressive patriarchal ideology that renders women more vulnerable to food insecurity, climate change and pandemics. For example, the multi-faith Circle of Concerned African Women Theologians (hereafter, the Circle) has been consistent in challenging the systematic exclusion and oppression of women within organised religion and in society.

For me, this investment by the Circle is extremely strategic as it unlocks African women's capabilities and promotes women's empowerment (see, e.g., Kuhumba 2022). The Circle has propounded life giving theologies that seek the full participation and flourishing of women and the environment in Africa (Matholeni et al. 2020; Chirongoma and Mombo 2021). It rejects patriarchal, androcentric and death-dealing theologies, such as those that condemn women to early deaths during pandemics such as HIV and AIDS (see, e.g., Dube 2008). Mercy Amba Oduyoye, arguably the foremost Circle activist, yearns for "…a world of women and men in which women's wisdom, work and words are placed as worthy of being an integral part of what constitutes being human" (Oduyoye 2019: 171).

Women in ATRs, Christianity and Islam are defying the restrictive and hunger intensifying patriarchal ideologies and lifting up liberating and satiating theologies found in their religions to take up leadership positions within their faith communities in Africa. They are providing hope that the SDGs and Agenda2063 can be achieved, as they bring their gifts to the struggles against food insecurity, climate change and pandemics. Religion is providing avenues for women's leadership in different African contexts. With passion and compassion (in faith settings, non-governmental organisations/civil society, in politics and government), they are "challenging and unseating globalised, structural, deeply gendered privilege" (Gaynor 2022: 181). This is consistent with the recognition that women have always enjoyed leadership in African settings (Achebe 2020), but colonialism tended to rigidify what used to be fluid gender identities.

Second, religious African women are actively involved in agriculture and improving the food security of their families, communities, nations and the continent. With religion as a ubiquitous feature of African life (Oduyoye 1995, 2019), it is clear that the majority of women who produce food in Africa are religious. Their religious commitment galvanises them to continue to work against formidable odds. In an informative article, Linda Naicker (2021) draws attention to a dimension that is often overlooked in discourses on food security, namely, the struggles of many women in urban areas in Africa to put food on the table. She reflects on the inspirational story of women urban community food gardeners, firmly committed to their Christian faith, in Pietermaritzburg, South Africa. She contends that their theory and practice embodied and expressed African Women's Theologies (AWT). She writes as follows:

> As I journeyed with the group of women gardeners, I became deeply mindful of the embodiment of AWT in their lives—they embraced it, and through it, found ways to overcome the harsh conditions they faced. While the group of women were not aware of the theological, academic, and activist nature of AWT, they embodied its tenets in their day-to-day struggle for survival, their resistance to systemic injustice, and their agency in the face of overwhelming disparity. In an environment where under-resourced households struggled to put food on the table, these women used urban community food gardens as an important safe space to mobilize and draw on communal strength as they worked in solidarity to overcome poverty. (Naicker 2021: 31–32)

Third, in the case of climate change, African women's indigenous spirituality (Siwila 2014) is a powerful resource in mitigation and adaptation. Across the region, women's indigenous knowledge systems are assisting families to improve their food security and to respond to the climate emergency by engaging in climate-smart agriculture (CSA) (see, e.g., Davies et al. 2019 for a general discussion on religion and CSA in Namibia). Although they face the threat of climate change, women in Africa are demonstrating high levels of creativity and adaptability, recovering the practice of planting traditional crops that are more drought resistant in some areas. In Nigeria, some churches actively support women's access to land, although patriarchal beliefs and practices continue to hold sway (Amusan et al. 2021).

Fourth, in the spirit of Ubuntu, namely, "I am because you are; because you are, therefore I am" in ATRs (Chisale 2018), charity in Christianity and almsgiving (*zakat*) in Islam, religious African women have been actively involved in supporting vulnerable members of society. Whereas the headlines tend to cover the once-off mega donations by multinational organisations and non-governmental organisations (NGOs), religious African women tend to work quietly but effectively. These include Catholic nuns, who were among the pioneers in supporting people living with HIV when the stigma was extremely high in the late 1980s and 1990s. They also rose to the occasion in the context of COVID-19 (see, e.g., Mwale 2022 for a discussion of their role in Zambia). This was also replicated by other African women's religious groups. In Uganda, for example, during the pandemic, the Church of Uganda's Mothers' Union was distributing food parcels to families and individuals that were affected by the strict lockdowns.

Fifth, women in religion in Africa have provided holistic health and healing in the face of pandemics and other challenges. Through their groups or individually, they have offered mental health support to many who would have otherwise fallen apart in the face of the pressing existential challenges that have been outlined above. Although women in Africa are not served well by the current health delivery systems (WHO Regional Office for Africa 2012), religious women have stepped in to accompany mostly fellow women, but some. For example, women constitute the majority of healthcare workers in Africa. Most of them are religious, as are the African Independent Church prophetesses and women traditional healers who provide support to women in the face of pandemics.

Sixth, and finally, African women in religion are forming robust interfaith networks (McGarvey 2022) that challenge the competitive religiosity championed by some male religious leaders. While competitive religiosity promoted by some male leaders is responsible for religious violence in Africa, interfaith networks led by women are generating new models of peacebuilding and integration. These women-led interfaith networks in Africa are providing a wide range of services, including addressing pandemics and climate change/food security, responding to sexual and gender-based violence, young women's leadership training and others. They are found on the frontlines in many of the continent's trouble spots, providing leadership beyond the patriarchal leadership found in the dominant religions.

Recommendations for Enhancing Religious African Women's Effectiveness to Respond to the Challenges Discussed

Having explored African women's vulnerability to food insecurity, climate change and pandemics, as well as having summarised some of the key interventions by women in religion in Africa in the foregoing sections, in this section I seek to make a few recommendations on enhancing the effectiveness of African religious women in the face of pandemics:

- There must be a more deliberate investment in enhancing African women's capacities to respond to multiple shocks. In particular, women in African faith communities can benefit greatly from disaster preparedness training. Women's role in responding to disasters in Africa is quite strategic, hence the need to bolster their knowledge and skills, even as they are already on the frontline
- Greater networking between African academic activist groups in the field of religion, such as the Circle, and women in African faith communities is required in order to initiate joint practical activities, as well as advocacy for women's human rights in their totality
- Programmes to further equip African religious women's active, meaningful and effective participation in African food systems are required
- African women's religious groups need to roll out training for young women leaders. Given the reality that adolescents and young people constitute the majority of Africa's citizens, such empowerment programmes are crucial
- Sensitising men and boys for solidarity with women and girls in pandemics.

While African religious women are playing a significant leadership role in the face of food insecurity and climate change, there is a need for men in Africa to stand and act in solidarity with them. Such men will join the struggle, not to take over, but to be mentored by women, learn from them and become allies. One of the leading African male theologians with a keen interest in gender justice and women's full liberation, Tinyiko S. Maluleke (see, e.g., Maluleke 2018) has called for liberating masculinities in Africa. These masculinities must become African ecological masculinities (see, e.g., Hultman and Pulé 2018). Such masculinities will mobilise men and boys to act in solidarity with African women and girls in their struggles against food insecurity, climate change and pandemics.

Conclusion

African women's vulnerability to food insecurity, climate change and pandemics has been well noted. Indeed, this story, urgent in its own right, runs the risk of becoming the "single story" that demobilises African women. In this chapter, I have sought to highlight the accompanying story of resilience, creativity and leadership. I have summarised how religious African women have negotiated the multiple shocks and existential threats by appropriating and deploying the social capital they derive from religion. I have illustrated how they are responding to food insecurity, climate change and pandemics by subverting patriarchy. In contexts where critics and prophets of doom have predicted death, African religious women have chosen life. Life in abundance. Their initiatives, activism and leadership provide a firm foundation for Africa to succeed in achieving the SDGs that are relevant to the themes discussed herein, namely, 2 ("Zero hunger"), 3 ("healthy lives for all") and 13 ("climate action"). The (religious) women of Africa are troubled, but not destroyed; in fact, they are providing effective leadership to a very resilient, rich and robust continent.

References

Achebe, Nwando. 2020. *Female Monarchs and Merchant Queens in Africa*. Athens, OH: Ohio University Press.

Adeleye-Fayemi, Bisi and Jessica Horn. Eds. 2009. *Voice, Power and Soul: Portraits of African Feminists*. Accra: African Women's Development Fund.

Adichie, Chimamanda Ngozi. 2009. "The Danger of a Single Story." *TED: Ideas worth Spreading*, July. http://www.ted.com/talks/chimamanda_adichie_the_danger_of_a_single_story, accessed 19 July 2022.

African Union Commission (AUC), UN Women, OHCHR and UNFPA. 2020. Gender-Based Violence in Africa during COVID-19. Addis Ababa: African Union Commission (AUC).

Ahmad, Nafees et al. 2021. Food Insecurity: Concept, Causes, Effects and Possible Solutions. *IAR Journal of Humanities and Social Science* 2(1): 105–113.

Alliance for Food Sovereignty in Africa (AFSA). 2019. African Food Systems and the SDGs. Saly, Senegal, 3–5 November 2018: Conference Report. Kampala: AFSA. Available at: https://afsafrica.org/wp-content/uploads/2019/05/

afsa-african-food-systems-and-the-sdgs-conference-report-final-en.pdf, accessed on 15 July 2022.

Amusan, L., M. E. Okokuwebe and G. Odularu. 2021. Women Development in Agriculture as Agency for Fostering Innovative Agricultural Financing in Nigeria. *African Journal of Food, Agriculture and Nutrition and Development* 21(6): 18279–18299.

CARE. 2020. Policy Report: Left Out and Left Behind: Ignoring Women Will Prevent Us From Solving the Hunger Crisis. Available at: https://www.care-international.org/files/files/LeftOutandLeftBehind.pdf, accessed on 15 July 2022

Chauke, Tinyiko and Puleng Segalo. 2021. Exploring women's gendered experiences of sexual reproductive health during a pandemic: Intergenerational reflections. *Agenda* 35(4): 151–160.

Chirongoma, Sophia and Esther Mombo. Eds. 2021. *Mother Earth, Postcolonial and Liberation Theologies*. London: Lexington Books.

Chisale, Sinenhlanhla S. 2018. Ubuntu as care: Deconstructing the gendered Ubuntu. *Verbum et Ecclesia* 39(1), a1790. https://doi.org/10.4102/ve.v39i1.1790

Chitando, Anna. Ed. 2020. *Women and Peacebuilding in Africa*. London: Routledge.

Chitando, Ezra, Masiiwa Ragies Gunda and Lovemore Togarasei. Eds. 2020. *Religion and Development in Africa*. Bamberg: University of Bamberg Press.

Chitando, Ezra, Ernst M. Conradie and Susan M. Kilonzo. Eds. 2022. *African Perspectives on Religion and Climate Change*. London: Routledge.

Cullinun, Kerry. 2012. In Africa, AIDS often has a woman's face. *Africa Renewal*: Special Edition on Women. Available at: https://www.un.org/africarenewal/magazine/special-edition-women-2012/africa-aids-often-has-woman's-face, accessed 19 July 2022.

Davies, Julia et al. 2019. Considering Religion and Tradition in Climate Smart Agriculture: Insights from Namibia. In Todd S. Rosenstock, Andreea Nowak and Evan Girvetz, Eds., *The Climate-Smart Agriculture Papers Investigating the Business of a Productive, Resilient and Low Emission Future*. Cham: Springer, 187–197.

Dube, Musa W. 2008. *The HIV and AIDS Bible: Selected Essays*. Scranton: University of Scranton Press.

Dube, Musa W. 2016. The Subaltern Can Speak: Reading the Mmutle (Hare) Way. *Journal of Africana Religions* 4(1): 54–75.

Ericksen, Polly J. 2008. Conceptualizing Food Systems for Global Environmental Change Research. *Global Environmental Change* 18: 234–245.

FAO, IFAD, UNICEF, WFP and WHO. 2022. *The State of Food Security and Nutrition in the World 2022. Repurposing food and agricultural policies to make healthy diets more affordable*. Rome, FAO.

Gaynor, Niamh. 2022. *Engendering Democracy in Africa: Women, Politics and Development.* London: Routledge.

Geiss, Imanuel. 1967. Notes on the Development of Pan-Africanism. *Journal of the Historical Society of Nigeria* 3(4): 719–740.

Hultman, Martin and Paul M. Pulé. 2018. *Ecological Masculinities: Theoretical Foundations and Practical Guidance.* London: Routledge.

International Food Policy Research Institute. 2022. *2022 Global Food Policy Report: Climate Change and Food Systems.* Washington, DC: International Food Policy Research Institute.

Joint Learning Initiative on Faith and Local Communities. 2022. Ed. *The State of the Evidence in Religions and Development.* Washington, DC: JLI.

Kezi-Nwoha, Helen. 2020. *Feminist Peace and Security in Africa.* Oxford: Oxfam.

Kobo, Fundiswa. 2021. Decentring the West: Womanism and the Circle of Mercy Oduyoye. In Lilian Cheelo Siwila and Fundiswa Amanda Kobo, eds., *Religion, Patriarchy and Empire: Festschrift in Honour of Mercy Amba Oduyoye.* Pietermaritzburg: Cluster Publications, 261–282.

Kuhumba, Kevin S. 2022. Towards Women's Empowerment in Africa: Insights from the Capability Approach. In Ezra Chitando, Sophia Chirongoma and Kudzai Biri, eds., *Women and Religion in Zimbabwe: Strides and Struggles.* London: Lexington Books, 237–254.

Labeodan, Helen et al. Eds. 2021. *COVID-19: African Women and the Will to Survive.* Bamberg: University of Bamberg Press.

Maino, Rodolfo and Drilona Emrullahu. 2022. Climate Change in Sub-Saharan Africa's Fragile States: Evidence from Panel Estimations. International Monetary Fund Working Paper No. 2022/054.

Malisa, Mark and Phillippa Nhengeze. 2018. Pan-Africanism: A Quest for Liberation and the Pursuit of a United Africa. *Genealogy* 2(3), 28; https://doi.org/10.3390/genealogy2030028

Maluleke, Tinyiko S. 2018. Of Wounded Killers and "Failed Men": Broadening the Quest for Liberating African Masculinities. *Journal of Gender and Religion in Africa* 24(1): 33–78.

Marindo, Ravayi. 2017. Gendered Epidemics and Systems of Power in Africa: A Feminist Perspective on Public Health Governance. *Africa Development* 42(1): 199–219.

Masenya, Madipoane (ngwan'a Mphahlele). 2006. Killed by Aids and Buried by Religion: African Female Bodies in Crisis. *Old Testament Essays* 19(2): 486–499.

Masuku, Mfundo Mandla and Tendayi Garutsa. 2021. Enhancing Food Security Through Systematic African Women Empowerment in South Africa. *Mediterranean Journal of Social Sciences* 12(6): 141–150.

Matholeni, Nobuntu P., Boateng, Georgina K. & Manyonganise, Molly. (eds.) 2020. *Mother Earth, Mother Africa & African Indigenous Religions.* Stellenbosch: SUN PReSS, https://doi.org/10.18820/9781928480730

McGarvey, Kathleen. 2022. Women of Faith Working Together as Mothers of a Culture of Peace: The Women's Interfaith Council in Northern Nigeria. In Ezra Chitando and Ishanesu S. Gusha, eds., *Interfaith Networks and Development: Case Studies from Africa*. Cham: Palgrave Macmillan, 99–115.

Mtata, Kenneth. Ed. 2013. *Religion: Help or Hindrance to Development?* Leipzig: Evangelische Verlagsanstalt.

Mwale, Nelly. 2022. Representations of Roman Catholic religious sisters' responses to COVID-19 in the Zambian media. *HTS Teologiese Studies/Theological Studies* 78(2), a6580. https://doi.org/10.4102/hts.v78i2.6580

Naicker, Linda. 2021. Sisters in Solidarity: Resistance and Agency through Urban Community Food Gardens in Pietermaritzburg. *Journal of Gender and Religion in Africa* 27(1): 24–41.

Ndikumana, Léonce. 2015. Integrated Yet Marginalized: Implications of Globalization for African Development. *African Studies Review* 58(2): 7–28.

Oduyoye, Mercy A. 1995. *Daughters of Anowa: African Women and Patriarchy*. Maryknoll, NY: Orbis Books.

Oduyoye, Mercy A. 2019. *Re-membering Me: Memoirs of Mercy Amba Oduyoye*. Ibadan: SEFER.

Onwutuebe, Chidiebere J. 2019. Patriarchy and Women Vulnerability to Adverse Climate Change in Nigeria. *SAGE Open* January–March: 1–7.

Platvoet, Jan G. 1996. The Religions of Africa in their Historical Order. In, Jan Platvoet, James Cox & Jacob Olupona (eds.), *The Study of Religions in Africa: Past, Present and Prospects*. Cambridge: Roots and Branches, 46–102.

Rakoczy, Susan. 2004. Religion and violence: the suffering of women. *Agenda* 18(61): 29–35.

Rodney, Walter. 1972. *How Europe Underdeveloped Africa*. London: Bogle-L'Ouverture Publications.

Schüssler-Fiorenza, Elisabeth. 1983. In *Memory of Her: A Feminist Theological Reconstruction of Christian Origins*. New York: Crossroad.

Siwila, Lilian C. 2014. "Tracing the Ecological Footsteps of our Foremothers": Towards an African Feminist Approach to Women's Connectedness with Nature. *Studia Historiae Ecclesiasticae* 40(2): 131–147.

Smith, Lica C., Amani E. El Obeid and Helen H. Jensen. 2000. The Geography and Causes of Food Insecurity in Developing Countries. *Agricultural Economics* 22: 199–215.

Steady, Filomina C. 2014. Women, Climate Change and Liberation in Africa. *Race, Gender and Class* 21 (1 and 2): 312–333.

UNDP. 2012. Policy Brief 4 Africa: Gender, Climate Change and Food Security. New York: UNDP.

USAID. n.d. Fact Sheet: Food Security and Gender. Available at: https://www.oecd.org/dac/gender-development/46460857.pdf, accessed 15 July 2022.

Valentini, Riccardo et al. Eds. 2019. *Achieving the Sustainable Development Goals Through Sustainable Food Systems.* Cham: Springer Nature Switzerland.

Van Dijk, H. G. and H. M. Nkwana. 2021. The Face of Food Insecurity is Female: A PostColonial Feminist Argument for Rural Women. *African Journal of Gender, Society and Development* 10(1): 99–116.

Vuola, Elina. 2017. Religion, Intersectionality, and Epistemic Habits of Academic Feminism. Perspectives from Global Feminist Theology. *Feminist Encounters: A Journal of Critical Studies in Culture and Politics* 1(1): 04.

West, Gerald O. and Beverley G. Haddad. 2016. Boaz as 'Sugar Daddy': Re-Reading Ruth in the Context of HIV. *Journal of Theology for Southern Africa* 155(July 2016 special Issue): 137–156.

WHO Regional Office for Africa. 2012. Addressing the Challenge of Women's Health in Africa: Report of the Commission on Women's Health in the African Region. Brazzaville: WHO Regional Office for Africa.

World Meteorological Organization. 2020. *State of the Climate in Africa 2019.* Geneva: World Meteorological Organization.

PART IV

Religion, Climate Change, Food Security and Shifting Vulnerabilities in Africa

CHAPTER 14

Farming God's Way to Avert Crop Failure and Pests in Malawi

Gertrude A. Kapuma and Mwawi Chilongozi

Introduction

Relying on subsistence farming is becoming a big problem for some people living in the rural areas. The land is so exhausted that it no longer produces enough harvest to sustain the farmers and their families for the whole year. The land has been overused over the years and it has become unproductive. Population growth has also made it difficult for people to own enough land to grow food and cash crops. The land problem is more pronounced in the southern region of Malawi because it is overcrowded. Families share very small plots to cultivate, and, in most cases, the size of the plot is not enough to bring good yields for the family. For the farmers to get better yields, they feel they must use chemical fertilizers. Malawi is now experiencing high costs of fertilizers, which is a big problem for most of the people in the rural areas who depend only on farming. People spend

G. A. Kapuma (✉)
Department of Practical Theology, Zomba Theological University, Zomba, Malawi

M. Chilongozi
CCAP General Assembly, Lilongwe, Malawi

L. Maseno et al. (eds.), *Religion, Climate Change, and Food Security in Africa*, Sustainable Development Goals Series,
https://doi.org/10.1007/978-3-031-50392-4_14

a lot of their time cultivating in the gardens expecting to have enough food in return. However, because the soil is overstretched and almost bare, most crops are not able to grow and produce the expected yield. As such, it is important as a country to introduce Farming God's Way methods. Farming God's Way is a type of conservation agriculture that promotes no tilling, mulching, and crop rotation. This method of farming is based on biblical metaphors such as "God doesn't plow, God's blanket, and the Garden of Eden" (Spaling and Kooy 2019: 1). Farming God's Way is "a development intervention for improving food security, adapting to climate change, and restoring soil productivity for resource-poor farming households" (Spaling and Kooy 2019: 1). Farming God's Way may help such people to improve their livelihood. This is a possibility because the method does not require chemical fertilizers.

Farmers can use natural local manure to improve the soil. These local manures include animal droppings, the tree leaves that are normally swept and burnt in our surroundings, and household wastes. When these are properly collected and handled, they turn into manure that can be used in gardens. By so doing, the soil would be enriched with proper nutrients resulting in good yields. The church can take a leading role in helping people to improve their ways of farming to harvest more. The leadership of the churches should be the ones to understand the advantages of this method and make it a priority to engage members in ways of improving their lifestyle of farming in new and less expensive ways. The focus in the following section is on some of the challenges that need to be considered to help people to understand and accept the new method of Farming God's Way (Shuwali, 2021).

Challenges that Farmers Encounter in Malawi

There are many challenges that farmers encounter in practicing agriculture in Malawi. Among many, soil erosion tops the list. Management of soil is a problem in many villages of Malawi. The soil's texture is eroded through unsuitable ways of making ridges. When heavy rains come, the ridges are swept together with the soil leaving behind unproductive and unfertile soils that cannot produce much yield. Growing up, we saw our parents making box ridges in their gardens to hold water. Today we do not see these in people's gardens. In a situation like this year (2023), following Cyclone Freddy, in Malawi, many gardens experienced water floods from the hills into the gardens. Cyclones are a new phenomenon in Malawi because Malawi had never experienced before such a disastrous cyclone as Freddy. The first Cyclone was Ana in 2022. In March 2023, Malawi had the worst experience from Cyclone Freddy. According to meteorologists, Cyclone Freddy was the most long-lasting and most traveled tropical cyclone in the Southern Hemisphere. The cyclone affected Madagascar, Mozambique, and Malawi. However, Malawi was the hardest hit by the cyclone than the other countries. The cyclone caused flash floods and mudslides that damaged infrastructures such as houses, roads, schools, and health facilities. Apart from these, crops and livestock were washed away. The impact of Cyclone Freddy was very devastating as more than 1000 people lost their lives and more than 500,000 people were rendered homeless. The crops that were washed away were almost ready for harvest, leading to food insecurity for most households in Southern Malawi.

Tropical cyclones such as Freddy are becoming common due to climate change. The other effects of climate change include increasing temperatures and increasing delays and inconsistencies in the onset of rainfall (Hunter et al. 2020: 2). Consequently, crop productivity is decreasing, resulting in food insecurity for many households. The areas that were affected by Cyclone Freddy and experienced mudslides had soil erosions. Such land will have to be cultivated differently to bring back its nutrients for plants to grow well. In this case, Farming God's Way should be encouraged to understand organic ways of taking care of the small piece of land so that it can be more productive.

As alluded to early on, the bulk of farmers in Malawi mostly practice subsistence farming and most farmers use chemical fertilizers. The government of Malawi has a subsidy program[1] for farm inputs, that is, chemical fertilizers and seeds. The types of fertilizers that are commonly used are NPK and Urea and these come mostly in 50 kg sacks and are distributed by private traders who are hired by the government to distribute fertilizer in the rural areas. It is important to note that in Malawi, the land is overused and, in most areas, there are no new methods of preserving the soil for future crops. To address this challenge, many farmers tend to use chemical fertilizers to improve the food nutrients in the soil. However, fertilizers must be used properly. One may have the capacity to buy fertilizer and fail to use it properly. For instance, in the year 2023, farmers had two challenges. The first one was the very high cost of fertilizer on the

[1] The Subsidy Program helps selected farmers buy fertilizer and seeds at reasonable prices. Fertilizer is very expensive in Malawi because it is imported from the Middle East or Europe.

market from 32,000 Malawi Kwacha to 72,000 (from US $32 to US $72), which most farmers in the village could not afford.

The second problem was that though the fertilizers were expensive, a delay in the distribution of fertilizers countrywide affected productivity. By the time the fertilizers were available, most crops were past the time of applying fertilizer. This caused most households to fail to harvest enough food. One male respondent, aged 42, who comes from Bvumbwe in Thyolo District, shared the following:

> In the past years, my harvest was usually between 12 to 15 bags of 50 kilograms of maize every year. However, in 2023 the same garden produced only 2 bags of maize. Even after applying the fertilizer, the crops did not look any better. I suspect that I bought fake fertilizer. … Only the small portion of the garden where I applied chicken manure for vegetable growing had a better yield because of the manure that was already in the soil. (Matebule, OI, 26 April 2023)

He may not be blamed for this understanding because he may have applied the fertilizer late and at the peak of the rains which eventually may have washed away the fertilizer. However, it is also possible that the fertilizer was fake because he bought it from people selling it on the streets.

The Malawi Fertilizer Input Subsidy Program (FISP) has not worked as expected. Whereas this program is for local small farmers to access and buy chemicalized fertilizer and seeds to grow to improve harvest and food security, it has faced many challenges and some local farmers who are supposed to benefit from the programs have not. Andy Currier (23 January 2020) argues, "The FISP program has failed to meaningfully reduce food insecurity and the high volatility of maize prices in Malawi. This volatility in maize prices harms farmers primarily dependent on maize. In years with good harvests, the prices of maize fall and incomes are stunted, while poor production years result in not having enough surplus maize to sell. Either way, volatile prices within a maize-dominant system keep Malawian farmers from getting out of poverty." With this challenge that Malawi is facing, there must be another farming method that can help farmers to have enough produce for consumption and surplus to sell in order to have the income to buy other basic necessities in their households. This includes alternative ways of boosting the land's fertility, such as using chicken manure, as mentioned by Matebule James (OI: 26 April 2023), which helped him to harvest four bags of maize on a small plot of land.

Another challenge that small farmers encounter in Malawi is the quality of the seed. Since quality seed is one of the important inputs in agricultural production, disease-resistant and increased yield potential seed is crucial for a farmer in terms of productivity and the value of the commodity at the market (Setimela 2006: 9). In pre-modern times, farmers used to save their own seed for replanting in the next growing season. The seed was preserved using indigenous methods of storage and the good seed was selected for good germination and yield. However, in Malawi, as elsewhere in Southern Africa, there is a paradigm shift in terms of seed selection and preservation. The shift is the result of the introduction of hybrid seed which smallholder farmers cannot preserve to grow in the next growing season because if replanted, the yield is very low. The farmers buy seed every year and the prices, like the price of fertilizer, have become exorbitant that most farmers cannot afford.

Thus, the FISP subsidizes the price of seed, fertilizer, and pesticides so that smallholder farmers can buy them at an affordable price (Chibwana and Fisher 2010: 1). However, maize seed and legume seed such as beans, groundnuts, cowpeas, and pigeon peas are free, but they go together with a voucher for fertilizer. Nevertheless, the FISP has got its own challenges in distribution and availability because sometimes seeds and fertilizer are not distributed at the right time. For instance, during the growing season for 2022/2023, seeds and fertilizer were delivered to the selling point a month after the rains had started. On the other hand, the selling points in the rural areas are limited, and farmers travel long distances to access seed and fertilizers. The other challenges are that the fertilizers and seed quality control system are weak and that monitoring, and evaluation, of the whole program is non-existent.

Nonetheless, not all farmers benefit from the FISP. The program is for the selected few, especially those who are poor and vulnerable such as people living with HIV, people living with disabilities, and child-headed households. The other farmers buy seeds and fertilizers at exorbitant prices. The high prices of seed make farmers plant seed from the previous growing season, and this results in low yield. The same is true with the high price of fertilizers alluded to in the previous section. The result is food insecurity for most households in the country.

Pests and disease control present another challenge that smallholder farmers face. Most hybrid seeds are prone to pests and diseases, unlike the indigenous varieties. Farming without mulching exposes the soil and plants that are under moisture and nutrient stress tend to attract infestation of pests and diseases. Farmers need to find a way of controlling the pests to have good yields. Thus, farmers end up using expensive pesticides to control pests.

The high poverty levels (HPL) among the local people are yet another cause of food insecurity in Malawi. Poverty is understood differently by different people depending on where they live. According to August (2014: 1), poverty can be classified into two depending on the level of disadvantage one experiences, that is, relative and absolute poverty. Relative poverty is where the income of a household is "so low that a minimum standard of nutrition, shelter, and personal necessities cannot be maintained" (August 2014: 1). Absolute poverty is a condition where an individual is so poor that he does not have basic human needs such as food and shelter. With this understanding, most of Malawi's population lives below the poverty line[2] and the required standard to live a comfortable life.

Poverty is a gendered phenomenon that significantly affects women more than men, hence the feminization of poverty. It is the power relations between men and women and gender inequality that put women at a disadvantage in terms of poverty (Chant 2008: 167). Although women play an important role in agriculture and food production, they are not given equal access to land. In some cases, they do not have access to subsidized farm inputs or loans. As a result, women from female-headed households are poorer than women from male-headed households.

The other factor that makes women poorer than men is unequal access to education. Most women in Malawi have low levels of education or are illiterate. In addition, most women are deprived of decision-making power. This makes women to be vulnerable, making it difficult for them to escape the poverty trap (Chilongozi 2022: 26). This is justified by UNICEF which avers that over 70 percent of Malawi's population lives below the income poverty line (https://www.unicef.org.Malawi).

Both in urban and rural areas, people live in poverty though the situations differ. In all these places, women and children are the most affected. Poverty is a serious concern among women in Malawi. Women become poor due to many factors such as being deprived of necessities of life and being disadvantaged because of their marital or social status. Whereas it is fact that poverty is a serious problem, it is also important to note that it

[2] Poverty line is the minimum amount of money a person requires to fulfill the necessities of life such as food and shelter. According to the World Bank, people who live below the poverty line do not have enough for their necessities.

affects more women in the rural areas. They are affected because of the responsibilities they have of looking after their families. Thus, Kapuma further states,

> Poverty is a serious concern among women in Malawi. Women become poor due to many reasons such as being deprived of the basic necessities of life and being disadvantaged because of their marital or social status. … In Malawi, the condition of poverty is manifested in the lack of a productive means to access basic needs of life such as food, water, shelter, health, and education. (Kapuma 2021: 48)

How Farming God's Way Can Help Farmers to Achieve Food Security

Farming God's Way is a farming method that can help local farmers to yield more from the piece of land, where they have been failing for years. As noted in the introductory section above, Farming God's Way is also known as conservation agriculture. It is an approach to agriculture that emphasizes sustainable and environmentally friendly practices, while it also improves crop yield and reduces the use of expensive inputs like fertilizer and pesticides (Spaling and Kooy 2019: 2). There is a need for a change of the mindset, so that we can revert to using what our parents used before. People tend to be attracted to simple ways of farming like the use of chemical fertilizer which is proven to be expensive and over time spoils the soil. They think that the other way of using natural manures is difficult because it takes a long period to prepare them (interview with Mphatso Sulumba on 23 April 2023). According to Bread and Water for Africa, "Farming God's way increases production, conserves precious topsoil, and transforms lives in Malawi. When awareness is raised and people see the advantages, they will go for it. Farmers would like to harvest more and at the same time be able to conserve the soil that is at risk now" (27 June 2023).

Many case studies show the advantages of Farming God's Way. According to Humphrey Kapito, a farm manager at Naming'azi in Zomba, Southern Malawi,

> Farming God's way has many advantages. There is soil improvement as there is minimum soil disturbance. It is low cost, hence affordable. It yields organic crops (No chemicals used) which adds value to good food and good

> health. By using organic manure, it lowers costs for the farmer. There is also optimum use of land through integration which promotes soil fertility, water conservation, and animal habitats. By so doing, this reduces waste as all residue is returned as manure. (Interview on 21 May 2023)

Naming'azi Farm which Kapito manages belongs to the Church of Central Africa Presbyterian, whose main purpose is to help the local people to gain farming information. It started as a training center in 1992 for young people in entrepreneurship. Its emphasis is to help young people in their communities to farm using organic manure so that they can have good harvests. It grows a variety of crops, including maize, vegetables, soya beans, pigeon peas, groundnuts, fruits, bananas, fish farming, piggery, and chicken. The surrounding villagers at this farm are the ones who benefit from the farming demonstrations that are done. Kapito, in his explanation, was very positive that this is the best way to go with the current high cost of farm inputs. He said that results may not be good the first year, it needs gradual adoption over some time, and then thereafter a farmer will be able to yield more. Subsistence farmers can benefit from this because of improved soil and water systems, and crop diversity that will bring food security to family units.

Malawians have been used to the system of farming that was brought by the European missionaries which is difficult to sustain and maintain because of climate change and the different weather systems. "Britain brought many practices from Europe. Unfortunately, many of these practices were not well suited to Africa. Others evolved from long-standing cultural practices powered by the lack of modern tools, knowledge, and products" (Y-Malawi, November 2020).

One of the benefits of adopting the method of Farming God's Way is that it "allows for a healthy balanced soil ecosystem, which is the safest environment to prevent an infestation of soil pests, through predation" (Dryden 2022: 29). In addition, "the practice of crop rotations also breaks pest and disease lifecycles and significantly reduces crop losses through their infestations" (Dryden 2022: 29). This is beneficial to the farmers because they do not spend money on pesticides and, at the same time, it is labor-saving as they do not spend time weeding in the garden as they have minimal weeds.

It is a sad reality of life that poverty makes women who are vulnerable to continue living in abusive relationships or to engage in transactional relations with men who will end up deserting them, leaving them with the responsibility of solely caring for and raising children from such relationships. Chebet and Cherop (2015: 195) further comment,

> Gendered poverty and the right to human dignity have been recognised as the central challenge to the development of humanity. Poverty is a dehumanising condition for everyone. It violates the human rights of the affected, whether women or men. Poverty subjects an individual to a state of powerlessness, hopelessness, and a lack of self-esteem, confidence, and integrity.

Since most farming in Malawi is done by women, the lack of involvement of women in conservation agriculture such as Farming God's Way exacerbates the situation. Hence, involving women in the Farming God's Way method will assist in alleviating poverty and making their households food secure. At the same time, it can assist in empowering women to have income if they harvest enough maize for consumption and the surplus maize for sale. When women are self-reliant, they cannot continue to live in abusive relationships and, at the same time, they are able to take care of their children and families. This empowers them to make decisions in the community.

The Role of the Church in Actualizing Farming God's Way

The introduction of Farming God's Way has two purposes: the first one is to help people to increase production and the second one is to let people know about Jesus. It is a holistic approach and helps people to grow both physically and spiritually.

In this case, the church has a bigger role to play. It must help the community to manage to grow their food and stop the mindset of always wanting handouts when there is insufficient food. God is not poor, and his people should not be treated as such. Most of the people in rural areas have two major problems. The first is that they cannot afford to buy fertilizer with the continued increase in prices. The second problem is that the land they use is very small for the family. This means that they ought to use a method that will yield them more. Farming God's Way can help to pull people from poverty to produce sufficient food for domestic use, as well as being able to sell the excess. For this to happen, the church should do the following:

1. Biblical teaching. "My people are perishing because of a lack of knowledge" (Hosea 4:6). Knowledge about God and what he is able to do is important. Apart from making disciples for Jesus, Farming God's Way is about stewardship. God entrusted human beings with the responsibility to take care of God's creation and to make people live happily and in return to thank God for his love and presence. This is echoed by Jesus, who says in John 10v10 that his coming was to give people abundant life. Mission2malawi.com says, "The parable of the talents—if you were faithful with the little God will add to you, but if you are not faithful with little, He takes away. If you are faithful with the land God has given you, even if it is little, then God will add to it" (accessed 30 June 2023). God wants his people to live happily with enough food security. This is the starting point of the church in advocating for the program of Farming God's Way.
2. Training. The church should engage itself in training church leaders to understand Farming God's Way methods and how these can change the stories of many people. The training should help leaders to embrace a shift in their mindset, changing from the normal way of farming that was adopted by the European missionaries to understanding that it is possible to farm and produce enough food which ushers positive changes in people's life situations. This will transform their lives from being riddled with poverty, which reduces them to a status of always waiting for handouts, and transitioning to self-sufficiency in food production. When people have food, it gives them self-esteem, but when they lack food, they become beggars, bereft of dignity and self-worth. It is therefore the church's responsibility to train leaders that can assist the communities in food production.
3. Church action. The church should take advantage of using the small plots around their churches as demonstration gardens. This will help members to learn and see for themselves the processes needed and the outcomes.
4. Encouraging the community to have rubbish pits in the home where they can throw away all the bio-degradable litter from household daily use so that these can decompose into manure. Those keeping animals like chickens and goats should be encouraged to keep the droppings—keeping them in such a way that they turn into manure. In these days, some villages have commercial farmers who are keep-

ing chickens and they throw away the droppings for people to collect them. The church should encourage people living in such communities to take advantage of the chicken manure to use for their farming.

Conclusion

Drawing insights from Malawi, our chapter has revealed that Farming God's Way is an immensely beneficial method, especially for smallholder farmers. It promotes the use of organic manure, unlike the farming methods that encourage the use of chemical fertilizer. This method is advantageous in view of climate change because it requires no tilling and mulching, while crop rotation helps to keep moisture in the gardens and minimizes soil erosion and maintains healthy soil and crops. At the same time, it increases the yields, making the smallholders' households food secure throughout the year.

The chapter has also emphasized the fact that the church has a crucial role to play in promoting Farming God's Way methods. The church should provide training and encourage its members to practice Farming God's Way in order to have food security and boost the amount of income which comes from farming. The Malawian case study has been used to illustrate the importance for the rest of the African continent to seriously consider embracing the method of Farming God's Way. The cost of chemical fertilizer is getting higher and higher, and ordinary people cannot afford it. The beautiful land of God has been misused. There is a need to return its beauty and glory, to bring back the nutrients it used to have. To allow the soil to be covered by God's blanket of nutrients for better yields.

References

August, K.T. 2014. *Equipping the Saints: God's Measure for Development.* 2nd Edition. Bellville, The Print-Man.

Bread and Water for Africa 2023 (https://africanrelief.org)

Chibwana, C. & Fisher, M. 2010. The Impact of Agricultural Input Subsidies in Malawi, *International Food Policy Research Institute.* https://www.aliment-erre.org/system/files/ressources/pdf/45_ifri_impacts_of_agricultural_input_subsidies_in_malawi.pdf [Accessed 16 July 2023]

Chilongozi, M.N. 2022. *Microfinance as a Tool of Socio-economic Empowerment of rural women in Northern Malawi: A Practical Theological Reflection.* Unpublished PhD Thesis, Stellenbosch University.

Currier, A. 2020. The Failure of Input Subsidies and a New Path Forward to Fight Hunger in Malawi. (www.oaklandinstitute.org)

Chant, S. 2008. The 'Feminisation of Poverty' and the 'feminisation' of Anti-Poverty Programmes: Room for Revision? *Journal of Development Studies.* 44(2), 165–197.

Chebet, D. & Cherop, B. 2015. Gender and Poverty: Rereading Proverbs 31 in Pursuit of Social-economic Justice for Women in the Reformed Church of East Africa. In Mouton, E., Kapuma, G., Hansen, L., & Togon, T. (Eds.), *Living with Dignity: African Perspectives on Gender Equality.* Stellenbosch: Sun Press. pp. 193–218.

Dryden, G.W. 2022. *Farming God's Way Field Guide.* https://www.farming-gods-way.org/Resources/FGW_Field_Guide.pdf [Accessed on 18 July 2023].

Hunter. R., Crespo. O., Coldrey, K., Cronin, K., & New, M. 2020. Research Highlights—Climate Change and Future Crop Suitability in Malawi. University of Cape Town, South Africa, undertaken in support of *Adaptation for Smallholder Agriculture Programme' (ASAP) Phase 2. International Fund for Agricultural Development (IFAD), Rome.*

Kapuma, G. 2021. Gender and Access to Land Ownership: Christian Reflection on the Experience of Malawian Widows. *African Journal of Gender and Religion.* 27(1): 42–64.

Setimela, P.S. 2006. Analysis of Community-based Maize Seed Production and Supply Schemes in Sub-Saharan Africa. In P.S. Setimela & P. Kosina (Eds.), *Strengthening and Scaling Up Community-based Production.* Mexico, D.F.: CIMMYT.

Shuwali, Dickson. 2021. Farming God's Way, A Missionary's Tale, Mission2Malawi (https://mission2malawi.com)

Spaling, H. & Kooy, K.V. 2019. Farming God's Way: Agronomy and faith contested. *Agriculture and Human Values.* 36, 411–426. https://doi.org/10.1007/s10460-019-09925-2 [accessed on 15 July 2023].

Y-Malawi. November 2020. (https://y-malawi.org)

CHAPTER 15

Religious and Cultural Drivers of the Feminization of Food Insecurity in South Africa: Contextualizing Ruth

Linda Naicker

Introduction

This chapter argues that food insecurity in South Africa, by virtue of its effects and long-standing pervasiveness, has reached epidemic proportions, particularly among women living in under-resourced areas, both rural and urban. Many scholars have highlighted the dire state of food insecurity in South Africa from the perspective of structural and systemic oppression. Battersby and Haysom (2016), for example, argue that to curb the perpetuation of food insecurity in South Africa, the state must address its systemic drivers. Lesetedi (2018), Govender (2018), Naicker (2021), and Kruger (2017) confirm that women's struggles are

L. Naicker (✉)
Research Institute for Theology and Religion, University of South Africa, Pretoria, South Africa
e-mail: naicklw@unisa.ac.za

L. Maseno et al. (eds.), *Religion, Climate Change, and Food Security in Africa*, Sustainable Development Goals Series,
https://doi.org/10.1007/978-3-031-50392-4_15

structurally driven and that male hegemony[1] plays a key role in the food insecurity status of women. By and large, in South Africa, men control the means of production and women, to a far greater extent than men, experience unemployment, poverty, and food insecurity. Rudolf et al. (2012: 15) posit that women are most affected by high food prices and limited purchasing power, while Battersby (2015: 1) elucidates that rampant corruption in South Africa, coupled with governmental mismanagement, drives poverty and food insecurity. I argue that the problem of food insecurity, particularly as it relates to women, must be comprehensively addressed. Its causes and consequences must be analysed from a myriad of perspectives. Essentially, I posit that religion and culture play a very damaging role in the perpetuation of female food insecurity in South Africa. Working from the perspective of Christianity, I employ the biblical text of Ruth in the Hebrew Bible to unpack the ubiquitous nature of the causes and consequences of food insecurity in South Africa, paying particular attention to the role of biblical interpretation in the disenfranchisement and food insecurity of women. Reinterpreting the Book of Ruth through the lens of African Women's Theology, I contextualize the context of food insecurity in the biblical narrative with the context of food insecurity in South Africa today.

The Dire State of Food Insecurity in South Africa

In South Africa, food insecurity was declared a national crisis in 2013 with over 1 million households experiencing hunger by 2018 (Statistics South Africa 2019: 18). Scholars argue that food insecurity in South Africa persists because of economic and spatial disparities (Pulker 2016: 4; Zager 2011: 2). As a predominantly cash-based society, income reliability is key to obtaining food and maintaining food security. However, income reliability is severely affected due to high unemployment. Low-income neighbourhoods are extremely under-resourced, and infrastructure is poor or non-existent. Access to land presents a mammoth problem as people are often unable to grow their own food and live in a humane way. Thousands of South Africans have little or no access to amenities such as electricity,

[1] Hegemony occurs through the subordination of particular groups based on dominant ideologies and enforced through maintaining and sustaining levels of hierarchy over subordinate groups, resulting in adverse conditions for the subordinate. "As such, hegemony is intricately related to power, ideology and struggle." See Herrmann (2017: 1).

water, and drainage, making the preparation of food difficult. Much of South Africa's urban low-income areas are dormitory townships created under apartheid and informal or squatter settlements. In rural South Africa, many people are farm tenants, farm labourers, or farm dwellers earning meagre wages and having little access to decent housing and arable land (Mthimkhulu 2018). Spatial access is a crucial factor in determining the food security of a large percentage of the population as people in rural, semi-urban, and even urban areas need to travel long distances to access food and transportation is costly. These levels of poverty and inequality put undue strain on penurious households rendering them food insecure (Naicker and Molobi 2022: 8). Women bear the brunt of food insecurity, because in the male-dominated context of South Africa, the burden of accessing and preparing food rests predominantly on the shoulders of women. Indubitably, systemic inequalities and marginalization, particularly of under-resourced black women, result in the feminization of poverty and the feminization of food insecurity (Battersby 2015: 3; Abrahams 2020: 7).

The Feminization of Food Insecurity in South Africa

The feminization of poverty, simply put, is that women experience poverty to a far greater extent than men and poverty amongst women is increasing in comparison to poverty amongst men (Uken 2018: 3). This concept frames the struggle of food-insecure women in South Africa. Moreover, race and class disparity and hegemonic gender configurations of South African society lead to the feminization of poverty and food insecurity. Women's poverty and food insecurity is a direct result of the uneven power dynamics inherent in patriarchal arrangements. As bell hooks (2000: 41) contends, the systemic control and disenfranchisement of black women is ideologically, socio-economically, socio-culturally, and politically constructed to maintain women in positions of subjugation.

Ortner (2006: 6) appropriately asserts that the ways in which society is constructed determine hegemonic configurations of gender and have a direct impact on the poverty status of particular communities and demographics. In South Africa, the economic status and disproportionate structural dimensions of power constrain women at the level of available options, hindering purchasing power and access to food. This results in sustained and perpetuated levels of poverty and food insecurity among women. Significantly, in South Africa there is sufficient availability of food.

The problem is the inability to access food (Haysom 2017: 145). A comprehensive exploration of the structures of power that render women food insecure necessitates an exploration of how various forms of institutionalized power operate in the lives of women. Uken (2018: 25–26) maintains that analysing monetary poverty alone is inadequate to understand the magnitude of the feminization of poverty. Gender privilege, women's subjectivities, the social exclusion of women, and a clear understanding of how poverty and gender equality conflate will facilitate a far-deeper understanding of women's poverty and food insecurity. Landman (2009: 40) notes that religion and culture control all areas of women's lives. It is therefore my inference that patriarchal biblical interpretation plays a major role in the oppression and food-insecure position of women.

Christianity and Patriarchy

In South Africa, where religiosity, in particular Christianity, is deeply entrenched, socio-cultural and religious arrangements are deeply connected to women's economic viability and food security. Eighty-six per cent of South Africans pledge allegiance to the Christian faith (Statistics South Africa 2016). According to Pillay (2015: 560), there is a deep connection between Christianity and the perpetuation of unequal gender relations in South Africa. Tonono (2018: 6) elucidates that patriarchal societies generally produce patriarchal churches. He expounds that "the stench of patriarchy permeates the pews in a similar manner that it does society at large". Patriarchy is an ideology that upholds male superiority over women and all the earth (Wood 2019: 2). Tonono (2018: 6) illuminates that patriarchy is a combination of ideologies that sustains male dominance, entrenches itself in all spheres of interaction, the church included, and is often posited as a god-ordained hierarchy sanctioned by biblical texts. Tonono (2018: 5) emphasizes that the interpretation of biblical texts and teachings about what the model Christian woman should be perpetuates the oppression and subjugation of women. Biblical texts are often used to affirm the superiority of men, enforce female subordination and submission, and perpetuate female dependence and the feminization of poverty and food insecurity.

Ecclesial Biblical Interpretation

In my own experience, I often observe the unquestioning trust towards local ministers and the biblical interpretations they put forward from the pulpit. This is a result of what I term, "patriarchal ecclesial enculturation". Patriarchal ecclesial enculturation is a process whereby local churches promote the oppression of women through the formation of socio-cultural and socio-religious environments that entrench male hegemony and enable patriarchal conceptions of women to become the taken-for-granted norm. The ecclesial enculturation of male hegemony and patriarchy is propagated through biblical interpretation and deliberately developed by ecclesial culture that is designed to subjugate women (Naicker 2021: i).

Ecclesial influences impact interpretation. Often, male power and control over women is valorised and hegemonic masculinities are lauded as God-given mandate and a taken-for-granted norm. In the story of Ruth for example, Ruth is portrayed as sexually naïve, chaste, and morally upright. By casting her in this light, Ruth is depicted as an example for all women to emulate. Concomitantly, a dichotomy is created between the representation of the character of Ruth and other women described as evil tempters bent on sexually luring men and engineering the downfall of men. Such interpretations produce patterns of domination, control, and hegemony at ecclesial, societal, and cultural levels and directly contribute to the subjugation, marginalization, and food insecurity status of women (Naicker 2021: 104). African women theologians decry such interpretations as oppressive to women. They further contend that such interpretations do not give theological weight to the contextual experiences of suffering and injustice endured by women. African women theologians choose to give voice to the specificity of African women's lived experiences (Nkiruka 2016).

Interpreting the Story of Ruth Through the Lens of African Women's Theology

Many womanist theologians find the text of Ruth replete with themes of patriarchy, male hegemony, gendered economic injustice, and food insecurity. For example, drawing on events portrayed in the story of Ruth, Kanyoro (2002) and Masenya (1998) demonstrate how women are often disenfranchised in contemporary African cultures because of patriarchal and androcentric captivity. Interpreting the story against the backdrop of women's migration due to poverty, Yeong-mee brings to light the struggle for survival in dangerous, hostile, and foreign environments. In Moyo's (2010) rendition, she retells the story of Ruth against the backdrop of widowhood, human trafficking, desperation, and survival. The parallels drawn between food insecurity in the text and food insecurity in the context of women's present-day realities are stark. Moyo (2010: 3) argues that for justice to be attained, social systems that commoditize women's bodies and leave them vulnerable to food insecurity and poverty must be challenged.

The Story of Ruth

The story of Ruth unfolds within the harsh realities of famine (hunger) in the land of Bethlehem (house of bread). Elimelech takes his wife Naomi and two sons to Moab where there is plenty of food. Unfortunately, after some time, a terrible misfortune befalls the family; Elimelech and their now married sons die in Moab. Naomi sets out back to Bethlehem with her daughters-in-law, Orpah and Ruth. At some point in the journey, Naomi urges her daughters-in-law to return to their homes where there is the possibility of them re-marrying and thereby securing their financial and food security in a patriarchal world where women were totally dependent on men for sustenance and survival. Orpah returns to Moab but Ruth refuses to leave her mother-in-law, thereby taking on the role of a breadwinner and provider for herself and her mother-in-law in Bethlehem. They return to Bethlehem during the barley harvest season where God once again blessed the land with plenty of food for all, yet the two women find themselves destitute because in the patriarchal world of the text, Naomi as a woman could not inherit her husband's property; she needed to have a male heir to inherit the property of her husband. So, Ruth went out into the fields of Boaz, a rich landowner to glean, thereby providing sustenance for herself and her mother-in-law. Gleaning was a means instituted by law whereby those who were poor and live on the margins of society were able to survive. As harvesters would reap the crop, workers were instructed to leave the corners of their fields un-harvested and anything that fell to the ground while reaping was left for the poor. However, Ruth alone gleaning in the fields since Naomi was too old to work made survival and the struggle to attain enough food difficult. Ruth as the breadwinner had to find another solution to their problem of hunger and poverty. She therefore brazenly asked Boaz, the rich landowner, directly for permission to glean among the reapers as opposed to gleaning behind them, thereby catching the attention of Boaz. This was considered a rather daring and innovative act on the part of Ruth since Ruth was also a foreigner. This act set the scene for another even more daring act later in the story. Naomi deemed Boaz to be a suitable partner for Ruth and one that will ensure their survival and food security, so she developed a strategy to permanently solve their problem of destitution and food insecurity. She instructed Ruth to wash and perfume herself and go to the threshing floor where Boaz will be, guarding the harvested crop. After Boaz had eaten and drunk and was asleep, Ruth was to offer herself to Boaz sexually. Since Ruth's primary concern was to provide for Naomi and herself, she does as her mother-in-law instructed (Nadar 2007).

An African Womanist Contextualization of Ruth Against the Backdrop of the Feminization of Food Insecurity

Androcentric Interpretations

African Women's Theology upholds the centrality of the Bible in the liberative work of addressing and combatting gender-based oppression. The use of biblical texts and stories forms a key component in the interpretation of theological arguments that advocate for equality, justice, and hope. Oduyoye (2001: 32) opines that the church must pursue justice by approving the full participation of women in the household of God. In doing so, the church will address the injustices of patriarchal oppression and demonstrate a commitment to the pursuance of justice. Adopting this positionality will ensure that the church becomes a place where life is nurtured to its fullest potential. Dube (2016: 2) argues that theological engagement with gendered oppression must acknowledge how patriarchy created men's claim to power and supremacy over women. Phiri and Nadar (2006) hold that theological engagement must confront ideological oppression by directly addressing social justice concerns. The church has an ethical responsibility to create and advocate for equitable gender relations through addressing cultural and ecclesial oppression and dominance of women. In sum, African Women's Theology is socially engaged theology, which is praxis-driven, placing emphasis on social analysis from the perspective of theological engagement.

Most ecclesial renderings of the Book of Ruth promote patriarchy and male hegemony. This is the patriarchal ecclesial enculturation that I highlighted earlier which enables dominance and cements women's vulnerabilities. Such dominance negatively affects every developmental trajectory in the lives of women. For African womanist theologians, the Book of Ruth presents an opportunity to confront patriarchal oppression and destabilize its ravaging effects. Working from the standpoint that culture is fluid and never static, African women theologians address how religion and culture can and must be re-envisioned and reinterpreted from the perspective of women's oppression. The story of Ruth is often used to critique oppressive ideologies in light of justice, hope, and redemption of women from all forms of oppression. African Women's Theology argues that to maintain a culture of male dominance is to perpetuate female oppression.

I use the biblical text of Ruth as a catalyst to decipher the religious and cultural bedrock that endorses women's subjugation, poverty, and food insecurity. Much like the South African context of food insecurity, in the patriarchal world of the text, Ruth and Naomi were food insecure in the context of abundance. As widows within a religious and cultural patriarchal context, Naomi and Ruth were food insecure and destitute because they had no male heirs to inherit the property of Naomi's dead husband, Elimelech. In this context, without the financial and material protection of men, women are unable to survive. The need for male protection in the text highlights the hierarchical gendered order of Ancient Israel and is used in patriarchal ecclesial interpretations to propagate and affirm male headship and superiority.

The story of Ruth and others in the Hebrew Bible are used at an ecclesial level as a spiritual basis for the submission of women and as a declaration of male power over women (Tonono 2018: 7). In encouraging such interpretations, male power is valorised and becomes the taken-for-granted norm with devastating effects on the financial, material, and food security status of women and children. Women are led to believe that without the protection of men, they would not survive. The normalization of such beliefs impacts all social spheres of women's existence, contributing to their subjugation, inequality, and financial insufficiency. Ecclesial interpretations of male superiority and headship foreground androcentricity and prevent women from resisting oppression and developing their own skills and abilities. Seen in this light, and from the perspective of food insecurity, the church can be seen as complicit in the perpetuation of the feminization of food insecurity. Theological incongruities that cement male headship, authority, and supremacy must be challenged through redefining biblical interpretation. Kanyoro (2002: 80) opines that the dignity and preservation of women is often trampled on through biblical interpretation.

Patriarchy and Male Hegemony

Essentially, the interpretation of biblical texts has the power to affect and shape the life of the individual whose belief system is anchored upon Christianity (Tonono 2018: 4). From an African womanist perspective, due consideration must be given to how female submission and subjugation is developed and explained and how male hegemony and authority is developed and explained. As noted earlier, in the patriarchal world of the text, Ruth and Naomi were vulnerable widows, dependent on the men in their lives to provide them with sustenance and security. Upon the death of Elimelech, Naomi depended on her two sons to carry on her husband's lineage and take ownership of all his property and wealth. With the death of Naomi's sons, Ruth, Orpah, and Naomi had no claim to the property, land, and wealth of Elimelech. While Orpah decided not to go to Bethlehem, Ruth decided to take on the role of the breadwinner and accompany the ageing Naomi back to Bethlehem knowing full well that as women they would be destitute, even in the land of bread (Bethlehem).

In making the journey to Bethlehem, Ruth and Naomi demonstrated a defiance towards the system of oppression of women. In Naomi's decision to return to the land of her birth rested the hope that the requirements necessary for their survival and food security would be realized. In Bethlehem, Ruth sets out to engineer initiatives that would ensure their food security and wellbeing. This act of courage is demonstrative of the agency of women in the face of sheer destitution. Ruth and Naomi's refusal to lay down and die in the face of overwhelming odds is reminiscent of the universality of women's suffering and agency. Their experiences poignantly represent the lived reality of so many food-insecure women in South Africa who find themselves vulnerable in the face of hegemonic and systemic inequalities.

The ideological vice of patriarchy as it intersects with religion, culture, economics, and politics affects women differently compared to the ways in which it affects even subjugated men. Research revealed that even in contexts where both men and women suffer systemic constraints, men still hold the economic upper hand (Hunter 2010; Naicker 2021). Male hegemony and dependence on men leads to the underdevelopment of women and the feminization of food insecurity. African Women's Theology as both contextual and communal offers theological engagement which challenges systemic oppressions propagated through patriarchy (Phiri 2004: 15).

Male headship in work environments and households creates conditions in which male power over women is exerted and control over decision-making and resources falls in the hands of men. Women are resultantly denied the right to supervise men and manage or own property. In some instances, women are even denied the right to meaningful employment. Moreover, in the South African workforce, women still earn decidedly less than men even when their duties are the same. These inequitable employment situations marginalize women who are often relegated to the informal sectors of employment. Furthermore, while women's participation in leadership roles at policy making and management levels in the workforce is disproportionately lower than that of men, on the domestic front (unpaid labour), their participation is disproportionately higher (Kruger 2017: 13). Notwithstanding this anomaly, women function within a multiplicity of roles especially in environments where men do not participate in family responsibilities such as caring for children and the elderly, housework, and purchasing and preparing of family meals. This level of patriarchal culture unjustly affects women, especially in times of financial crisis (Kruger 2017: 24), and has a direct impact on their food security status.

African Women's Theology values the interconnectedness of life and seeks to empower women in instances where inequality negatively affects their lived realities (Tarus 2014: 4). Their theological engagement therefore serves to give weight and substance to the contextual experiences of the suffering they endure. In listening to and giving voice to the specificity of African women's struggles, African women theologians confront patriarchy and critique exclusion and domination (Nkiruka 2016). In the South African context, the basis for women's food security is empowerment[2] through access to education and skills development. The marginalization of women in the public and domestic spheres perpetuates gender stratification and women's vulnerabilities. Yet, there is a scarcity of development and empowerment opportunities for women, which hinders their access to meaningful employment and economic viability. Additionally, sexual and reproductive health services for women in South Africa are

[2] I am aware that the word "empowerment" in the discourses on women and black people is a contested term (Nazneen et al. 2014). I use "empower" to denote increasing the spiritual, political, social, cultural, and economic strength of individuals and communities.

inadequate, further entrenching women's vulnerabilities (Kruger 2017: 23–24). The exclusion of women from full participation in all aspects of social life directly contributes to the feminization of food insecurity. The valorisation of patriarchal biblical interpretations and the failure of the church to empower women are what Wood (2019: 2) refers to as "modern day slavery". Patriarchy and male hegemony have devastating personal implications for the food security status of vulnerable women and are indicative of how power and control shape the lived experiences of women in situations of limited access (Naicker 2021).

Vulnerabilities and Agency

As the story of Ruth and Naomi unfolds, Ruth goes into the field of Boaz to glean as a means of acquiring food, but this initiative was clearly inadequate to meet the needs of the two women. Ruth therefore decided to address Boaz boldly and ask for further assistance. This was an extremely dangerous move, but Ruth's only other choice was bringing home an inadequate amount of food which would not cater to the family's dietary needs. Ruth as both woman and foreigner gleaning in the fields made her extremely vulnerable to both physical and sexual assault. Moreover, as a young woman, she ran the risk of reprisal from both women and men if they perceived any misconduct on her part. This level of abuse and subjugation of women is an inherent part of patriarchal societies. Characteristically, women's bodies have always been linked to immorality and inferiority. Theological and ecclesial engagement must therefore acknowledge this historic misdeed and affirm to the full humanity, full participation, and equality of women in both church and society (Oduyoye 2001: 73).

Likewise in the South African context, the marginalization and discrimination of women in terms of economic opportunities and access, coupled with the financial dependency of women, increases their vulnerability to food insecurity. Resultantly, women are at the mercy of patriarchal systems of power and control. In situations of powerlessness, where the basis of control lies in who has access to food, without the protection of men, the situation is often dire for under-resourced women. Women's unequal position in the workforce, lack of access to land and resources, and unequal rights and access within family structures entrench poverty and food insecurity (Mthimkhulu 2018: 16).

Studies detailing the correlation between food insecurity and intimate partner violence (IPV) revealed that within environments of socio-economic constraints in South Africa, there exists a direct link between men's perpetration of IPV and food insecurity (Hatcher et al. 2019: 769; Chop et al. 2017: 935). Reliance on male support and food supply results in festering depression and anger among men who feel pressurized to provide food within the context of economic constraints. Women in these situations find themselves at the mercy of the men in their lives who control not only access to food, accommodation, and basic subsistence needs but also the negotiation of safer sex. Dependence on their partners renders women powerless to negotiate condom use, making them susceptible to a myriad of health issues including HIV and AIDS as well as unwelcome pregnancies.

Notwithstanding these extreme levels of vulnerability, both Ruth and Naomi point us towards a way that defies patriarchy and traditional perspectives of how women should be in the world. In defying patriarchy, they point us towards social change that is life affirming and life giving. In leaving Moab and returning to Bethlehem even though in Bethlehem too they were at the mercy of the good graces of men to survive must be seen as agential. The actions of Ruth and Naomi present a challenge to men and women to contest ideological, religious, and cultural dominations that oppress women and rob them of their full humanity. Ruth's innovative initiatives to attain food security within a context of extreme social, cultural, political, and economic constraints must also be seen as agential.

From the perspective of African Women's Theology, there must be continuous dialogue with socio-economic, socio-cultural, and socio-political influences that cement women's oppression. Oppressive elements must be critiqued and liberation from all forms of oppression must be advocated for (Oduyoye 2001: 17).

Conclusion

In this chapter, I highlighted the dire state of food insecurity in South Africa and argued that male hegemony, instituted through the vice of patriarchy, plays a galvanizing role in the feminization of food insecurity in South Africa. Focusing predominantly on the impact of ecclesial interpretations of the story of Ruth in the Hebrew Bible, I contended that ecclesial interpretations of biblical texts are often used to affirm the superiority of men and the subjugation of women, thereby entrenching food insecurity and its effects among women. Through the lens of African Women's Theology, I put forward a life-affirming interpretation, calling for the liberation of women from all forms of oppression that lead to life-denying situations and the feminization of food insecurity.

References

Abrahams, M. 2020. Economic justice through the lens of human dignity: food and corona virus. Pietermaritzburg Economic justice and dignity. http://pmbejd.org.za/index.php/household-affordibility-index.

Battersby, J. 2015. A call for a food systems approach to alleviating food insecurity. *Input for Province*, 5 April.

Battersby, J. and Haysom, G. 2016. "Urban agriculture: The answer to Africa's food crisis." *Quest.* 12 (2). 8–9. https://www.researchgate.net/publication/320805362.

Chop, E., Duggaraju, A., Malley, A., Burke, V., Caldas, S., Yeh, P., Narasimhan, M., Amin, A. and Kennedy, C. 2017. Food insecurity, sexual risk behaviour and adherence to antiretroviral therapy among women living with HIV: A systematic review. *Health Care Women Int.* 38(9). 927–944. https://doi.org/10.1080/07399332.2017.1337774.

Dube, M. (ed.). (2016). *Other ways of reading: African women and the Bible. Society of Biblical Literature.* Atlanta. 237–251.

Govender, V. 2018. *Local Economic Development and Rural Women: A Case Study of the Vulindlela Region*, KwaZulu-Natal. Masters, University of KwaZulu-Natal. Pietermaritzburg.

Haysom, G. 2017. Climate change, food and the city: Agency and urban scale food system networks. In: Thomas-Hope, E. (ed). *Climate Change and Food Security: Africa and the Caribbean.* Routledge, London. 145–155.

Hatcher, A. Stockl, H., McBride, R., Khumalo, M., and Christofides, N. 2019. *Pathways from Food Insecurity to Intimate Partner Violence Perpetration Among Peri-Urban Men in South Africa.* Prev Med, 56(5). 765-772. https://pubmed.ncbi.nlm.nih.gov/30905482.

Herrmann, A. 2017. Hegemony. International Encyclopaedia of Organizational Communication. https://doi.org/10.1002/9781118955567.wbieoc094.

hooks, b. 2000. *Feminism is for everybody: Passionate politics.* Pluto Press, London.

Hunter, M. 2010. *Love in the Time of AIDS: Inequality, Gender and Rights in South Africa.*

Kanyoro, M. 2002. *Introducing feminist cultural hermeneutics: an African perspective.* New York, Sheffield.

Kruger, T. 2017. Msunduzi Municipality Strategy for Economic Empowerment. SIVEST South Africa. http://www.msunduzi.gov.za.

Landman, C. 2009. *Religious Discourse Supporting Patriarchy: The Evil of Patriarchy in Church, Society and Politics.* A Consultation Held at Mount Fleur Conference Centre.

Lesetedi, G. 2018. A theoretical perspective on women and poverty in Botswana. *Journal of International Women's Studies*, 19(5). 193–208.

Masenya, M. 1998. Ngwetsi (bride): the Naomi-Ruth story from an African-South African woman's perspective. *Journal of Feminist Studies in Religion*, 14/2, 81–90.

Moyo, F. 2010. Widowhood and Desperation for Food: Retelling Ruth in the Context of Human Trafficking. www.overcomingviolence.org/fileadmin/dov/images.

Mthimkhulu, S. 2018. *Reading Ruth in Light of the Environmental Food Insecurity and Landlessness Faced by Rural Female-Headed Households in South Africa.* Honours Dissertation, UKZN

Nadar, S. 2007. *Theology in the HIV & AIDS ERA: Studying the Hebrew Bible in HIV & AIDS contexts.* Ed. M. Dube. Geneva: World Council of Churches.

Naicker, L, and Molobi, V. 2022. "A Socially-Engaged Theological Response to the Historic and Structural Nature of Food Insecurity in South Africa". *Studia Historiae Ecclesiasticae.* https://doi.org/10.25159/2412-4265/10656.

Naicker, L. 2021. *Food, Sex and Text: Exploring Survival Sex in the Context of Food Insecurity Through Communal Readings of the Book of Ruth.* PhD Dissertation, University of the Western Cape.

Nazneen, S., Darkwah, A. and Sultan, M. 2014. 'Researching Women's Empowerment: Reflections on Methodology by Southern Feminists.' Centre for Gender Studies and Advocacy. (4). http://ugspace.ug.edu.gh/handle/123456789/25759.

Nkiruka, O. 2016. *Transforming Narratives through "LIVE": A Pastoral Model of Care for Women in Sex Work.* PhD Dissertation, Paper 2. http://digitalcommons.luthersem.edu/phd.

Oduyoye, M. 2001. *Introducing African Women's Theology.* Sheffield: Sheffield Academic Press.

Ortner, S. 2006. *Anthropology and Social Theory: Culture, Power and the Acting Subject.* UCLA University Press, Los Angeles.

Pillay, M. 2015 "The Church, Gender and AIDS What's Wrong with Patriarchy?", *Missionalia* 43(3). 558–567.

Phiri, I. 2004. Southern Africa. In: Parratt, J. (ed.). *An introduction to Third World Theologies.* Cambridge University Press, Cambridge. 137–162.

Phiri, I. and Nadar, S. 2006. Introduction. In Phiri, I., and Nadar, S. (eds.), *African Women, Religion and Health. Essays in Honor of Mercy Amba Ewudziwa Oduyoye.* Cluster Publications, Pietermaritzburg.

Pulker, A. 2016. *The Relationship between Urban Food Security, Supermarket Expansion and Urban Planning and Policy in the City of Cape Town: A Case of the Langa Junction Mini Mall.* Master's dissertation. University of Cape Town.

Rudolf, M., Kroll, F., Ruysenaar, S. and Dlamini, T. 2012. *The State of Food Insecurity in Johannesburg.* Urban Food Security Series. 12. AFSUN. Cape Town.

Statistics South Africa. 2019. Towards Measuring the Extent of Food Security in South Africa: An Examination of Hunger and Food Inadequacy. www.statssa.gov.za/publications/03-00- 14/03-00-142017.pdf.

Statistics South Africa. 2016. General Household Survey. Retrieved from https://www.statssa.gov.za/publications/P0318/P03182015.pdf.

Tarus, D. 2014. Theology of Social Transformation in the Theological Works of the Circle of Concerned African Women Theologians. *African Journal of Evangelical Theology*. 33 (1).

Tonono, S. 2018. *The Christian Church and Intimate Partner Violence: Patriarchy, Gains and Opportunities*. Paper presented at the National Research Symposium, Department of Social Development. https://methodist.org.za/wp-content/uploads/2020/.

Uken, B. 2018. *The feminization of poverty: a gendered underclass*. North Carolina State University. http://www.researchgate/net.publications/325229586.

Wood, J. 2019. Gender inequality: The Problem of Harmful, Patriarchal, Traditional and Cultural Gender Practices in the Church. *HTS Teologiese Studies/Theological Studies*. 75(1). 1–8. https://doi.org/10.4102/hts.

Zager, K. 2011. *Commutes, Constraints, and Food: The Geography of Choice*. Honours, University of Cape Town, Cape Town.

CHAPTER 16

Faith-Based Organisations and Food Security and Sovereignty in Africa: A Critical Review

Tshenolo J. Madigele and Ezra Chitando

Introduction

Without food security and food sovereignty, there is no development. A continent that relies on outside players to feed its citizens is ill-prepared to drive its development agenda and achieve prosperity for all. Thus, the African Union's Agenda 2063, Goal 3 within Aspiration 1 ("A prosperous Africa based on inclusive growth and sustainable development") says that by 2063 Africa will be a continent where "Citizens are healthy, well-nourished and have long life spans…" (African Union 2015: 30). The United Nations (UN) Sustainable Development Goals (SDGs) also address the theme of food security. Sustainable Development Goal

T. J. Madigele (✉)
University of Botswana, Gaborone, Botswana

E. Chitando
University of Zimbabwe, World Council of Churches, Harare, Zimbabwe

L. Maseno et al. (eds.), *Religion, Climate Change, and Food Security in Africa*, Sustainable Development Goals Series,
https://doi.org/10.1007/978-3-031-50392-4_16

Number 2 focuses on, "Zero Hunger."[1] Addressing hunger is, therefore, a challenge at the global, continental, regional, national, and household levels. Religion and spirituality have a key role to play in addressing Africa's food security and food sovereignty challenges. As Nyakecho et al. (2022) argue, African religions and spirituality are strategically placed to promote and secure the continent's food security and sovereignty. This becomes particularly urgent in the wake of climate change and the threat it poses in Africa (Chitando et al. 2022), the impact of COVID-19 (see e.g. Mukiibi 2020 and La Via Campesina 2022), and (at the time of writing) Russia's invasion of Ukraine.

This chapter argues that there is an intersection between religion and development in Africa, with special reference to the role of faith-based organisations (FBOs) and food security and sovereignty. It acknowledges that while the relationship between religion and development in Africa is a complex and multifaceted topic that has been studied by scholars and researchers for decades (Schuurman 2011), there has been far less attention to how the theme of food security and sovereignty is relevant to reflections on religion and development. Religion has played a significant role in shaping the social, cultural, and economic landscape of Africa. Consequently, its impact on development in the wake of climate change, food security and sovereignty cannot be overlooked. Faith-based organisations also played a significant role in the overall response to the HIV epidemic in Africa (see e.g. Schmid et al. 2008), while Muslim FBOs are engaged in social welfare in Africa (see e.g. Weiss 2020).

One of the ways in which religion intersects with development in Africa is through the provision of social services. Many religious organizations, such as churches and mosques, have established schools, hospitals, and other social welfare institutions that provide services to the community. These institutions often serve as a critical source of support for marginalized communities that lack access to basic services (Kunhiyop 2008: 226). Religion also plays a role in shaping attitudes towards development. In some cases, religious beliefs may promote values that are conducive to development, such as hard work, honesty, and community service. However, in other cases, religious beliefs may hinder development by promoting fatalism or discouraging innovation. Religion can be a source of conflict in some African societies. Religious differences have been a

[1] https://www.un.org/sustainabledevelopment/hunger/ (accessed on 26 June 2023).

contributing factor to many conflicts on the continent (Agbiji and Swart 2015).

Despite the potential challenges posed by religion, many scholars argue that it can also be a catalyst for development (Osagie-Usman 1994; Salih 2002). For example, some religious organizations have been at the forefront of efforts to promote peace building and conflict transformation in Africa. Nonetheless, religious organizations can play an essential role in promoting development by providing social services and promoting values that are conducive to progress. On their part, some FBOs have been playing a notable role in mobilizing communities in Africa to become more effective in responding to climate change, food security and sovereignty. This chapter reviews and critiques these efforts by FBOs in Africa. In the section that follows, the chapter attends to definitional matters. This is followed by an overview of food security and sovereignty in Africa. The chapter then summarizes the activities by FBOs. The ensuing section provides a critical analysis of the role of FBOs in promoting food security and sovereignty in the context of climate change in Africa. In conclusion, the chapter calls for more liberative approaches to this highly strategic issue if the continent is to meet its development agenda.

Understanding the Concepts of Food Security and Sovereignty in Africa Over the Years

In order to put the discussion into its proper context, it is important to begin by providing some background details relating to food security and sovereignty. Food security in Africa is a complex issue that is influenced by various factors such as climate change, population growth, poverty, political instability, and conflicts. The definition of food security has been developed over the years in the continent of Africa. In the year 1974, food security was defined as the availability, accessibility, utilization, and stability of food supply within African countries (FAO 1974). This definition was established during the World Food Conference held in Rome in 1974. The conference was organized by the United Nations (UN) to address global food shortages and to develop strategies to ensure food security for all nations (FAO 1974).

During this conference, it was recognized that Africa faced significant challenges in achieving food security due to a combination of factors, including population growth, poverty, political instability, and environmental degradation. As a result, the definition of food security in Africa focused on not only the availability of food but also on its accessibility and utilization. Availability refers to the physical presence of food within a country or region. Accessibility refers to the ability of individuals or households to obtain food through means such as purchase, production, or aid. Utilization refers to the ability of individuals to use and benefit from the nutrients provided by food. Stability refers to the ability of a country or region to maintain consistent levels of food supply over time (FAO 2008). These dimensions of food security are entirely dependent on each other. This means that any approach which addresses one dimension and leaves the other one would be inefficient. Food may be available, but not all would be able to access it. On the other hand, food may be easily accessible, but would not guarantee utilization and sustainability (Arene and Anyaeji 2010). In this regard, people need to be food secure but would not be food secure if they do not have sufficient funds to access food. People also need to correctly utilize the food they would have accessed. This means that they should be able to know and enjoy the nutritional benefits of food (ECI 2002). Addressing the dimensions of food security in Africa requires a multifaceted approach that considers all four dimensions of availability, access, utilization, and stability. From the understanding of these dimensions, food security has come to be understood as food availability, acceptability, and affordability (Enilolobo et al. 2022). The following citation also brings out the dynamics and concerns in a clear way:

> The differences in conceptualizing food security notwithstanding, there is a somewhat unanimity among scholars that quality (what kind of food?), quantity (how much food?), universality (who should get the food?), stability (when?) and dignity (how?) are universal components of the concept of food and nutrition security while availability (Household food production and crop diversity), access (Percentage food expenditure to total household expenditures), utilization (Degree of access to services-water, health and sanitation), and Stability (Stability of food prices and supply) are the set of indicators to be considered in any analysis of food and nutritional security…(Ogujuiba et al. 2012: 91)

Food security in Africa has remained a major concern due to various factors such as droughts, conflicts, and economic instability. In 1983, the World Food Programme (WFP) estimated that over 30 million people in Africa were at risk of starvation. The WFP launched several emergency relief efforts to provide food aid to affected populations (FAO 1983). To address the issue of food security in Africa, several initiatives were launched, such as the Comprehensive Africa Agriculture Development Programme (CAADP). The CAADP aims to increase agricultural productivity and promote sustainable agriculture in Africa. Another initiative is the African Union's Malabo Declaration on Accelerated Agricultural Growth and Transformation for Shared Prosperity and Improved Livelihoods, which was adopted in 2014. The declaration sets targets for increasing agricultural productivity and reducing poverty and malnutrition in Africa (World Bank 1986). The United Nations' Sustainable Development Goals (SDGs) which aim to end hunger by 2030 are also aimed at addressing food security in Africa.

Food security in Africa in 1990 was defined as the state in which all people, at all times, have physical, social, and economic access to sufficient, safe, and nutritious food that meets their dietary needs and food preferences for an active and healthy life (FAO 1996). This definition was developed by the Food and Agriculture Organization (FAO) of the United Nations in 1996 and is still widely used today (FAO 1996). This definition was adopted by the World Food Summit in 1996, which aimed to reduce the number of undernourished people in the world by half by 2015. As with many other global declarations and initiatives, this goal is yet to be realized.

In 1990, many African countries were facing severe food insecurity due to a combination of factors such as droughts, conflicts, economic instability, and poor governance. The situation was particularly dire in sub-Saharan Africa where more than 30% of the population was undernourished (FAO 1996). To address this challenge, various initiatives were launched by international organizations such as the FAO, the World Food Programme (WFP), and the International Fund for Agricultural Development (IFAD). These initiatives aimed to improve agricultural productivity, promote rural development, enhance food distribution systems, and strengthen national food policies (Pinstrup-Andersen et al. 1997, cited in Lado 2001).

However, the most widely accepted definition of food security is that developed by the Food and Agriculture Organization (FAO) of the United Nations. According to the FAO, food security exists "when all people, at all times, have physical, social and economic access to sufficient, safe and nutritious food that meets their dietary needs and food preferences for an active and healthy life" (FAO 1996). This definition emphasizes the importance of access to food as well as its quality and safety. It is important to recognize that achieving food security requires not just addressing immediate hunger but also addressing underlying issues such as poverty and inequality. As a result, definitions of food security should include broader issues such as social justice and sustainability. This means that any attempt to address food security in Africa should be intentional in making sure that food is produced and consumed in a way that is environmentally sustainable and socially just. This understanding highlights the importance of addressing both immediate and underlying issues to achieve long-term food security (FAO 1996). Food insecurity, therefore, has common and interrelated contributing factors (Bjornlund et al. 2022). Having reflected on food security, it is important to pay attention to another, equally important concept, namely, food sovereignty.

Food Sovereignty

1. Focuses on food for the people by: (a) placing people's need for food at the centre of policies; and (b) insisting that food is more than just a commodity.
2. Values food providers by: (a) supporting sustainable livelihoods; and (b) respecting the work of all food providers.
3. Localizes food systems by: (a) reducing the distance between suppliers and consumers; (b) rejecting dumping and inappropriate food aid; and c) resisting dependence on remote and unaccountable corporations.
4. Places control at a local level by: (a) placing control in the hands of local food suppliers; (b) recognizing the need to inhabit and share territories; and (c) rejecting the privatization of natural resources.
5. Promotes knowledge and skills by: (a) building on traditional knowledge; (b) using research to support and pass on this knowledge to future generations; and (c) rejecting technologies that undermine local food systems.
6. Works with nature by: (a) maximizing the contributions of ecosystems; (b) improving resilience; and (c) rejecting energy intensive, monocultural, industrialized, and destructive production methods (Gordillo and Jerónimo 2013: v).

Food sovereignty is particularly important for Africa. At the height of the COVID-19 pandemic, it became clear that Africa needs to invest in food sovereignty, as countries in the Global North shamelessly engaged in vaccine nationalism (see e.g. Benyera 2021). Many countries and pharmaceutical companies were hoarding vaccines when Africa was in desperate need. It was by divine providence that Africa did not have very high death rates, but the message was clear: the continent needs to ensure that it can support itself in all the key areas of life. As outlined in the citation above, food sovereignty ensures that food becomes central in policy formulation and implementation. As we argue below, this is quite consistent with the teachings found in the religions of Africa. Food sovereignty implies investing in food producers, as well as ceding power to local food suppliers. This implies critiquing the current model where global players tend to receive priority.

Food sovereignty also respects indigenous knowledge systems. It is not as if Africa began producing food with the advent of more intense interaction with Westerners. Africa must utilize indigenous knowledge systems to respond to climate change and enhance food sovereignty (see e.g. Mafongoya and Ajayi 2017). Food sovereignty is also highly conscious of gender. It recognizes the critical role that women play in food production, as well as their strategic place in household food security and consumption. Thus:

> Food sovereignty is also becoming increasingly influential, offering what more mainstream food and nutrition security approaches lack—a chance for people to self-determine what they produce and eat through the creation of local, environmentally sustainable food systems. The food sovereignty movement has also made considerable progress in promoting a more gender-equitable approach, although little has been done to challenge inequitable intra-household distribution of food. At a theoretical level the right to food and food sovereignty approaches offer potential opportunities for strengthening the way food insecurity is understood and addressed. (Ake et al. 2019: 32)

Having attended to clarifying the key conceptual issues, in the following section, the chapter examines causes of food insecurity and struggle for food sovereignty in Africa. Although there is merit in considering food sovereignty as more strategic than food security (see e.g. Tambi et al. 2014), in this chapter we generally regard the two concepts as complementary. Both are driven by the need to see Africa being assured of food and determining how this food is accessed and consumed. However, for us food sovereignty is more appealing because of its emphasis on autonomy and liberation, with special emphasis on Africa utilizing African seeds to grow food for Africa.

Causes of Food Insecurity and the Struggle for Food Sovereignty in Africa

Food insecurity in Africa is a complex issue with various causes, ranging from environmental factors to political and economic instability. One of the main causes of food insecurity in Africa is poverty. Many people in Africa live below the poverty line, which means they cannot afford to buy enough food to meet their basic nutritional needs. Poverty is often linked to low levels of education, limited access to healthcare, and lack of employment opportunities (Mbaya 2003). Another major cause of food insecurity in Africa is climate change. The continent is highly vulnerable to the effects of climate change, including droughts, floods, and other extreme weather events. These weather patterns can have a devastating impact on agriculture, which is the primary source of livelihood for many Africans. In addition, climate change can lead to the spread of pests and diseases that can destroy crops and reduce yields (Drimie and Mini 2003).

Political instability and conflict are also major drivers of food insecurity and struggle for food sovereignty in Africa. Wars and civil unrest can disrupt food production and distribution systems, leading to shortages and price spikes. In some cases, conflict can also lead to the displacement of populations, making it difficult for people to access food and other basic necessities (Van Rooyen 2000). Other factors that contribute to food insecurity in Africa include poor infrastructure, inadequate storage facilities, and limited access to markets. These challenges make it difficult for farmers to get their products to market and for consumers to access affordable and nutritious food (Drimie and Mini 2003).

One of the main challenges to achieving food security and sovereignty in Africa is the limited availability of arable land, water resources, and other agricultural inputs. Many African countries also face frequent droughts, floods, pests, and diseases that can damage crops and reduce yields. In addition, climate change is exacerbating these challenges by altering weather patterns and increasing the frequency and intensity of extreme weather events (FAO 2016). Food insecurity therefore has significant social, economic, and environmental consequences. It can lead to malnutrition, stunted growth, poor health outcomes, reduced educational attainment, and lower productivity. It can also exacerbate social inequalities, increase political instability, and contribute to environmental degradation (Wheeler and von Braun 2013; Parker et al. 2019).

In order to address food insecurity and achieve food sovereignty in Africa, there is a need for comprehensive strategies that tackle these underlying causes. This may involve investing in education and healthcare programmes that help lift people out of poverty, as well as implementing policies that promote sustainable agriculture practices and mitigate the impact of climate change. It may also involve addressing political instability and conflict through diplomatic efforts and peace building initiatives. Overall, addressing food insecurity and achieving food sovereignty in Africa will require a multifaceted approach that tackles these underlying causes in a coordinated manner.

The Importance of Food Security and Sovereignty in Africa

Food security and sovereignty is a critical issue in Africa due to its impact on the economic, social, and political stability of the continent. Firstly, food security and sovereignty is crucial for sustainable development in Africa. The continent has vast agricultural resources that can be harnessed to meet its food needs and generate income for farmers. However, the lack of access to inputs such as seeds, fertilizers, and irrigation facilities limits production. It is therefore important to increase various means such as providing farmers with access to improved seeds, fertilizers, and other inputs. It is important to invest in irrigation systems, research and development, and infrastructure development to support agriculture (Ghenna 2022). As noted earlier, however, food sovereignty demands that these should not be imported and that indigenous varieties be both promoted and protected.

Additionally, poor infrastructure and inadequate storage facilities lead to post-harvest losses, reducing the availability of food. Therefore, investing in agriculture and improving infrastructure can enhance food security and promote sustainable development in Africa. Investing in transportation infrastructure such as roads and railways to improve access to markets. It is also important to implement policies that support small-scale farmers and local food systems (Limão and Venables 2001).

Secondly, food insecurity and lack of food sovereignty contributes to poverty in Africa. The continent has the highest number of people living in extreme poverty globally, with over 400 million people living below the poverty line (FAO 2009). Food insecurity exacerbates poverty by limiting access to nutritious food and increasing vulnerability to diseases. Malnutrition affects children's growth and development, leading to stunted growth and cognitive impairment. Therefore, addressing food insecurity and promoting food sovereignty is crucial for reducing poverty levels in Africa. Addressing poverty through various means such as education, job creation, and social safety nets can help reduce food insecurity (UNFAO 2018).

Thirdly, food security is essential for human well-being in Africa. The continent faces several health challenges such as HIV/AIDS, malaria, and tuberculosis that require proper nutrition for effective treatment. Malnutrition weakens the immune system, making individuals more susceptible to infections. Additionally, hunger causes stress and anxiety that affect mental health. Therefore, ensuring access to nutritious food is critical for promoting good health and well-being in Africa. Improving food distribution systems is another critical solution to addressing food insecurity. This includes investing in transportation infrastructure such as roads and railways to improve access to markets. It is also important to implement policies that support small-scale farmers and local food systems, as promoted by the principles of food sovereignty.

Lastly, it is important to promote peace by resolving conflicts and addressing the root causes of conflicts. This will create a conducive environment for food production and distribution (Food and Agriculture Organization of the United Nations 2021). In summary, food security and sovereignty is a critical issue in Africa that affects sustainable development, poverty reduction, and human well-being. Addressing this challenge therefore requires investment in agriculture, improving infrastructure, and promoting access to nutritious food. The lack of food security and struggle for food sovereignty in Africa could cause political instability, societal disintegration, economic downturn, scramble for resources, high crimes and exploitation of vulnerable populations among others. It is in this context that FBOs have been actively involved in seeking to secure mostly food security, but sometimes also, food sovereignty.

Faith-Based Organisations and Food Security in Africa

As noted earlier, according to the Food and Agriculture Organization (FAO), Africa is the continent with the highest prevalence of undernourishment, with an estimated 250 million people suffering from hunger (FAO 2009). This is due to various factors such as low agricultural productivity, inadequate infrastructure, limited access to markets, and poor governance. Food security in Africa remains a critical challenge that requires concerted efforts from governments, civil society organizations, private sector actors, and development partners. It is essential to adopt a holistic approach that addresses the root causes of food insecurity while promoting sustainable development. Achieving food security in Africa requires a multifaceted approach that includes improving agricultural productivity, promoting sustainable farming practices (particularly indigenous ones), enhancing market access, strengthening social protection programmes, and addressing gender inequalities. It also involves ensuring that food systems are resilient to shocks and stresses such as droughts, floods, and pandemics (as became clear during the COVID-19 pandemic).

In addressing food insecurity and the quest for food sovereignty in Africa, FBOs have played a crucial role in providing relief and support to affected communities. It is beyond the scope of this chapter to provide a detailed definition of FBOs. The following dimensions of FBOs have been identified:

> ... such organisations typically have one or more of the following characteristics: funding by a religious body, direct denominational affiliation, governance by a board of religious leaders, not-for-profit status, a religious mission statement, religiously affiliated personnel, and/or projects directly or indirectly influenced by religious beliefs...(Welty 2014: 66)

Faith-based organisations (associated with different religions) have been actively involved in virtually all the regions/countries where drought, famine, and other threats to food security have been experienced. For the most part, FBOs have a religious mission and work to promote social justice and welfare. They have a presence in many African countries and are involved in various activities such as education, healthcare, and disaster relief. Thus, many FBOs have been involved in addressing food insecurity through various initiatives such as food aid programmes, agricultural training, advocacy, and community development. Food aid programmes involve distributing food to vulnerable populations such as refugees, internally displaced persons, and those affected by natural disasters. Agricultural training involves teaching farmers sustainable farming practices that can improve their yields and incomes. Advocacy involves raising awareness about the causes of food insecurity and advocating for policies that can address the issue. Community development involves empowering communities to take charge of their food security by providing them with resources and skills. It is significant to note that FBOs from different faith traditions, including World Vision (Evangelical Christian), Catholic Relief Services (Catholic), and Islamic Relief Worldwide (Muslim) have been responding to the food crisis/insecurity in Africa.

In an informative article, Lingrani and Niewolny identify three key reasons why some FBOs have been involved in food systems work. First, they identify the historical connection between faith and food across diverse religions, as food serves different functions in these religions. Second, they maintain that many religions have an in-built commitment towards caring for the poor. Third, they observe that most FBOs have the requisite social and cultural capital that enables them to pursue social justice issues. For Lingrani and Niewlny (2017), these factors have enabled FBOs to contribute to the quest for food security. In the specific case of Africa, global ecumenical bodies such as the World Council of Churches (WCC) have been consistent in challenging the emphasis on profits ahead of human lives through economic systems that condemn the poor to die of hunger. Through the Ecumenical Advocacy Alliance's (EAA) Food for Life Campaign, the WCC has underscored the urgency of recognizing the right to food for all. The WCC's Ten Commandments of Food[2] are anchored on the principles of justice and are particularly relevant to an Africa that is disadvantaged by the skewed global economic systems.

One strategic intervention to promote food security and sovereignty by FBOs in Africa is through faith-based land restoration. For example, World Vision Niger has sought to restore 40,000 hectares and improve 90,000 households' incomes through tree-based restoration; the Interfaith Mediation Centre and EverGreening Network for Landscape Restoration is working to restore the environment and promote peaceful co-existence between pastoralists and farmers in Nigeria; Catholic Relief Service Madagascar has sought to improve rural livelihoods and increase landscape restoration on land owned by the church and the Minda Trust/MIKA Initiative working on land rights with Muslims in the coastal areas of Kenya (Regreening Africa 2022: 5).

On its part, KAICIID reports on faith actors preventing food loss in Rwanda, highlighting the work of the Rwanda Interfaith Council on Health (RICH). RICH works with farmers to reduce post-harvest losses

[2] https://www.oikoumene.org/sites/default/files/Document/TenCommandments Food.pdf, accessed 3 July 2023.

by providing training on proper storage, processing, and packaging techniques.[3] Staying in the Great Lakes Region, Food for the Hungry Uganda focuses on empowering communities to identify and solve food security challenges through capacity building, education, and innovative agricultural practices.[4]

The Contribution of Faith-Based Organisations to Food Security and Sovereignty in Africa: A Critical Review

While there are challenges that accompany the interventions by FBOs, it is fair to concede that they have made significant strides in promoting food security, and to some extent, sovereignty, in Africa. They have empowered small-scale farmers, increased agricultural productivity, promoted nutrition, and improved access to markets. They have refused to operate with a limited understanding of religion, initiating inter-religious engagement in the context of the World Food Programme's Hero Hunger initiative.[5] On its part, the Catholic development agency, the Diocesan Development Services (DDS) has operated in the Catholic Diocese of Idah (Diocese) in Nigeria from 1970 to the time of writing. It has sought to promote environmental sustainability and food security among the disadvantaged members of the community (see e.g. Morse and McNamara 2021). In other instances, FBOs in Chad and Malawi are strengthening health systems and promoting food security. They are empowering communities, generating innovative practices and interacting with other stakeholders in the quest to promote food security and sovereignty in Africa.

Overall, we contend that through capacity building and education (e.g. Food for the Hungry Uganda[6]), collaborating with other stakeholders, developing sustainable agricultural practices and providing food relief (e.g. Buddhist Global Relief in the Horn of Africa[7]), tackling food loss, bolstering healthcare facilities and other engagements, FBOs are contributing to

[3] https://www.kaiciid.org/news-events/news/how-faith-actors-are-preventing-food-loss-rwanda, accessed 3 July 2023.

[4] https://www.christianitytoday.com/ct/2022/october/africa-christian-aid-agencies-have-new-approach-to-famine.html.

[5] https://documents.wfp.org/stellent/groups/public/documents/reports/wfp284537.pdf, accessed 3 July 2023.

[6] https://reliefweb.int/report/uganda/food-hungry-nutrition-training-saving-lives-uganda, accessed 17 July 2023.

[7] https://www.facebook.com/BuddhistGblRlf/, accessed 17 July 2023.

food security and sovereignty in Africa. However, they face a number of challenges. First, the charity model tends to dominate the FBO engagement in food security and sovereignty in Africa. There is a tendency to respond to famine without embracing a long-term liberating vision. As Naicker and Molobi (2022) rightly argue, food insecurity is structural violence. Unless FBOs creatively and prophetically challenge global and national capitalistic systems of oppression, food sovereignty will remain a forlorn ideal in Africa. Thus, there is an urgent need for FBOs and activists in Africa to adopt a revolutionary approach when tackling food security and sovereignty. The following observation is, therefore, an apt one:

> Until there is a fair and transparent world trading system and western governments change their current agricultural policy approaches, there is no reason why African countries should not undertake reforms in the agricultural sector that entail[s] a central and dominant role for the state and which can contribute to food sovereignty. (Arthur 2012: 17)

Second, and closely related to the foregoing, is that most FBOs have not prioritized Africa's food sovereignty in their short- to long-term planning. The image of the continent as home to emaciated, dependent, and desperate beings continues to dominate. There can be no development built on dependency. Consequently, FBOs must invest in partnering with Africans to work for a new world order in which capitalist exploitation is overcome and a more just world is attained. This is not a utopian ideology, but a reality that can be achieved in the contemporary period.

Third, most FBOs have not maximized on the resources that they possess to contribute towards Africa's food security and sovereignty. Although a few FBOs have demonstrated willingness to let go of their land, or to ensure maximum productivity through sustainable agricultural practices, most are holding on to the land. The land question remains a deeply political and spiritual issue, particularly in Southern Africa where the mission churches became beneficiaries of land stolen from the indigenous people (see e.g. Chitando 1998). A serious commitment to relinquishment, where the churches give up the land they got unjustly, is required (see e.g. Gundani 2002).

Fourth, and finally for this chapter, there is a need for FBOs to invest in addressing the serious gender imbalances that characterize African women's lack of access to land. It is a travesty of justice for those who are responsible for food production not to own the land. Due to a multiplicity of factors, including historical, religio-cultural, political, economic and others, many women do not own land. It is critical for FBOs to work with governments, other civil society actors, as well as other stakeholders, to promote women's ownership of land. Such an intervention will contribute towards food security and sovereignty in Africa.

Conclusion

No meaningful and sustainable development can emerge from stomachs that are rumbling due to hunger. The African Union, regional economic communities and national governments have proclaimed the need for Africa to ensure it meets and surpasses its food requirements. To go on to global platforms and declare Africa's territorial integrity, autonomy and right to self-determination is illusory when the continent continues to rely on external players to rescue it when famine strikes. In this chapter, we have reviewed Africa's food security and sovereignty concerns. We have assessed the strengths and challenges of FBOs as they seek to meet these challenges. We conclude by reiterating the point that only when the continent wins the battle for food security and sovereignty can it be said to be on a more certain trajectory towards sustainable development.

References

African Union. (2015). *Agenda 2063: The Africa We Want: Framework Document.* Addis Ababa: African Union.

Agbiji, Obaji M, & Swart, Ignatius. (2015). Religion and social transformation in Africa: A critical and appreciative perspective. *Scriptura 114*: 1-20.

Ake, M. et al. (2019). Food Sovereignty: An Enhancer to Women Contributions to Food Security in Nigeria. *Journal of Social Sciences and Public Policy* 11(1): 14–36.

Arene, CJ, Anyaeji, RC. (2010). Determinants of Food Security among Households in Nsukka Metropolis of Enugu State of Nigeria. *Pakistan Journal of Social Sciences (PJSS)* 30(1): 9–16.

Arthur, P. K. (2012). Food Security and Sovereignty in Africa: Issues, Policy Challenges and Opportunities. ACBF Occasional Paper No. 17. Available at https://elibrary.acbfpact.org/acbf/collect/acbf/index/assoc/HASH0111/89f5f679/634201b8/79f1.dir/Occa17.pdf, accessed 30 June 2023.

Benyera, E. (2021). *The fourth industrial revolution and the recolonisation of Africa: The coloniality of data* (p. 200). Taylor & Francis.

Bjornlund, V., Bjornlund, H. & van Rooyen, A. (2022). Why food insecurity persists in sub-Saharan Africa: A review of existing evidence. *Food Sec.* 14: 845–864. https://doi.org/10.1007/s12571-022-01256-1.

Chitando, E. (1998). Towards Land Answers? Spirituality, Reconciliation and Resettlement in Zimbabwe. *Bulletin for Contextual Theology in Africa* 5(3): 55–60.

Chitando, E. et al. Eds. (2022). *African Perspectives on Religion and Climate Change*. London: Routledge.

Drimie, S. & Mini, S. (2003). Food Security and Sustainable Development in Southern Africa. Human Sciences Research Council (HSRC). Available: www.hsrcpress.ac.za. Accessed 22 February 2023.

ECI (Ebony Consulting International (Pty) Ltd.) (2002). The food security effects of the deregulation of agricultural marketing in South Africa. A report by ECI to the National Agricultural Marketing Council, 18 February.

Enilolobo, O. S., Nnoli, T. I., Olowo, S. O., Aderemi, T. A., Adewole, A. O., Olapade, V. O., & Esedeke, J. F. (2022). Determinants of Food Security in Nigeria. *Acta Universitatis Danubius: Oeconomica,* 18(3).

FAO. 1974. World Food Summit. Rome, Italy.

FAO. 1983. World food security: A reappraisal of the concepts and approaches, Director General's report, FAO, Rome, available at: www.fao.org.

FAO. 1996. Rome declaration on world food security and world food summit plan of action. World Food Summit, November 13–17, 1996, Rome. https://www.fao.org/3/w3548e/w3548e00.htm.

FAO. 2008. An introduction to basic concepts of food security, EC-FAO Food security Programme, available at: http://www.foodsec.org/docs/concepts_guide.pdf.

FAO. 2009. (Food and Agriculture Organization of the United Nations). One sixth of humanity undernourished – more than ever before." https://www.sciencedaily.com/releases/2009/06/090619121443.htm. Retrieved on 22 February 2023.

FAO. 2016. Climate change and food security risks and responses.PP.1–46 (available on the FAO website (www.fao.org/publications).

Food and Agriculture Organization of the United Nations. (2021).

Ghenna, K. (2022). Impact of Ukraine war on commodities and Africa: possible policy responses and opportunities. Pan African Chamber of Commerce and Industry.

Gordillo, G. and O. M. Jerónimo. (2013). Food Security and Sovereignty: A Base Document for Discussion. Rome: FAO. Available at https://www.fao.org/3/ax736e/ax736e.pdf, accessed 26 June 2023.

Gundani, P. H. (2002). The Land Crisis in Zimbabwe and the Role of the Churches towards its Resolution. *Studia Historiae Ecclesiasticae* 28(2): 122–169.

Kunhiyop, S. W. 2008. The Challenge of African Christian Morality. In Chinne, Z (ed.), *Service with Integrity: The Christian in the Nigerian Project,* 211–244, Kaduna: The ECWA Goodnews Church.

La Via Campesina. International Peasants' Movement. (2022). COVID-19 and Conflicts: Impact on Africa's Food Sovereignty. https://viacampesina.org/en/grassroots-impacts-of-covid19-conflicts-and-crises-on-the-right-to-food-and-food-sovereignty/.

Mafongoya, P. L. and O. C. Ajayi. Eds. (2017). *Indigenous Knowledge Systems and Climate Change Management in Africa*. Wageningen, Netherlands: CTA.

Mbaya, S. (2003). The Southern African food security crisis, causes and responses: A Regional overview, in S. Drimie & M. Lafon (eds.), *Food Security in Southern Africa: Causes and Responses from the Region*. Johannesburg: Institut Francais d'Afrique du Sud.

Morse, S. and N. McNamara. (2021). Role of Faith-Based Social Groups in Promoting Sustainable Food Security in Nigeria. In S. E. Silvern and E. H. Davis (eds.), *Religion, Sustainability, and Place: Moral Geographies of the Anthropocene*. Singapore: Springer Nature Singapore: 173–198.

Mukiibi, E. (2020). COVID-19 and the State of Food Security in Africa. *Agriculture and Human Values* 37: 627–628.

Naicker, L. and V. S. Molobi. (2022). A Socially-engaged Theological Response to the Historic and Structural Nature of Food Insecurity in South Africa. *Studia Historiae Ecclesiasticae* 48(1), 17 pages. https://doi.org/10.25159/2412-4265/10656.

Nyakecho, O. H. et al. (2022). A review of African spirituality and food sovereignty and the gender aspects along the staple food value chain among smallholder farmers in western Kenya. *International Journal of Scientific and Research Publications* 12(8): 106–114.

Lado, C. (2001). Environmental and socio-economic factors behind food security policy strategies in Botswana. *Development Southern Africa* 18(2), 141–168.

Limão, N., Venables, A. J., (2001). Infrastructure, geographical disadvantage, transport costs, and trade. *The World Bank Economic Review* 15: 451–479.

Ogujuiba, K., U. M. Ogbonnaya and O. E. Omoju. (2012). Food Security, Policies and Institutions in Africa: Prospects for a Revolution. *Journal of Economic Cooperation and Development* 33(4): 87–115.

Osagie-Usman, C. (1994). Mission of Revenge. West Africa, 3984 p.205.

Parker, L., Bourgoin, C., Martinez-Valle, A., & Laderach, P. (2019). Vulnerability of the agricultural sector to climate change: The development of a pan-tropical climate risk vulnerability assessment to inform sub-national decision making. *PLoS ONE*, 14(3), e0213641. https://doi.org/10.1371/journal.pone.0213641eology.

Regreening Africa. (2022). Engaging faith-based organisations in land restoration across Africa: Insights and opportunities. Nairobi: World Agroforestry.

Salih, M.A. M. (2002). Islamic NGOs in Africa: The Promise and Peril of Islamic Voluntarism. Copenhagen: University of Copenhagen, Centre of African Studies.

Schmid, B. et al. (2008). The contribution of religious entities to health in sub-Saharan Africa. Study commissioned by B & M Gates Foundation. Unpublished report. ARHAP.

Schuurman, E. (2011). Technology and Religion: Islam, Christianity and Materialism. *Koers: Bulletin for Christian Scholarship* 76(2):373–386. (https://doi.org/10.4102/koers.v76i2.21).

Tambi, E. et al. (2014). Food sovereignty and food security: Where does Africa stand? Accra: Forum for Agricultural Research in Africa.

UNFAO. (2018). United National Food and Agriculture Organization – Economic and Social Development Department, 2008. "The State of Food Insecurity in the World, 2008: High food prices and food security – threats and opportunities." Retrieved on 17 July 2023.

Van Rooyen, J. (2000). Regional food security and agricultural policy in Southern Africa: A challenge of policy conversion in diverse settings. Development Southern Africa 17(1): 7-22.

Welty, E. (2014). Faith-based Peacebuilding and Development. *Journal of Peacebuilding & Development* 9(2): 65–70.

Weiss, H. (Ed.). (2020). *Muslim Faith-based Organisations and Social Welfare in Africa*. Cham: Palgrave Macmillan.

Wheeler, T. & von Braun, J. (2013). Climate change impacts on global food security. *Science* 341(6145): 508–513.

World Bank. (1986). *World development report*. Oxford University Press.

PART V

Religion, Climate Change, Food Security and Resilience

CHAPTER 17

Pentecostalism, Theology of Survival and Food Security in Zimbabwe

Kudzai Biri

Introduction

The United Nations reported on 7 June 2023 that Africa imports 85 per cent of its food. This is disturbing news, given that Africa is a vast continent that is endowed with resources and a large labor force. It is therefore important to examine how Pentecostals navigate the terrain of hard work and survival in their theology to establish how it affects food security or alleviating poverty and deprivation. There is no doubt that Pentecostalism has remained one of the most vibrant forms of Christianity in Africa. Biri (2013) has praised the ethos of hard work that most Pentecostals teach. In Zimbabwe, the Zimbabwe Assemblies of God Africa is one of the oldest Pentecostal churches that has consistently taught believers to work

K. Biri (✉)
Religion Department, Hope College, Holland, MI, USA

L. Maseno et al. (eds.), *Religion, Climate Change, and Food Security in Africa*, Sustainable Development Goals Series,
https://doi.org/10.1007/978-3-031-50392-4_17

Talents.[1] The church has also regularly encouraged members to accumulate as much food as possible, so that in times of need and crisis, they have enough food to sustain themselves. During the decades of crisis in Zimbabwe, especially the years 2000, 2008, and just before the Covid 19 lockdown era, church members were encouraged to store food, especially the basic commodities. While this is noble given the unpredictable economic challenges in Zimbabwe, the practice has multiple challenges that it causes. Many poor people cannot afford to buy and stock basic commodities. Only the medium income people and the rich can manage to accumulate large quantities of food. This is mainly because the majority of civil servants and the jobless live on a hand to mouth basis. Accumulating food in bulk to cater for crisis moments remains farfetched for the majority. This is one practice that divides the rich and the poor and it becomes manifest even in the church. As a movement that thrives on theology and testimonies on progress and prosperity, the poor have nothing to testify since most of them survive on a hand to mouth basis. Ruth Marshall Fratani and Kalu aptly point that Pentecostalism has failed to alleviate pauperism. The vibrant theology of dominion, survival and prosperity in Africa does not match with the levels of poverty and deprivation. What then is at stake? The Pentecostal theology of dominion and survival and the cycle of poverty deserve scrutiny. Vengeyi (2012), with reference to Zimbabwe, argues that the cycles of poverty and deprivation are human made. He argues that Zimbabwe is endowed with natural resources that enable wealth creation and subsequently abundant food production that can suffice to feed the whole nation. However, the majority swim in poverty while the elite live excessively luxurious lives. How does the church respond to the issues of corruption, mismanaged, misplaced priorities and the heavy taxes imposed on citizens who earn little wages and subsequently affects production, whilst creating a food crisis? Zimbabwe imports almost all agricultural and industrial goods especially from South Africa. The proliferation of retail shops popularly known as 'matakishopu' derived from tuck-shops is an example of open markets for smuggled goods from neighboring countries that are sold at cheaper prices, compared to those in the regular supermarkets. However, the strict usage of the United States

[1] Talents in ZAOGA refer to different entrepreneurship projects that individuals can undertake to raise money for their families or for the church. The teaching is based on Matthew 25. The idea behind Talents is to generate wealth outside the pay check and for those who are not employed to empower themselves economically.

dollar makes it difficult to buy the basic food commodities. This contrasts with 2008 where shops were empty, and people had nowhere to buy basic commodities except from neighboring countries. The section below examines the nature of Pentecostalism that infringe with wealth creation and food security. It probes the nexus between religiosity, poverty, deprivation, and pauperism in Zimbabwe.

The Theology of Election and Dominion

2 Thessalonians 2:17 admonishes that, if any man is in Christ, he is a new creation. Most Pentecostal churches uphold the doctrine that those who make confession to invite Jesus Christ into their lives become born again and constitute the elect of God. They therefore envisage a unique life of victory, abundance over, above, and against those who are outside the Pentecostal fraternity. As such, they believe that Pentecostals should enjoy privileges. These privileges translate to the ability to accumulate wealth and enjoy material blessings. While the study does not condemn the desire to get rich, the quest to accumulate wealth has witnessed oppressive and exploitative practices in most Pentecostal churches. I have witnessed poor people being pressured to give what they have as a prerequisite for receiving divine blessings. This has worsened many and immersed them in poverty. The economic state in Zimbabwe defies claims of operating in faith because people must be employed to get money, they have to work and be productive. An economy that is on its knees for decades exposes the weakness of the Pentecostal theology of giving by faith and miraculous provision and blessings. It also presents a paradox and ambivalence that prevents practical economic empowerment of the majority. The realistic harsh economic environment must be acknowledged and the prevailing Pentecostal theology in Zimbabwe should be reformulated to combat hunger through personal savings and investments and not sowing to the ministry or to the man/woman of God. The prevailing practice seems to glorify working towards giving to the man/woman of God. While ZAOGA for example has trained people to work Talents, with the goal of empowering adherents to become economically sound, the ever- consistent demand on money even where resources are scarce is extreme. The researcher has also carried out research on the Apostolic Faith Mission in Zimbabwe before and after the split of the church. Most congregants, even in leadership positions complained about how leaders ignore the biting economy and insist that congregants should give to the church.

The elites in leadership hierarchies usurp the resources of innocent believers. The principle of sowing and reaping are popular, and congregants are expected to sow into the church, ministry, or life of the man/woman of God consistently. For example, Makandiwa of the United Families International Church had sitting arrangements according to how congregants give in the church and he had many investments that he defended (Biri 2013). On the other hand, Prophetess Memory Matimbire of Daughters of Virtue Ministries urged single women to sow in order to unlock doors for finding a marital partner. The best to be sown was pegged at least 100 USD not in local currency, the Zimbabwean bond notes.[2] While the researcher protested at the teachings, as they are not based on biblical truth, the impact of the teaching is what is significant. Marriage is revered in most African societies as single motherhood is looked down upon (Biri 2021). Poor single, and desperate women give money and tokens in the hope of getting married, to the detriment of resources that can be channeled towards themselves or their children. Thus, enriching the leadership. Makandiwa castigated poor congregants by saying, 'you eat what you get, but I eat what I want'.[3] What is significant in the statement is the acknowledgement of his church member's poverty. They cannot afford to eat what they want, but only what they can get. In contrast, he eats whatever he wants because according to him, God has blessed him. But the source of their (Pentecostal leaders) wealth is derived from the same poor congregants. This raises the question of inequality and social justice. Does Pentecostalism suffice to stand in the gap for the poor, widows and needy, the groups that constitute God's special people? These groups have suffered more during the decades of the economic crisis, with many people including civil servants struggling to access basic commodities.

In addition, Makandiwa interpreted the story of the widow of Zarephath by claiming that the widows must provide for the man of God, even if it is the last dime in their pockets. The wrong interpretation and deployment of scriptures in contexts of poverty continue to impoverish congregants. Congregants are brainwashed and often threatened with 'touch not God's

[2] The author was a member of a social media group of single mothers where funds were mobilized for deliverance regardless of your denominational affiliation.

[3] Emmanuel Makandiwa is a controversial Pentecostal leader that broke away from the Apostolic Faith Mission in Zimbabwe. The orientation towards sowing and church practices such as anointed oil, apples, bricks etc. has deprived congregants who have remained poor.

anointed scripture' to silence voices that protest against the exploitative practices. The believers are told to continue sowing and that they are not bound by economic situation.[4] The result is that most leaders and their families and those that surround them benefit and live lavishly while the majority live in poverty. Paul Gifford made this observation earlier when he made the claim that Pentecostals in Southern Africa lack sociological awareness. They defy prevailing the socio- economic situation by over emphasizing the prosperity gospel and that individuals should prosper as children of God. Gifford's claim makes sense because the economic principles that can prosper individuals or a nation to combat poverty or hunger and have enough food reserves are ignored. One can therefore safely conclude that Pentecostals lack the foresight and stamina to effectively fight hunger and boost food security because of a weak theology of dominion and prosperity.

The claims of dominion while people do not work has also produced lazy people who claim material things by faith. This is one seamy side of Pentecostalism. This is in spite of entrepreneurship programs because the capitalist outlook and orientation of Pentecostalism has torrid results on believers. Most Pentecostal leaders tap into the corrupt political practices of the ruling regime (Biri 2013) that promote the politics of the belly (Bayart 1989). The politics of the belly runs in tandem with the Pentecostal theology of dominion and survival. It focuses on the individual first as one elected by God and dominion, with privileges above others. It does not promote the spirit of being 'my brother's keeper'. In addition, there is no accountability because everything belongs to the men/women of God who cannot be questioned or interrogated. This inevitably disempowers believers economically because they contribute and at the end their individual economic needs are not met in times of crisis and in alleviating their poverty. There are few exceptions in ZAOGA. For example, the working of Talents, popularly known as Matarenda has economically empowered believers by teaching and promoting the ethos of hard work. Church members through Talents have built a big hospital, Mbuya Dorcas hospital, several schools such as Ezekiel Guti primary school, Ezekiel Guti high school in Belvedere and Zimbabwe Ezekiel Guti University in Bindura. However, these investments do not accrue a direct financial benefit for

[4] Eunor Guti teaching at ZAOGA Grange, Harare, 07-08-2015.

most church members because there is no accountability on proceeds from the properties since they are solely managed by the Guti family.[5]

The usurping of congregants' resources and money creates poverty in individual lives and the families, while creating generational wealth for the elite. This is manifest in the economic sector where the elite continue to enjoy material wealth and get an advantage because they can manage food reserves while the poor remain economically marginalized. Hence, this impacts heavily on food security because the majority are vulnerable. The vibrancy of Pentecostalism, not only in Zimbabwe has drawn at the center the nexus between religiosity and poverty. It is baffling that a country such as Zimbabwe is endowed with vast natural resources and vibrant churches and yet it remains enmeshed in poverty. Kalu and Marshall Fratani laments how religion has failed to alleviate poverty in Africa and Zimbabwe is not an exception. While the theologies of survival and hard work are crucial, they have failed to meaningfully translate them practically to empower people to take charge and have dominion over poverty.

The traditional Shona society had means and ways to provide checks and balances to prevent excessive accumulation. The practice of *zunde ramambo* was also a way in which the grain that was given to the king would be later distributed in times of need, especially to the needy, poor widows and orphans (Gombe 1985). The ethos of hard work, sowing to the church and accumulation is commendable. This provided checks and balances and prevented people to be extremely rich. In addition, it promoted social justice as leaders ensured that every member of the community had enough food for survival. However, the big men/women should have a sense of responsibility and accountability to share resources that can boost food security for the congregants, given that Pentecostal churches have amassed wealth. There are no democratic structures that guarantee that the same members of the church that consistently contribute are also taken care of during crisis moments. Hence, in a way, it destroys the ability to sustain oneself in the name of giving to God. Crisis moments are litmus tests for Pentecostal members and their churches. First, Pentecostal churches operate in a capitalist mode of orientation. They are ever ready

[5] The author is a member of the ZAOGA church. She has witnessed fundraising projects in the name of the church, e.g. hospital, university, schools etc. but the church members do not benefit in any way from the proceeds of the investments. Those who question and protest do so behind the scenes. They lack the courage to do so because of the authoritarian structures of the church.

to get but not to give away what has been accumulated. If there are donations, in rare incidents, they do not match the needs of the crisis against the backdrop of what they have.

In 2019, Cyclone Idai exposed the weak structures of Pentecostalism in relation to food security and managing crisis. Zimbabwe was hit hard by the cyclone and this saw many people becoming homeless, in need of shelter and food. Some few Pentecostals donated with cameras around and some boasted that they were not left behind in donating clothes and food. While it is commendable, there are no regular donations that are made to the poor or extremely vulnerable groups within the communities, when many people consistently give, the virtues of care, and love become questionable. In this regard, Pentecostalism lacks the practice of managing food security.

Climate Change, Economic Challenges and Alternatives

The ravaging climate change and its adverse effects require strategizing in organizational structures. The so-called charity ministries should be effective in mapping ways to alleviate poverty and hunger because Zimbabwe has the potential of being a breadbasket of Southern Africa once again. Economic and political instability affects the economy, especially the production and preservation of food. Land that belongs to many Pentecostal churches is not used for agricultural purposes. Instead, there are halls, schools, and business offices. The debilitating climate change and the threat to food security should push Pentecostals who own vast tracts of land to invest in agricultural activities in Zimbabwe. Zimbabweans have experienced harsh economic meltdown and political turmoil. The political theology that they espouse is clearly partisan. The 2012-edited volume on Religion and Politics by Chitando examines the role of churches in the midst of crisis. In 2008, many Pentecostal churches led by ZAOGA accumulated food staffs because of the deepening crisis. It created shortages and shops ran out of food since there were already some shortages. There was panic buying and those who did not have money suffered a lot.

Church owned properties should sustain congregants in times of crisis, if they are established by the church. However, this does not seem to happen because the majority have no control of the church's resources and they do not decide what to do with the proceeds from the properties. The former minister of Tourism, Walter Mzembi proposed accountability and taxing churches. There was fierce resistance from Pentecostal churches as they claimed that the state could not interfere with matters of spirituality. This gave the late Robert Mugabe, who was then President of Zimbabwe, the opportunity to warn and threaten the church not to interfere with politics if activities of the church could not be monitored by the state. These developments expose the weakness of Pentecostalism that disempowers them to demand and confront brutal regimes and stand for the citizens. Furthermore, it confirms the observations of Ruth Marshal-Fratani and Andre Cortini (2001) that Pentecostals espouse a weak political theology. Failure to confront the government in terms of issues surrounding security feeds into the economy. Hence, Pentecostalism has not successfully redeemed the ethos of sheltering and providing the needy in times of crisis. This is one key issue that requires attention because Zimbabwe is hard hit by cycles of deprivation and poverty and the vibrancy of Pentecostalism and their entrepreneurship programs have the potential to cater to the poor and the needy. However, their oppressive and exploitative structures that lack democracy run parallel with the ruling ZANUPF regime (Biri 2013).

The World Food Program donates food for the needy in different parts of the country. Record has it that over the years, the powerful politicians consistently hijack the food for political purposes because they selectively parcel out food for their supporters.[6] For instance, on 27 August 2019, Emmanuel Makandiwa, the founder of United Families International church was denied access to donate food and blankets to the needy in Manicaland Province, Cyclone Idai hit area because he did not want to involve the ZANUPF regime in the exercise. If the food donations are successfully hijacked, they do not benefit the needy people that are intended, but the privileged. This is not different from the contributions in the church, which do not benefit the congregants but mostly the elite and the privileged. Membership in many churches comprises of women and children. The rise of single motherhood homes is a challenge alongside child headed homes. Several factors that range from divorce, deaths from HIV and AIDS have fueled the crisis leading to entrenching the already dire poverty levels. Research done by the writer as a member of a

[6] The researcher worked in Mashonaland West province that is consistently hit by drought and receives donations from the World Food Program and other charity organizations. Donations were taken by some members of the ruling regime and distributed to their supporters. In addition, there was also an outcry on how donated funds and food were misused by the elite, to the deprivation of the masses who needed help because of the effects of both Cyclone Idai and Covid 19.

Pentecostal denomination show that most of the churches do not have the facilities to cater to widows and orphans in need of food aid as well as other essentials. Likewise, the government of Zimbabwe does not offer any meaningful child support for single mothers. To what extend then can Pentecostalism guarantee food security in a continent and countries ravaged by many socio-economic and political woes? This question re-ignites the critique on the image of Pentecostalism worldwide.

Capitalists under Religious Guise?

Pentecostalism has been accused of promoting the worst forms of capitalism under the guise of religion. The hierarchy that usurps the resources of innocent desperate believers, as has been noted above, seem to justify criticisms that are levelled against it. The leadership that is decorated with unquestioned authority deploy evangelical power that disempowers believers. The practice of continuously making believers to contribute without matched reciprocity when the same contributors need support is a capitalist mode.

Pentecostalism promotes individualism, the moral principle typical of bourgeois ideology and morality. The theoretical foundation of individualism is the recognition of autonomy and the absolute rights of the individual in society (Salifun and Dixon 1984). The absence of democratic structures to provide checks and balances on the man/woman of God and the elite opens avenues of exploitation, oppression, and abuse. The abuse of vulnerable women that sought help from the church is high and there have not been any collective efforts to address that in many Pentecostal circles. The key defining mark has been to urge members to consistently sow in order to reap, sowing to the leadership and at the end, most of them (Pentecostal leaders), live obscene lavish lifestyles while the congregants languish in poverty. The wrong deployment of scriptures whereby people are encouraged to give the little that they have for miraculous provision has contributed largely to the impoverishment of believers. They become vulnerable and most women are exposed to abuse because they cannot feed their children.

Food security ropes in issues of gender justice. Violence is rampant in the political sphere in Zimbabwe and it is also endemic in the church (Kaulem 2007). The usurping of women's power affects the families and communities economically and leads to food shortages because most women sustain many households. They are innovative and cross border trading became a pillar for many families to survive as women embarked on businesses that include barter trade. If women are pushed to the periphery, they are disempowered and because they do not participate in production and management of food, it affects production and management of resources. Not all the negative aspects in Pentecostalism are meant to discredit the movement. The loopholes mark areas that Pentecostals need to improve, especially to become pro-active in crises that rage in the nation. It remains one of the leading forms of Christianity in Zimbabwe, which has the potential of transforming masses from poverty through increased food production. If it channels its energies in the following areas, first, towards engaging in political and economic stability because of the privileges that Pentecostals have with the state. This means pursuing a political theology that confronts evils of injustices in the land for economic and political stability. Second, establishing transparency and investing in agricultural activities with the aim of alleviating hunger through mastery of climate patterns that occur because of climate change. Most Pentecostals advocate social engagement as they seek to transform society, to be concerned with issues affecting the world. Hence, the issue of climate change and its threat to food production and security should be central to Pentecostal theology and practice.

A Critique on the Theology of Survival: Towards a Theology of Food Security and Holistic Flourishing in Zimbabwe's Pentecostalism

Most Zimbabwean Pentecostals have not emphasized agricultural activities as a viable means to guarantee food production and security. For example, food production is the focus on the Apostolic Pentecostal Church in Liberia. On 10 May 2022, Bishop Kartu argued that privileging private revelation and prophetic accumulation does not benefit people but only founders. In contrast, most women have embarked on cross border trading, which has sustained families during the decades of crisis. However, climate change requires not a shift in theology, but actualizing the theology of survival. Pentecostals need to focus on what survival means in the midst of climate change, poverty, and wealth creation that benefits masses and not the elite only. The accumulation of resources in Pentecostal churches is one key effective way to ensure food security by redistributing or sharing equally or giving where there is need. It is a sure way of overcoming the food crisis in Zimbabwe because rainfall patterns and food production are not uniform in the country. If the elite overcome and defeat greed by sharing with congregants and allowing them to benefit, families will flourish in every sector of life.

The theology of food security is key in sermons because abundant life translates to a life without lack of food and other necessities. The focus and emphasis on abundant life marks the key characteristic of Pentecostal theology that promotes holistic flourishing. The survival strategies in the wake of climate change require a focus on production of food more than the faith for, for example, tithing and miracle provisions and many more promises. Hence, the seamy side of the popular gospel of prosperity can be turned around to execute energies into food production through agricultural activities to boost food security. This approach ensures both material and spiritual prosperity and flourishing in Zimbabwe. David Maxwell (1998) has a good report on ZAOGA's entrepreneurship programs. To him, they have 'delivered people from the spirit of poverty' which is the spirit to cultivate and encourage believers to work hard for self-sustenance. During the times of climate change, the same spirit is required. It is an effective way of being pro-active through participation in global efforts to combat food shortages and boost food security in times of climate change and its harsh consequences in our contemporary times.

Conclusion

The Pentecostals are strategically positioned to deal with the issue of food security. The establishment of schools, colleges, and other properties is noble, but they (Pentecostal leaders), should see that they benefit the congregants and avoid monopoly of resources and profits by the elite or families of the leaders. The resources that are mobilized and invested should be channeled towards the needy families and communities in times of crisis. The country's food security is challenged because of decades of crisis and climate change. Hence, new strategies of survival are required. For example, there is need to invest more in food production and agricultural or farming activities. This can boost food reserves in the country and empower individuals and families to sustain themselves from local produce, which saves their income, instead of buying expensive imported foodstuffs. In addition to that, the mobilized resources should be directed to acquire basic food commodities to the less privileged and to those in arid areas where production of food is difficult. It is a form of justice to those who live in geographical areas whose climate is not friendly to food production. The contexts of 'human made or created poverty' should open way to re-think the prosperity gospel. Compelling believers to give under hardships is thuggery and promotes poverty and vulnerability. Hence, prosperity should also shield and provide to the poor and the needy and see to it that there is justice in distribution and access to food while guaranteeing food security for all.

References

Bayart, J. (1989). *The State in Africa: Politics of the Belly.* Cambridge, Polity.

Biri, K. (2021). Redeeming the Land ... Pentecostalism, Neo-colonialism and Reimagining Africa in *Mother Earth, Post Colonial and Liberation Theologies.* (eds), Chirongoma, S and Mombo E, Lanham, Lexington Books, 139–149.

Biri, K. (2013). Religion and Politics in Post-Colonial Zimbabwe: A Comparative Analysis of Robert Mugabe and Ezekiel Guti Leadership Styles, in Chitando, E. (ed). *Prayers and Payers: Religion in Post-Colonial Zimbabwe*, Harare: SAPES.

Gombe, J.M. (1985). *Tsika DzevaShona*, Gweru, Mambo Press.

Kaulem, D. (2007). *Ending Violence in Zimbabwe*, Harare, AFCAST.

Marshall-Fratani, R. and Cortini, A. (eds.). (2001). *Between Babel and Pentecost: Transnational Pentecostalism in Africa and Latin America*, Indiana, Indiana University Press.

Maxwell, D. (1998). "Delivered from the Spirit of Poverty"?: Pentecostalism, Prosperity and Modernity in Zimbabwe, *JRA*, 28, 3, 309–337.

Salifun, M and Dixon, R. (eds). (1984). *The Dictionary of Philosophy*, Moscow, Progress Publishers.

Vengeyi, O. (2012). Zimbabwean Poverty is man made! Demystifying Poverty by appealing to the peophetic book of Amos, Scriptura, *Journal for Biblical, Theological and Contextual Hermeneutics*, 107, 107–138.

CHAPTER 18

Religion, Climate Change, and Food Security in Africa: Religion, Food Security, and Resilience of Rural People in Ghana

Sylvia Owusu-Ansah, Beatrice-Joy Owusu-Ansah, and Appiah Kubi

Introduction

Food security, religion, and resilience are vital issues for rural communities in Ghana. This chapter begins by examining existing literature on food security, religion, and resilience of rural people in Ghana. This review identifies the key drivers of food insecurity and the role of religion in shaping the resilience of rural communities. Various factors drive food security

S. Owusu-Ansah (✉)
Perez University College, Accra, Ghana

B.-J. Owusu-Ansah
West Africa Centre for Crop Improvement, University of Ghana (WACCI), Accra, Ghana

A. Kubi
Korle-bu Teaching Hospital, Accra, Ghana

L. Maseno et al. (eds.), *Religion, Climate Change, and Food Security in Africa*, Sustainable Development Goals Series,
https://doi.org/10.1007/978-3-031-50392-4_18

in Ghana, including poverty, low agricultural productivity, climate change, and limited access to markets and infrastructure (Yaro 2013). Over 20% of the population in Ghana experiences food insecurity, with the northern region being the most affected (FAO, IFAD, UNICEF, WFP, & WHO 2021). The inadequate access to markets and infrastructure is often compounded by inadequate storage facilities, limited access to credit, and inadequate extension services (Adu et al. 2018). These factors limit food production among rural farmers, preventing them from meeting their food consumption needs and selling the surplus to generate income. Religion is very significant in shaping the resilience of rural communities in Ghana. Religion is a source of hope, comfort, and support during adversity. In Ghana, Christianity and Islam are the dominant religions, providing a framework for social organization, moral values, and community cohesion. Religious institutions also play a critical role in providing social services such as healthcare, education, and food assistance to vulnerable groups (Abunyewah 2020). Policy and community-level interventions can enhance food security and resilience in rural communities in Ghana. These interventions include improving access to markets and infrastructure, promoting climate-smart agriculture, enhancing extension services, and providing financial and technical support to smallholder farmers (Asenso-Okyere et al. 2019). Community-level interventions include promoting community-based natural resource management, strengthening social networks, and promoting local food systems (Asenso-Okyere et al. 2019).

Additionally, integrating religious institutions in designing and implementing these interventions can enhance their effectiveness and sustainability (Abunyewah 2020). Food security, religion, and resilience are critical issues for rural communities in Ghana. Addressing food insecurity requires a multifaceted approach that tackles the root causes of poverty, low agricultural productivity, and limited access to markets and infrastructure (Yilmaz and Ahmet 2023). Religion plays a vital role in shaping the resilience of rural communities, and its integration into policy and community-level interventions can enhance their effectiveness and sustainability.

As part of the study, fifty (50) rural people were interviewed across five regions in Ghana to find out the extent of their knowledge of climate change, its adverse effects on their daily lives, how they are coping with the challenges, and the role their religion plays in their resilience. We interviewed men and women under three categories: farmers, market women/

retailers, and consumers. Of the 50 respondents, 35 are farmers, 28 are market women/retailers, and 25 are consumers. Some respondents fell under more than one category. We used these three categories to capture the disruptions in the food value chain caused by climate change factors and how the main actors of the food system respond to them in Ghana. To find out about the level of food security in rural areas in Ghana, we asked questions covering the availability, access, utilization, and stability of food in various regions of Ghana, especially the northern region. We sampled rural people from Bepong (eastern region), Tolon (northern region), Tampulla (upper east region), Kaleo (upper west region), and Techiman North (Bono east region).

None of the respondents practiced traditional religion. Twenty-eight (28) respondents identified as Christians (56%) while 22 respondents (44%) were Muslim (see Table 18.1). There were 33 male participants (66%) and 17 female participants (34%).

The Concept of Food Security

Food security is the availability, access, utilization, and stability of safe and nutritious food that meets dietary needs and preferences for an active and healthy life (Brown et al. 2015). Food insecurity occurs when people have restricted access to safe and nutritious foods and cannot access enough food to meet dietary energy requirements. Food security is a complex issue that cannot be limited by geography or well-defined by a single grouping, demography, education, geographic location, or income. Achieving food security for everyone everywhere is a widely agreed upon international objective, most recently codified in the United Nations Sustainable Development Goals for 2030 (Brown et al. 2015). Food security's four key dimensions are availability, access, utilization, and stability (Food and Agriculture Organisation of the United Nations 2015). Climate

Table 18.1 Characteristics of rural people interviewed

Characteristics	*Percentage (%)*
Gender	
Male	66
Female	34
Religion	
Christian	56
Muslim	44
Traditional	–

change exacerbates the issue of food security, particularly in low-income and middle-income countries, as it affects the availability, access, and utilization of safe and nutritious food (Jarvis et al. 2008). The world's food challenge is ensuring food security, as more than 811 million people, mainly from low-income and middle-income countries, faced hunger in 2020. McCarthy et al. (2018) state that one billion people, which is 16% of the global population, suffer from lasting hunger at a time when there is more than enough food for everyone to feed on this planet. Climate change affects food stability, which is the absence of significant fluctuation in food availability, access, and utilization. The connection between food security and climate change is complex and influenced by various environmental and socioeconomic factors (McCarthy et al. 2018).

During our interviews, we asked all 50 respondents about the factors that influenced their access to the food of their choice. The majority of them (39 respondents, i.e., 84.8%) mentioned low income as a limiting factor to accessing their preferred food. Twenty-three respondents (50%) reported the non-availability of food caused by flooding, droughts, poor infrastructure, and others as the second most important factor. Poor health was the third factor that influenced their access to food reported by 18 respondents (39.1%) (see Fig. 18.1). Thirteen of the respondents (28.3%) mentioned limited access to markets as the next factor influencing their access to food, while one respondent (2.2%) mentioned old age as a limiting factor.

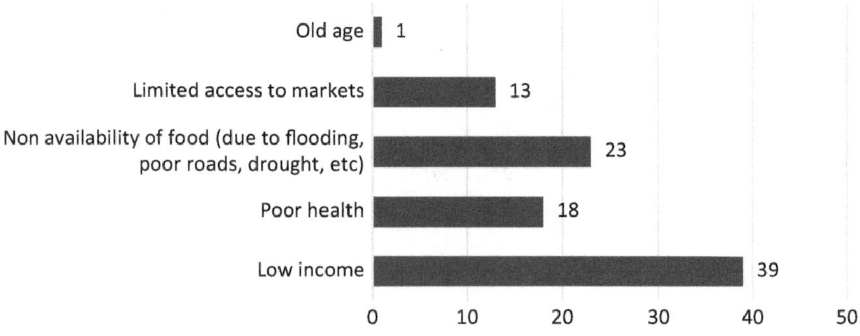

Fig. 18.1 Factors that influence food security in rural areas

Resilience

Resilience is the dimensions of social, economic, and environmental systems to cope with a hazardous event or trend, or disturbance, responding or reorganizing in ways that maintain their essential function, identity, and structure while also maintaining the capacity for adaptation, learning, and transformation (IPCC 2018). Ansah et al. (2019) define resilience as the ability of a household to cope with and recover from shocks and stresses that may affect their access to food, such as climate change, natural disasters, and economic downturns. Household food security, on the other hand, refers to the ability of a household to access sufficient, safe, and nutritious food to meet their dietary needs and preferences. Ansah et al. (2019) review the relationship between household food security and resilience, including various concepts, methodological approaches, and empirical evidence. The authors argue that household food security and resilience are closely intertwined, and Ghanaian communities and beyond should consider them together when designing policies and programs to address food insecurity. They also identify gaps in the existing research and suggest areas for future studies, including the need for a more nuanced understanding of the various dimensions of resilience and the development of more robust measures of food security and resilience (Ansah et al. 2019). The article provides a comprehensive overview of the current state of research on the topic. It highlights the importance of addressing household food security and resilience to achieve sustainable food systems (Ansah et al. 2019).

Ansah et al. (2019) argue that household food security and resilience are mutually reinforcing, as more resilient households are better able to maintain food security in the face of shocks and stresses, while households that are food-secure are better able to invest in activities that build resilience, such as diversifying income sources, building social networks, and investing in natural resource management. Therefore, the authors suggest that policies and programs addressing food insecurity should also consider household resilience, and vice versa, to achieve sustainable and resilient food systems.

Ansah et al. (2021, p. 414) indicate:

> The 2017 Sustainable Development Goals (SDGs) report highlights that economic shocks, natural disasters, conflict, and wars, among other shocks, create economic losses of more than USD 250 billion annually. These

> various shocks usually affect both farm and off-farm activities of the rural people and the vulnerable in developing countries, thus creating significant threats to their food security.

Ansah et al. (2021, p. 414) define "a shock as any event which may disrupt the normal functions of socioeconomic agents and their activities, impose challenges and threaten household food security." They identify four main shocks:

> climate, pest, health, and price. Climate shocks include droughts, floods, and storms. Pest shocks include crop pests, diseases, and livestock pests or diseases. Health shocks include illness and the death of a household or family member. Price shocks include large increases in input or food prices and significant dips in crop sale prices. (Ansah et al. 2021, p. 417)

Drought will likely cause low farm product yield, eventually decreasing household income. However, an increase in prices because of scarcity in the product supply mitigates the situation (Ansah et al. 2021, p. 416). Coping mechanisms include using one's cash savings, asset depletion, social networks with family and friends, and reducing expenditure or change of lifestyle support. Other ways identified include support from the government, NGOs, or religious groups and labour deployment (Ansah et al. 2021, p. 417).

Ansah et al. (2019) note that communities understand resilience in various ways, including ecological, social-ecological, and livelihood. Ecological resilience focuses on the ability of ecosystems to withstand and recover from disturbances. In contrast, social-ecological resilience emphasizes the relationship between human communities and the ecosystems on which communities depend. Livelihood resilience focuses on the ability of households to adapt to changing conditions and maintain their livelihoods over time. Ansah et al. (2019) discuss different definitions of household food security, including availability, access, utilization, and stability. Availability refers to the physical availability of food in a given area, while access refers to the ability of households to obtain food through production or purchase (Opare 2013). Utilization refers to the nutritional value of the food available, and stability refers to the ability of households to maintain food security over time. Ansah et al. (2019) argue that these concepts and definitions of resilience and household food security are essential when designing policies and programs to address and build resilience. A more nuanced understanding of these concepts' various dimensions can help identify the most appropriate interventions for different contexts and populations.

Tables 18.2a, 18.2b, and 18.2c show the response given by the fifty respondents when asked to highlight the challenges they face daily as a result of climate change factors. For each challenge mentioned, we asked how they adapted, managed, or coped with them. It can be observed from the various responses, that though all three categories (farmer, market woman/retailer, and consumer) have some overlapping elements, they each have peculiar challenges that are presented as a result of climate change. Moreover, each category has evolved ways of dealing with these challenges that directly or indirectly impact food security in the community.

From Tables 18.2a, 18.2b, and 18.2c, one can clearly see how climate change factors such as drought, and frequent flooding affect the availability, access, utilization, and stability of nutritious food that meets the preferences of individuals in the rural communities. The findings agree with Antwi-Agyei & Nyantakyi-Frimpong (2021) who expounded in their literature, the effect of external shocks such as climate change on household income, and the coping mechanisms employed by smallholder farmers.

Climate change factors have several adverse effects on farming activities and hence the food security of communities, more so in rural communities. The impact of climate change is complex and results in huge cascading effects throughout the food system. Farmers need to invest more into land preparation and crop/animal production to achieve higher yields. However, erratic rainfall patterns, flooding and other unfavourable factors affect the effectiveness of farm inputs such as fertilizers, and weedicides leading to lesser and lesser yields.

Aside losses incurred at the production stages in the food system, poor postharvest management, coupled with higher temperatures, poor road networks caused by flooding or heavy rains, to mention but a few, exacerbate the problem of food insecurity. Delayed transportation of farm produce not only leads to food losses but also reduces the aesthetic and nutritional quality of the produce. Losses lead to scarcity, which then lead to price surges, making it even harder for the rural poor to be food-secure.

In the long run, both the farmer and market woman/retailer are also consumers. No one in the value chain is spared from the debilitating effects of climate change factors. Nonetheless, human agency gives us the ability to adapt, transform, and persist in response to these changes.

Table 18.2a Effects of climate change factors on the livelihood and income of farmers and their coping strategies

Challenges caused by climate change factors	*Farmers affected*	*Coping strategies adopted by Farmers*	*Adoption rate*
Cost of crop/animal production			
The increased cost of production (animal feed, seedlings, certified seeds, irrigation, labor, and land preparation) is due to drought.	13* (37.1%)	Use of alternate water sources such as harvested rainwater, wells, and planting near streams.	6* (17.1%)
The increased cost of production (labor, harvesting, seedlings, and certified seeds) is due to erratic rainfall.	6* (17.1%)	Skip unfavorable planting seasons or plant according to weather conditions.	3* (8.6%)
The increased cost of production (fertilizer, labor, harvesting, seedlings, and certified seeds) due to flooding/heavy rainfall.	6* (17.1%)	Plant crops that require less investment and effort in production.	4* (11.4%)
The increased cost of production (labor, harvesting, seedlings, and certified seeds) is due to higher temperatures.	4* (11.4%)	Reduce the amount of fertilizer or supplement with manure.	3* (8.6%)
		Manual weeding instead of weedicides.	2* (5.7%)
		Save seeds from the previous harvest.	3* (8.6%)
		Reduce the number of laborers.	1* (2.9%)
		Plant early, use improved varieties, and follow recommended planting distance and practice.	2* (5.7%)
		Reduce farm sizes.	1* (2.9%)
		Apply for loans from financial institutions.	1* (2.9%)
Yield (animal/crop)			
Decrease in yield (number and size) due to drought.	11* (31.4%)	Planting of improved certified seeds.	6* (17.1%)
Decrease in yield (number and size) due to erratic rainfall.	10* (28.6%)	Storage of farm produce in tarpaulins to prevent moisture damage.	1* (2.9%)
Decrease in yield (number and size) due to flooding/heavy rainfall.	6* (17.1%)	Irrigation with stored water or river.	5* (14.3%)
Decrease in yield (number and size) due to higher temperatures.	4* (11.4%)	Timely harvesting and efficient storage.	6* (17.1%)

(*continued*)

Table 18.2a (continued)

Challenges caused by climate change factors	*Farmers affected*	*Coping strategies adopted by Farmers*	*Adoption rate*
Decrease in farmer seeds for next growing season due to decrease in yields.	2* (5.7%)	Frequent veterinarian visits.	1* (2.9%)
		Process farm produce to increase shelf life.	1* (2.9%)
		Tend a backyard garden to support your family.	1* (2.9%)
		Increase laborers and chemical use.	1* (2.9%)
		Supplement fertilizer with manure.	2* (5.7%)
Transportation (of farm inputs and produce)			
Increased transportation costs due to bad roads caused by flooding/heavy rainfall.	19* (54.3%)	Share the transportation cost with the retailer.	1* (2.9%)
Delayed transportation due to inaccessible roads caused by flooding/heavy rainfall.	10* (28.6%)	Use of tricycles or motorcycles instead of trucks to convey produce.	9* (25.7%)
Lack of transportation due to the unwillingness of drivers to use deplorable roads caused by flooding/heavy rainfall.	3* (8.6%)	Use of labourers and family to carry the goods to the market which is less expensive.	6* (17.1%)
Increase in cost of vehicle maintenance.	1* (2.9%)	Share transportation costs with other farmers.	1* (2.9%)
		Use of wheelbarrows.	2* (5.7%)
		Sell farm produce at the farm to avoid transportation cost.	4* (11.4%)
		Reduce the number of trips to the market.	2* (5.7%)
		Sell farm produce at the farm gate to avoid the high cost of transportation.	3* (8.6%)
		Invest in personal transportation (motorbike, truck)	3* (8.6%)
Marketing of farm produce			

(*continued*)

Table 18.2a (continued)

Challenges caused by climate change factors	*Farmers affected*	*Coping strategies adopted by Farmers*	*Adoption rate*
Low patronage at farm gate or market due to poor yield and/or quality caused by climate change factors (drought, flooding/heavy rainfall, erratic rainfall, and higher temperatures).	9* (25.7%)	Engage in contract farming to secure a ready market before I produce.	7* (20%)
Low patronage at the market due to poor yields, spoilage, or poor quality during transportation caused by higher temperatures.	4* (11.4%)	Sell on credit.	3* (8.6%)
Low patronage at the market due to spoilage and poor quality caused by delayed transportation from the farm as a result of flooding/heavy rainfall.	5* (14.3%)	Sell at markets in neighbouring towns.	4* (11.4%)
Low patronage due to increased cost of farm produce caused by an increase in production and transportation cost.	2* (5.7%)	Store produce till there is higher demand.	2* (5.7%)
Failure to market produce due to limited access to market (as a result of bad roads caused by flooding/heavy rainfall or insufficient market centres).	6* (17.1%)	Process farm produce to increase its value.	2* (5.7%)
Low patronage at farm gate or market due to lack of customers as a result of flooding/heavy rainfall or higher temperatures.	2* (5.7%)	Grade/sort farm produce and sell quality ones at a premium price.	5* (14.3%)
Income			
Low income due to poor yields caused by climate change factors.	12* (34.3%)	Increase price of farm produce.	1* (2.9%)
Low income due to increased production cost.	5* (14.3%)	Engage in multiple businesses.	5* (14.3%)
Low income due to low patronage and/or reduced cost of farm produce as a result of poor quality.	10* (28.6%)	Rely on farm produce for sustenance.	7* (20%)
Low income due to limited market acess as a result of flooding/heavy rainfall or insufficient markets.	3* (8.6%)	Apply for loans from financial institutions or ask family and friends.	6* (17.1%)

Numbers with * indicate the number of farmers. Numbers in brackets indicate the percentage of farmers

Table 18.2b Effects of climate change factors on the livelihood and income of market women/retailers and their coping strategies

Challenges caused by climate change factors	*Market women/ retailers affected*	*Coping strategies adopted by market women/retailers*	*Adoption rate*
Transportation to market			
Increased transportation cost due to bad roads caused by flooding/heavy rainfall.	15* (53.6%)	Carry the goods to the market.	5* (17.9%)
Delayed transportation due to inaccessible roads caused by flooding/ heavy rainfall.	6* (21.4%)	Use motorcycles or tricycles to convey goods.	3* (10.7%)
Lack of transportation due to unwillingness of drivers to use deplorable roads caused by flooding/ heavy rainfall.	3* (10.7%)	Buy from farmers/ wholesalers at the market to avoid transportation cost.	4* (14.3%)
		Transport in bulk.	2* (7.1%)
		Sell on demand to avoid unnecessary cost.	2* (7.1%)
		Sell in the neighbourhood.	3* (10.7%)
Marketing of foodstuff			
Low patronage at market due to poor quality of foodstuff from the farm caused by climate change factors (drought, flooding/heavy rainfall, erratic rainfall, and higher temperatures).	7* (25%)	I engage in contract marketing.	6* (21.4%)
Low patronage due to spoilage or poor quality of foodstuff caused by delayed transportation to market as a result of flooding/heavy rainfall.	6* (21.4%)	I sell in small quantities.	1* (3.6%)
Low patronage due to increased cost of foodstuff caused by increase in transportation cost.	4* (14.3%)	I reduce price of goods to increase demand.	2* (7.1%)
Low patronage due to lack of customers as a result of flooding/heavy rainfall or higher temperatures.	3* (10.7%)	I grade foodstuff according to quality to sell at premium prices.	5* (17.9%)

(*continued*)

Table 18.2b (continued)

Challenges caused by climate change factors	*Market women/ retailers affected*	*Coping strategies adopted by market women/retailers*	*Adoption rate*
Income			
Low income due to lack of customers during flooding/heavy rainfall and higher temperatures.	4* (14.3%)	Buy on credit and pay later.	2* (7.14%)
Low income due to low patronage as a result of increased cost of foodstuff.	6* (21.4%)	Run multiple businesses.	8* (28.6%)
Low income due to poor sales as a result of delayed transportation.	5* (17.9%)	Increase the price of foodstuffs accordingly.	2* (7.14%)
Low income due to food spoilage and poor quality during transportation or at market caused by higher temperatures or flooding/heavy rainfall during.	15* (53.6%)		

Numbers with * indicate the number of market women/retailers. Numbers in bracket indicate the percentage of market women/retailers

Table 18.2c Effects of climate change factors on the livelihood and income of consumers and their coping strategies

Challenges caused by climate change factors	*Consumers affected*	*Coping strategies adopted by consumers*	*Adoption rate*
Acess to market			
Increased transportation cost during heavy rainfalls.	12* (48%)	Go to the market after the rain has stopped.	2* (8%)
Increased transportation cost due to bad roads caused by flooding/ heavy rainfall.	3* (12%)	Use a bicycle.	1* (4%)
Unmotorable roads caused by flooding.	3* (12%)	Pay for delivery services from the market (Okada).	1* (4%)
Higher day temperatures or heavy rainfall discourage visits to market.	8* (32%)	I buy in bulk to reduce the number of market trips.	2* (8%)
Lack of transportation during flooding/heavy rainfall.	2* (8%)	Rely on others who have access to the market.	2* (8%)

(continued)

Table 18.2c (continued)

Challenges caused by climate change factors	*Consumers affected*	*Coping strategies adopted by consumers*	*Adoption rate*
		I mostly buy foodstuff in the neighborhood instead of going to the central market.	1*(4%)
Food choices			
Low quality foodstuff due to climate change factors (drought, flooding/heavy raingfall, higher temperatures, and erratic rainfall).	10* (40%)	Bulk purchasing of foodstuff when possible.	5* (20%)
Limited variety of foodstuff at market due to seasonal production of foodstuff as a result of climate change factors.	10* (40%)	I grow a backyard garden.	3* (12%)
Reduced quantity of marketable foodstuff due to spoilage caused by climate change factors.	3* (12%)		
Higher cost of foodstuff due to climate change factors.	5* (20%)		

Numbers with * indicate the number of consumers. Numbers in brackets indicate the percentage of consumers

Religion, Resilience, and Food Security

Ghana has a diverse religious landscape, with Christianity and Islam being the dominant religions and traditional beliefs also playing a significant role. The agricultural sector is the mainstay of the economy, with over 60% of the population engaging in farming. However, rural communities in Ghana face many challenges in achieving food security, such as climate change, poor infrastructure, and limited access to resources (Alves and Williams 2019).

Drought and less rainfall curtail farming activities, and the rural people in the farming communities must find a way to survive since farming is their only livelihood. Some depend on their saved income or find other ways of doing business. Such a situation is described in the Bible as follows: The waters of Nimrim are dried up and the grass is withered; the vegetation is gone. Furthermore, nothing green is left. So the wealth they have acquired and stored up they carry away over the Ravine of the Poplars (Isaiah 15:6–7).The Bible also describes a similar situation in the *Book of Jeremiah*: The nobles send their servants for water; they go to the cisterns, they find no water. They return with their jars unfilled; dismayed and despairing, they cover their heads. The ground is cracked because there is no rain on the land; the farmers are dismayed and cover their heads (Jeremiah 14:3–4) (Dunn n.d.).

There were various water sources for human and animal use in the Old Testament. These include rainfall that collects in valleys and cisterns and the Gihon Spring in the Kidron Valley. Ancient civilizations, such as Egypt and Mesopotamia, developed large canals and underground tunnel systems for water supply (Abdelkader & Mohamed 2016). The importance of storing water during the rainy season for use during the dry season is emphasized for agriculture in areas without adequate water supply systems. The water accumulating in valleys and cisterns due to rainfall is a crucial source for Ghanaians and their livestock. In 2 Kings 3:16–17 (NIV), God promised to miraculously fill a valley with pools of water to provide for the needs of the people and their animals. The promise that there shall not be rainfall but the valley shall be filled was the source of hope, comfort, and security. Religion becomes the epitome of confidence and resilience in a situation like the one described above. In Egypt and Mesopotamia, large canals and underground tunnel systems provided water for crop and human use. The Nile River and irrigation canals were the primary sources of water in Egypt, and the canal system allowed agricultural use of desert lands not covered by the annual flooding of the Nile. In Jerusalem, the Gihon Spring was the only source of water for the city, but its intermittent flow required water storage in a pool for times when the spring was not flowing (Geva 1998–2023). Water storage during the rainy season is crucial for agriculture in areas without adequate water supply systems.

The Old Testament patriarchs were engaged in farming, animal rearing, and the production of foodstuff. Wells were a vital source of water supply for humans and agricultural use. The Bible reports that Abraham and Isaac engaged in farming activities during famine. Genesis 26 declares that Isaac sowed in famine, and God blessed his labour with a 100-fold harvest the same year. There was a famine in the land could also mean there was drought. Very often, famine in the Old Testament was because of drought. Isaac possessed some wells his father Abraham had dug. He faced opposition. The Philistines stopped the wells, filling them with earth. He did not give up. He continued to dig more wells until there was no more opposition. That is resilience. The ability to keep going even in the face of opposition. The account of the entire story is in Genesis 26:12–22:

> Isaac planted crops in that land and, the same year, reaped a hundredfold because the LORD blessed him. The man became rich, and his wealth continued to grow until he became very wealthy. He had so many flocks, herds, and servants that the Philistines envied him. So all the wells that his father's servants had dug in the time of his father Abraham, the Philistines stopped up, filling them with earth. Abimelek told Isaac, "Move away from us; you have become too powerful for us." So Isaac moved away from there and encamped in the Valley of Gerar, where he settled. Isaac reopened the wells his father Abraham dug, which the Philistines had stopped up after Abraham died, and he gave them the same names his father had given them. Isaac's servants dug in the valley and discovered fresh, well water there. But the herders of Gerar quarreled with those of Isaac and said, "The water is ours!" So he named the well Esek because they disputed with him. They dug another well, but they quarreled over that one also, so he named it Sitnah. He moved on from there and dug another well, and no one quarreled over it. He named it Rehoboth, saying, "Now the LORD has given us room, and we will flourish in the land (Dunn n.d.).

Today, rural people in Ghana face obstacles and opposition concerning food security. These are classified into shocks due to climate change, inadequate resources, and limited road access. Religion is vital in shaping attitudes and practices towards food security in rural Ghana. Traditional beliefs, Christianity, and Islam are the dominant religions in Ghana, and they all have varying influences on food security. Traditional beliefs promote sustainable agricultural practices, respect for natural resources, and community cooperation. For example, traditional beliefs encourage using organic fertilizers, crop rotation, and planting drought-resistant crops, which enhances food security (Amponsah et al. 2019).

Several factors contribute to resilience in rural Ghana. These include access to information and technology, social networks, and financial resources. Religion also plays a role in building resilience. For example, traditional beliefs that emphasize community cooperation and solidarity may facilitate sharing of resources and knowledge during times of crisis. Similarly, Christian and Islamic teachings emphasizing the importance of helping the poor may encourage individuals and communities to support one another during difficult times (Antwi-Agyei & Nyantakyi-Frimpong 2021). Of the 50 rural people interviewed, 45 respondents (90%) mentioned that their faith in God/Allah gave them hope, and strength to go on with their lives regardless of their current situations. They believed that things were going to get better and they were determined not to give up. Five (10%) of the respondents stated that their religion taught them to be content in every situation, whether good or bad. Five of the respondents (10%) also mentioned that they received assistance from their brothers/ sisters in the faith.

Religious institutions, such as churches and mosques, can also play a critical role in addressing food insecurity through their charitable activities. For example, many religious organizations operate food banks, soup kitchens, and other food aid programs to provide food to those in need. Religious institutions can also use their influence to advocate for policies and programs that promote food security, such as supporting small-scale farmers and increasing access to nutritious foods (Brouwer et al. 2018). Interfaith collaboration can also be critical in addressing food insecurity by promoting cooperation and understanding between different religious groups. For example, the African Faith Leaders' Declaration on Food Security and Sovereignty calls for interfaith collaboration to address food insecurity in Africa by promoting sustainable agriculture and supporting small-scale farmers. Similarly, the Interfaith Food Bank Network in the United States brings together different religious organizations to coordinate their efforts to address food insecurity in their communities (Alves and Williams 2019).

The above table is a summary of the services received by some of the respondents over the past five years from religious groups, mostly churches is provided from the data collected (see Table 18.3). According to the beneficiaries, the assistance received from these religious groups provided immense support during difficult times. The respondents requested for more of such services as it helped to cushion the adverse impacts of climate change factors, especially during the Covid-19 pandemic.

Linking Rural Livelihood, Resilience, and Food Security

In many parts of the world, a high degree of vulnerability to climate change characterizes rural livelihood, market volatility, and political unrest. Ghana's economy primarily depends on agriculture, and rural areas are the main contributors to agricultural production. However, rural people in Ghana face numerous challenges that affect their food security and resilience, including low agricultural productivity, climate change, and poverty (Owusu et al. 2020). Increasingly, resilience informs development initiatives aimed at building the capacity of rural households and communities to cope, adapt, and transform in the face of diverse shocks and stressors (Addaney et al. 2021).

Incorporating resilience thinking into food and nutrition security policy and programming is not without its challenges. This is primarily

Table 18.3 Services provided by religious groups to rural people that promotes resilience in light of climate change

Services provided by religious groups to rural people	*Number of respondents/ beneficiaries*
Healthcare	(32%)
Free health screening by church.	12*
Free supply of drugs/medicine.	1*
Free vaccination.	3*
Education	(18%)
Education on personal hygiene and good home keeping.	2*
Provision of school stationery to children by church.	2*
Education on efficient postharvest handling of farm produce.	4*
Education on Covid-19 prevention and management.	1*
Food assistance	(14%)
Provision of foodstuff especially during drought and flooding by the church.	7*
Financial support	(12%)
Payment of hospital bills	2*
Financial support from church	2*
Payment of school fees	1*
Access to loans by the church	1*
Technical support	(24%)
Training on proper farm management and cultural practices.	8*
Training on efficient management of fires on farm and at home.	3*
Provision of farm tools by church.	1*

Numbers with * indicate the number of respondents. Numbers in bracket indicate the percentage of respondents

because the concept is best understood as embedded within dynamic and highly contextual processes that various parties can interpret differently (Pelletier et al. 2016). In many parts of the world, rural livelihoods consist primarily of family farms involved in agricultural, fisheries, community forestry, pastoral, or aquaculture production activities (Food and Agriculture Organization of United Nations, FAO 2014).

The Impact of Climate Change on Food Security

Mohammed (2021) examines the impacts of climate change on food security and smallholder livelihoods in northern Ghana. The author argues that climate change exacerbates existing challenges to food security and livelihoods in the region, including low agricultural productivity, poverty, and limited access to markets and credit. He suggests that these impacts are particularly severe for smallholder farmers disproportionately affected by climate change due to their reliance on rain-fed agriculture and limited resources to adapt to changing conditions. Mohammed (2021) further describes various strategies that smallholder farmers in northern Ghana use to adapt to climate change, including using drought-tolerant crop varieties, soil conservation practices, and alternative livelihoods such as beekeeping and livestock rearing. However, he notes that these strategies are often limited in their effectiveness and that more support is needed from government and development organizations to build the capacity of smallholder farmers to adapt to climate change. Mohammed (2021) and Warner et al. (2012) identify the potential role of policy interventions in addressing the impacts of climate change on food security and livelihoods in northern Ghana. They suggest that policy measures such as improved access to credit, markets, and extension services can help smallholder farmers to adapt to changing conditions and build more resilient livelihoods. They also argue that policies aimed at reducing greenhouse gas emissions and mitigating the impacts of climate change are essential to preventing further damage to food systems and livelihoods in the region (MENA Report 2023). Overall, the article highlights the urgent need for action to address the impacts of climate change on food security and smallholder livelihoods in northern Ghana and the potential role of policy interventions in building resilience and sustainability in the region's food systems (Warner et al., 2012). Bang et al. (2018) examine the relationship between frequent flooding and food security in Africa and suggest disaster management strategies to enhance local livelihood resilience. The article argues that climate change is causing more frequent and severe flooding, which disrupts local food systems and undermines food security. The authors suggest that disaster management strategies can help communities build resilience and adapt to the challenges posed by flooding (Bang et al., 2018; Anon. n.d.-a; Anon n.d.-b).

Bang et al. (2018) also describe various disaster management strategies implemented to enhance local livelihood resilience and food security in the face of frequent flooding. These strategies include early warning systems, flood risk mapping, community-based flood management, and disaster risk reduction programs. The authors argue that these strategies could help communities to mitigate the impacts of flooding on food systems, prevent loss of life and property, and build more resilient livelihoods

(Muhonda 2014). The authors also highlight the importance of stakeholder engagement in disaster management strategies. They suggest that effective disaster management requires the participation of local communities, government agencies, civil society organizations, and other stakeholders. Additionally, the authors emphasize the need for disaster management strategies to be tailored to local communities' specific needs and capacities (Bang et al. 2018).

Essegbey et al. (2015) present an overview of Ghana's National Climate-Smart Agriculture and Food Security Action Plan for 2016–2020. The plan aimed to increase Ghana's agricultural productivity and food security while reducing greenhouse gas emissions and building resilience to climate change. The plan includes several key strategies, including promoting climate-smart agricultural practices, developing sustainable land use and management, strengthening institutional capacity and governance, and promoting private-sector agricultural investment. The article discusses various components, highlighting the importance of stakeholder engagement, capacity building, and monitoring and evaluation. The article also highlights the role of international cooperation in supporting Ghana's efforts to address the challenges of climate change and food security. All the rural people interviewed had prior knowledge of climate change, though to various degrees. To find out the extent of their knowledge on climate change, we divided it into various factors: higher temperatures, frequent flooding, drought, and erratic rainfall patterns (see Fig. 18.2).

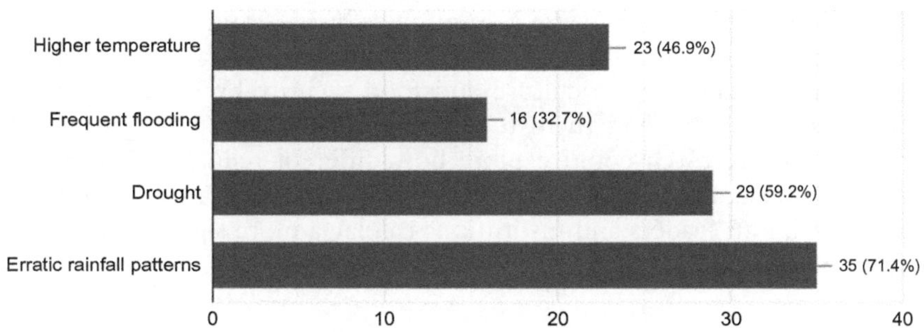

Fig. 18.2 The extent of knowledge on climate change of rural people

Conclusion

Although food insecurity can affect everyone, certain groups are more vulnerable. Women, particularly those in poor households, often bear the brunt of high food prices, forcing them to sacrifice their health, education, or food consumption. Small-scale farmers are also at risk since they usually need more resources to cope with climate variability and change. Climate change can, therefore, have a damaging impact on their crops and livelihoods. Finally, poor people in urban communities are particularly susceptible during economic crises, as they are often the most likely to suffer from unemployment, which can hinder their access to food. Religion can significantly impact food security by shaping people's attitudes, beliefs, and practices towards agriculture, influencing dietary choices, and impacting access to food aid. Religious institutions can be critical in addressing food insecurity through their charitable activities and advocacy efforts. Interfaith collaboration can also promote cooperation and understanding between different religious groups to address food insecurity at the local and global levels.

Food security and resilience of rural people in Ghana are critical for the country's development and growth. Agricultural productivity, climate change, poverty, and social networks are among the factors that affect food security and resilience in the rural areas of Ghana. Enhancing agricultural productivity, promoting climate-smart agriculture, reducing poverty, and strengthening social networks can enhance the food security and resilience of rural people in Ghana. Religion, food security, and resilience are interconnected issues in rural Ghana. Religious practices can promote sustainable agriculture and equitable access to food. Religion can also contribute to the resilience of rural communities by fostering social networks and community cooperation. Religious leadership plays a crucial role in promoting sustainable agriculture and food security by educating their followers and mobilizing them to support one another during times of crisis. Future research could explore how different religious beliefs and practices impact food security and resilience in rural Ghana and the potential for religious leaders and institutions to play a more significant role in promoting sustainable agriculture and food security.

References

Abunyewah, E., 2020. Role of religion in disaster management and risk reduction in Ghana: A case study of the 2015 Accra flood disaster. *Journal of Disaster Research,* 15(5), pp. 635–644.

Adu, K. A., Sarpong, D. A. & Amegashie, D., 2018. Food insecurity and coping strategies among farming households in the northern region of Ghana. *Journal of Development and Agricultural Economics,* 10(8), pp. 259–270.

Abdelkader T. Ahmed, Mohamed H. Elsanabary. (2016). The Water Supply History of Underground Aqueducts in Egypt from: Underground Aqueducts Handbook CRC Press [Accessed 22 Dec 2023] https://www.routledgehandbooks.com/doi/10.1201/9781315368566-8

Addaney, M, Sarpong, G.E. & Akudugu, J.A. (2021). "Climate Change Adaptation in Akropong, Ghana: Experiences of Female Smallholder Farmers." *Journal of Land and Rural Studies*, 9(2), 344–367. https://doi.org/10.1177/23210249211008537

Alves, F. M., & Williams, T. O. 2019. Religion, agriculture, and food security in sub-Saharan Africa. *Religion* 49(3), 469–490.

Amponsah, O. W., Darko, J., & Boakye, E. K. 2019. Exploring the link between religion and food security in Ghana: A qualitative study. *Journal of religion and health* 58(1), 50–61. https://doi.org/10.1007/s10943-018.

Anon., n.d.-a *'Cities Race to Resilience' Campaign Launched to Mobilize Action Ahead of COP26—Global Covenant of Mayors.* [Online] Available at: https://www.globalcovenantofmayors.org/press/cities-race-to-resilience-campaign-launched-to-mobilize-action-ahead-of-cop26 [Accessed 24 07 2023].

Anon., n.d.-b *EverGreen Agriculture: Rethinking Modern Farming Evergreen Agriculture.* [Online] Available at: http://evergreenagriculture.net/evergreen-agriculture-rethinking-modern-farming/ [Accessed 24 07 2023].

Antwi-Agyei, P., & Nyantakyi-Frimpong, H. 2021. Evidence of Climate Change Coping and Adaptation Practices by Smallholder Farmers in Northern Ghana Sustainability 13(3) 1308. https://doi.org/10.3390/su13031308

Antwi-Agyei, P., Stringer, L. C., Dougill, A. J., & Codjoe, S. N. A. 2016. Livelihood adaptations to climate variability: Insights from farming households in Ghana. *Regional Environmental Change* 16(7), 2015–2027.

Ansah, I. G. K., Gardebroek, C. & Ihle, R., 2021. Shock Interactions, Coping Strategy Choices and household Food Security. *Climate and Development,* 13(5), pp. 414–426.

Ansah, I. G. K., Gardebroek, C. & Ihle, R., 2019. Resilience and household food security: a review of concepts, methodological approaches and empirical evidence. *Springer,* Volume 11, pp. 1187–1203.

Asenso-Okyere, K., Asante, F. A. & Andam, K. S., 2019. *Policy and program options for improving food security in Ghana,* s.l.: International Food Research Institute (IFRI).

Bang, H. N., Miles, L. & Gordon, R., 2018. Enhancing local livelihoods resilience and food security in the face of frequent flooding in Africa: A disaster management perspective. *Journal of African Studies and Development,* 10 September, 10(7), pp. 85–100.

Brouwer, I. D., Williams, T. O., & Kyalo-Willson, B. 2018. Religion and food security: Examining the role of religious institutions in addressing food security.

Brown, M. E. et al., 2015. *Climate Change, Global Food Security, and the U.S. Food System.*, s.l.: The U.S. Department of Agriculture (USDA).

IPCC, 2018. *Centering Equity in Climate Resilience Planning and Action | CAKE: Climate Adaptation Knowledge Exchange.* [Online] Available at: https://www.cakex.org/documents/centering-equity-climate-resilience-planning-and-action [Accessed 24 July 2023].

Dunn, B. N., n.d. *Isaiah 15 New International V UK—The Bible Door.* [Online] Available at: http://www.bibledoor.com/isaiah-15-new-international-version-brother.neal-dunn-uk/ [Accessed 23 July 2023].

Essegbey, G. O., Nutsukpo, D., Karbo, N. & Zougmo, R., 2015. *National Climate-Smart Agriculture and Food Security Action Plan of Ghana (2016–2020),* Copenhagen: CCAFS Working Paper No 139.

FAO, IFAD, UNICEF, WFP and WHO, 2021. *United State of Food security and nutrition in the world 2021. Transforming food systems for food security, Improved nutrition and affordable healthy diets for all. The state of food security and nutrition in the world (SOFI) 2021. Rome,.* Rome: FAO.

Food and Agriculture Organisation of the United Nations, F., 2015. *The State of Food and Agriculture. Social protection and agriculture: breaking the cycle of rural poverty.* Rome: FAO.

Food and Agriculture Organization of United Nations, FAO, 2014. *The state of food and agriculture: Innovation in family farming.* Rome, Italy: s.n.

Geva, H., 1998–2023. *Jerusalem Archaeological Sites: Biblical Water Systems.* [Online] Available at: https://www.jewishvirtuallibrary.org/biblical-water-systems-in-jerusalem

IPCC, 2018. *Global Warming of 1.5 0C: An IPCC Special Report on the impacts of global warming of 1.5 0C above pre-industrial levels and related global greenhouse gas emission pathways, in the context of strengthening the global responds to the threat of climate change,* Geneva: World Meteorological Organisation.

Jarvis, A. et al., 2008. *Climate Change and its Effect on Conservation and Use of Plant Genetic Resources for Food and Agriculture and Associated Biodiversity for Food Security,* s.l.: s.n.

McCarthy, U. et al., 2018. Trends in food science and technology. *Global food security—Issues, challenges and technological solutions,* July, Volume 77, pp. 11–20.

MENA Report, 2023. *Brazil: Waldez Goes participates in the launch of the Parliamentary Front for the Strengthening of COP-30 in Brazil.,* Brazil: MENA.

Mohammed, K., 2021. *Impacts Of Climate Change On Food Security And Smallholder Livelihoods In Northern Ghana,* Ontario: Western University.

Muhonda, P., 2014. *"Analysis of Institutional Mechanisms That Support Community Response to Impacts of Floods in the Middle-Zambezi River Basin, Zimbabwe." Physics And Chemistry Of The Earth, *Parts A/B/C,.* [Online] Available at: https://doi.org/10.1016/j.pce.2014.11.013 [Accessed 24 07 2023].

Opare, U. L., 2013. *"Perspective: The Evolving Dimensions and Perspectives on Food Security—What Are the Implications for Postharvest Technology Research, Policy and Practice?" International Journal of Postharvest Technology and Innovation.* [Online] Available at: https://doi.org/10.1504/ijpti.2013.059340 [Accessed 24 07 2023].

Pelletier, B., Hickey, G. M., Bothi, K. L. & Mude, A., 2016. Linking rural livelihood resilience and food security:an international challenge. *CrossMark,* 6 May. p. 5.

Warner, K., Afifi, T., Henry, K., Rawe, T., Smith, C., de Sherbinin, A. (2012). Where the Rain Falls: Climate Change, Food and Livelihood Security, and Migration. Global Policy Report of the Where the Rain Falls Project. Bonn: CARE France and UNU-EHS.

Yaro, J.A., 2013. Building resilience and Reducing Vulnerability to climate change: Implications for food security in Ghana. FES Ghana.

Yilmaz, S. & Ahmet, G. M., 2023. *"Food Insecurity Indicators of 14 OECD Countries in a Health Economics Aspect: A Comparative Analysis." Frontiers in Public Health.* [Online] Available at: https://doi.org/10.3389/fpubh.2023.1122331 [Accessed 24 07 2023].

CHAPTER 19

Religion, Food Security, and Climate Change Mitigation: A Case of Luangwa Valley Women of Eastern Zambia

Nelly Mwale and Nicholas Miyoba Haambokoma

Introduction

Climate change and food security remain topical challenges for present-day societies. The United Nations observes that climate change was one of the greatest challenges of our time as increases in global temperature, sea level rise, ocean acidification and other climate change impacts were seriously affecting coastal areas and low-lying coastal countries, including many least developed countries and Small Island Developing States (United Nations Development Programme, 2019). The Government of the Republic of Zambia (2016) also acknowledges that climate change has significantly affected rainfall patterns, a situation that has resulted in crop failure and a further reduction in economic growth. As such, both the global Sustainable Development Goals and the national policies affirm the need for tackling

N. Mwale (✉) • N. M. Haambokoma
Department of Religious Studies, University of Zambia, Lusaka, Zambia
e-mail: nellymwale@unza.zm; nicholas.haambokoma@unza.zm

L. Maseno et al. (eds.), *Religion, Climate Change, and Food Security in Africa*, Sustainable Development Goals Series,
https://doi.org/10.1007/978-3-031-50392-4_19

climate change and food security. For example, Zambia's Vision 2030 seeks to, among other things, develop and/or advocate policies and programmes that will ensure food and nutrition security, food quality and safety at individual household, community and national level in the area of food and security (Government of the Republic of Zambia 2016).

The Zambia National Agricultural Policy also seeks to make agriculture more productive and sustainable through promoting environmentally friendly farming systems such as conservation farming, afforestation, and the use of green manure and lime; promoting expansion of production of oil seed crops (soybean, sunflower, groundnuts) in rotation with food grains as a way of reducing fertilizer costs on the one hand and increase farm yield, incomes and consumption of protein rich food crops on the other; and promoting production of fruits and vegetables as a way of encouraging consumption of balanced diet and increasing incomes and diversifying income sources (Ministry of Agriculture and Co-operatives 2011). The policy recognizes the interconnectedness of improvement in food insecurity and poverty reduction among poor small-scale farmers. The United Nations Development Programme (UNDP) (2019) acknowledges that while hunger and malnutrition constitute real and present risks in Zambia as approximately 60 per cent of people live below the poverty line, and 42 per cent are considered extremely poor, climate change is expected to worsen these impacts by 30 per cent by 2030.

This chapter explores how women are contributing towards climate change mitigation, resilience, and food security from their indigenous religious worldview perspective. The chapter approaches the concept of religion, as related to African Indigenous Religion in which case the interest was to explore aspects of African Indigenous Religion which were exploited for mitigating climate change and promoting food security among women in the Luangwa Valley as expressed in the notion of ubuntu. The understanding of the concept of food security is informed by the definition provided by the World Food summit (1996), and refers to a situation where all people, always, have physical and economic access to sufficient, safe, and nutritious food that meets their dietary needs and food preferences for an active and healthy life. It is acknowledged that more recently, the concept of food security has been conceptualized to encompass food access, availability, utilization, and sustainability. By this, food access is ensured when households and all individuals within them have adequate resources to obtain appropriate foods for a nutritious diet while food availability is achieved when enough food is consistently available to all individuals within a country. Food utilization and sustainability entails the consumption of food through adequate diet, clean water, sanitation, and health care to reach a state of nutritional well-being where all physiological needs are met, and a situation where the above three components of food security are fulfilled at any time, respectively (FAO 2006). In terms of climate change, the chapter acknowledges the debates surrounding the definition of climate change and takes the perspective provided by Gyampoh et al. (2009: 71) that, indigenous people may not understand the concept of global warming or climate change, but they observe and feel the effects of decreasing rainfall, increasing air temperature, increasing sunshine intensity, and seasonal changes in rainfall patterns. As such, climate change is related to some drastic changes in its weather patterns in the form of extreme temperatures, thunderstorms, floods, and droughts in certain instances.

Although climate change and food security have been addressed from different fronts, the role of religion and women remains cardinal. For example, it is well acknowledged that women constitute most food producers in the world and continue to account for a prominent role in food security. In the case of sub-Saharan Africa (FAO 2011): "most of the food production that is attributable to women makes them the principal agents of food security and household welfare" (IFAD 2009: 1). Oniang'o (2005) also affirms that women play a crucial role in meeting the food and nutrition needs of their families through all three pillars of food security (food production, economic access to food, and nutrition security). Similarly, women are acknowledged as tapping into their indigenous religious worldviews to mitigate climate change and foster food security, as revealed in a study by Ibnouf (2012) in Sudan. In Zambia, women make up more than 70 per cent of the agricultural labour force and play a critical role in enhancing food security and nutrition (Ministry of Agriculture and Co-operatives 2011).

Despite the foregoing acknowledgements, the contributions of women to climate change and food insecurity mitigation as informed by their religiosity have remained inadequately studied in Zambia. Additionally, the studies that focused on the intersection of religion and climate change have tended to dwell on Christianity (Mwale 2021, 2022). The studies which have dwelt on African Indigenous Religion have often lacked a gendered dimension as the focus has been for instance, on interrogating indigenous practices of environmental sustainability in the Tonga Community of Southern Zambia (Kanene 2016) and the general impact of climate change on African Indigenous Religion (Mwale 2016). The chapter takes the perspective that more could still be learnt about the ways in which women have tapped into African Indigenous Religion to mitigate climate change and foster food security. This view agrees with the observation by Tarusarira (2017) that religion and culture have been conspicuously missing as a factor that can help in mitigating the effects of global warming, as existing strategies and programmes have not considered the spiritual and cultural dimension to solve the environment crisis in Africa. Therefore, the chapter seeks to show the potentials of indigenous beliefs and practices as expressed in the notion of ubuntu to address climate change and food insecurity using the context of the Luangwa Valley in Eastern Zambia.

It purposively focuses on women's strides because the impact of climate change and food insecurity is gendered. Santisteban (2020: 7) observes that indigenous, peasant, and rural women, along with most women in the Global South, bear an even greater climate change burden due to the historic and on-going impacts of colonialism, racism, and inequality; above all, they are more dependent upon natural resources for their survival. The UNDP (2019) acknowledges that in most places with high prevalence of undernourishment, women farmers have significantly less access to land, information, finance, and agricultural inputs, a situation that makes them more vulnerable to climate shocks and affects their health and the food security and nutrition of the entire household. Santisteban (2020:7) also further notes that drought, flooding, and extreme and unpredictable climate patterns present life and death challenges for many women because it is they who are responsible for providing food, water, and energy to their families: they are both player in and manager of the family's food sovereignty. Hence, it became imperative to understand how women were contributing to the mitigation of climate change and food insecurity as informed by their indigenous religious worldview.

The chapter draws on a qualitative literature-based research approach in which data were collected through document analysis in the forms of reports and media features (videos, radio, and newspapers) on climate change and food security interventions in the Luangwa Valley. This is supplemented by narratives of three women farmers whose accounts were reported in the public sphere (daily tabloids) and informal conversations with one of the author's mother who grew up in the valley. The chapter unfolds by giving a brief context of climate change and food security and its link to religion. This is followed by a description of the theoretical lens, brief review of related literature and the ways in which women were utilizing aspects of indigenous religion to mitigate climate change and promote food security.

Locating the Context and the Reality of Climate Change and Food Security in Luangwa Valley

The Luangwa Valley in Eastern Zambia is well known for its game parks, namely, the North Luangwa National Park, the Lukusuzi National Park, and the South Luangwa National Park. Its history is often associated with the nineteenth century when a trading post was opened at Malambo, which became the main trade route from Tete to Lake Mweru (Carr 1996).

The Luangwa Valley is mostly populated by the Kunda speaking people who migrated from the Luba area in Congo in the early nineteenth century. The Kunda people are matrilineal and governed by six chiefs who are assisted by indunas. The indunas are appointed by the Chiefs as advisors, and each of the villages has a headman (or headwoman). In terms of the religious landscape, like many parts of the country that have been exposed to other religions, especially Christianity and Islam, Luangwa Valley was also Christianized. Despite this, African Indigenous Religion still exists as observed by Taylor (2006) that most Zambian communities continued to adhere to traditional practices alongside their new Christian faiths, and these practices continue to influence behaviour and cultural norms.

In the Luangwa Valley, sorghum, finger millet, and maize are the major starch food crops, while groundnuts, cowpeas, and pumpkins are also grown. Goats and chickens are commonly kept by farm households, and some farmers have a few cattle. Other areas of the region mainly produce bulrush millet, sorghum, and cassava. On the economic front, Luangwa Valley is one of the poorest districts in Zambia (Kunda Daily Mail, 19 October 2022). The valley has recorded its own experiences of climate change. Kunda (Daily Mail, 19 October 2022: nd) recounted that:

> For many years, Luangwa River has been a centre of economic activities for the majority of the people who live around it, but the effects of climate change have seen the river slowly drying up, thereby forcing those who depend on it to find alternative ways of survival. The economy of the Luangwa Valley was reliant on the river for food and water, making it the economic backbone of the area. Like in most rural areas in Zambia, women carry the bigger burden of looking after their families, hence they are the ones immensely hit by the effects of climatic changes that have resulted in

some parts of the river drying up. Some of the effects of climate change include prolonged droughts and excessive heat, which all result in water levels going down, thereby making it difficult for the inhabitants to easily access the commodity and practise irrigation-based farming.

Ubuntu as a Theoretical Framework

The chapter is conceptually anchored on women as agents of a healthy environment through the prism of indigenous knowledge systems, particularly ubuntu as understood in African indigenous thought. In this regard, ubuntu is related to a unified community in which it is believed that life belongs to a community (in which the value is strongly linked to the concept of wholeness). This entails that a human being finds fulfilment in the total community of their fellows in whose existence they participate. African thought emphasizes that individuals can only exist when they are part of a group: "they are because they belong" (Metz 2011: 532; Thorpe 1991: 120). This is extended to ubuntu's focus on the equality of all creation. Therefore, while it is acknowledged that the notion of ubuntu may have different words to describe it, it is used, as observed by Mugumbate and Chereni (2020), that it is a collection of values and practices that Africans view as making people authentic human beings. Although these values and practices vary across different ethnic groups, they all point to the idea that an authentic individual human being is part of a larger and more significant relational, communal, societal, environmental, and spiritual world.

The chapter uses the ubuntu lens because as observed by Chitando (2022), African perspectives on religion and climate change are heavily shaped by ubuntu and indigenous knowledge systems (IKS). Often associated with solidarity, ubuntu remains a resource for understanding how women have responded to climate change given that the understanding of solidarity and community in the African indigenous worldview is extended to nature too. Chibvongodze (2016) captures this perspective by stressing that ubuntu is as much about human relations as it is about humanity's interface with nature. This is because the African indigenous worldview is one that sees a connection between the physical and biological environments and the humans and non-humans.

Brief Review of Related Literature

Situated in the Zambian context, the chapter converses with similar studies that have focused on indigenous women and food security in the context of climate change. Globally, the World Bank (2022) observes that the shifting climatic conditions have disrupted traditional indigenous food systems and threatened many of the foundational elements that bind indigenous peoples together. The role of indigenous women who as traditional food providers are frequently at the forefront of efforts to adapt and mitigate the effects of climate change is also acknowledged. Additionally, some African specific studies have focused on indigenous women and their role in climate change mitigation. For example, Aluko (2018) explored women's use of indigenous knowledge for environmental security and sustainable development in Southwest Nigeria, while Kilonzo (2022) dwelt on women's use of indigenous knowledge systems in addressing the effects of climate change from the Kenyan context. This chapter extends the utilization of indigenous knowledge by women to food security.

The chapter is also situated in studies that have addressed African Indigenous Religions and climate change. For example, from the Zimbabwean context, different scholars have interrogated climate change and indigenous religions. Some of the notable studies include Taringa's (2006) focus on African Traditional ecology and Tarusarira (2017)'s study on African religion, climate change, and knowledge systems. The role of religious leaders in mitigating climate change has also been explored by Chitando (2017). More recently, Shoko (2022) showed from the Zimbabwean context that African Traditional Religion (ATR) had a valuable contribution towards mitigating effects of climate change in Zimbabwe through indigenous knowledge systems, values, and practices such as respect for land and environment, maintenance of sacred water bodies, trees, forests, and mountains, observing taboos, rituals, totems, and hunting. Nyawo (2022) also demonstrated that Eswatini, like many African and other Global South countries, bears the brunt of the impact of climate change, but indigenous ecological knowledge facilitated the mitigation and adaptation to different life situations.

The links between women and food security are also numerous in scholarship. For example, Oniang'o (2005) showed that women played a crucial role in meeting the food and nutrition needs of their families. Others like Ibnouf (2012) have linked the role of women in the promotion of indigenous foods to food security by observing that women in rural Sudan employed a variety of methods based on indigenous knowledge in the processing and preservation of agricultural and animal raw materials, wild food products to acquire alternative food sources, and to relieve the stress of seasonal food scarcity particularly during the shortage of rainfall or drought and thus achieving household food security. In this regard, Ibnouf has demonstrated that women in rural Sudan tend to use locally available resources effectively in achieving household food security, and since women were responsible for food processing and preserving, they contributed more to household survival. Similarly, Masekoameng and Molotja (2019) explored the role of indigenous foods and indigenous knowledge systems for rural households' food security in Sekhukhune district of the Limpopo province in South Africa.

Although notable studies have focused on the role of indigenous women in mitigating the climate change challenge and in promoting food security, the chapter agrees with Kilonzo (2022) that women and indigenous knowledge systems is an issue that has rarely been explored extensively in relation to climate change on the continent in general. As such, the chapter seeks to complement these studies and show the ways in which indigenous women have tapped into ubuntu for climate change mitigation and food security from the context of the Luangwa Valley in Eastern Zambia.

Women's Actions to Mitigate Climate Change and Promote Food Security

This chapter unpacks the contributions of women towards mitigating climate change ranging from themes centred on water harvesting, crop rotation and reclaiming the seeds of the former generations, returning to old ways of farming (conversation farming), nurturing trees and other forms of vegetation, livestock rearing and harvesting wild fruits to passing on the heritage to future generations.

Water Harvesting, Crop Rotation, and Reclaiming the Seeds of the Former Generations

In a context where climate change and deforestation have had a negative impact on the rains, it had become a challenge to have good yields. Breeze FM radio farmers' programme (2016) trailed a family in the semi-arid Luangwa Valley of Eastern Zambia which was using simple methods to capture the little water available so that it was available to their crops for a longer period:

> Climate change doesn't always mean poor rains. Good rain may fall at the right time, only to go away while the young crops still need water. So we use simple ways of farming that capture enough rainwater and keep the moisture in the soil for a longer time. (Radio Farmer, Breeze FM, June 2016)

This aspect of harvesting water is closely aligned to the sanctity that is attached to water as a source of life. Eneji et al. (2012: 50) observe that in the African indigenous worldview, water is sacred because it is life and an abode of spirits. As such, water is approached carefully alongside the observation of taboos. Thus, while water may be treated with reverence and harvested for sustenance of life, this also pointed to the quest to connect with nature in ways which renewed the earth and the women's livelihood through improved yields. FAO (2014: 4) affirmed that small-holder agriculture can be enhanced and transformed by increasing productivity of crops and livestock through application of sustainable land management approaches, notably using rainwater harvesting technologies, soil conservation measures, applying conservation agriculture principles and including agroforestry or afforestation in crop-livestock systems in the cultivated watersheds.

Besides harvesting water, women were also growing indigenous crops and practising crop rotation.

> The Luangwa Valley has always had poor rains. So in the early years, the people cultivated small fields of sorghum and millet. However, with the government introduction of maize and other crops among the Kunda, poor harvest became the order of the day. The traditional farming methods could not give good yields in the harsh environment of the Luangwa Valley. The changes in weather patterns have made us go back to the seeds of our forefathers....We practice crop rotation for two reasons. First, legumes like groundnuts and soybeans leave nitrogen in the soil. So the next crop that we plant after the legume crop usually benefits. Secondly, certain crops encourage bad weeds like witch weed, as well as pests. By rotating our crops, we reduce the bad impact of these weeds and pests. (Radio Farmer, Breeze FM, June 2016)

The use of traditional crops grown in rotation as a women's response to climate change and food insecurity is in line with the conclusion by Lukhele-Olorunju and Gwandure (2018) that indigenous knowledge systems used by women in rural subsistence farming in Africa encompassed seed and breed selection, seed protection and preservation, weeding and intercropping, free-range farming, and the use of indigenous technologies in livestock and poultry treatment. By going back to the indigenous crops which had originally been cultivated, the women not only reclaimed their identity but also connected with their ancestral heritage through the indigenous seeds. This resonates with the observation that numerous examples of women innovating at the community level to adapt and mitigate the effects of climate change include reviving indigenous seeds and setting up seedbanks, preserving food and rain-harvesting systems (The World Bank 2022). For example, Mai Charity, a farmer recounted that: "the primary reason for farming must always be to have good food before thinking about selling. So we kept one bag of soybeans and one of groundnuts for ourselves, and we sold the surplus. We do that for all our crops. The baby is enjoying soy porridge. Even we adults eat soy porridge for breakfast" (Breeze FM Radio programme, n.d June, 2016). Sithole and Lekorwe (2019) also observed that crop diversification was a practice employed by many indigenous women and in practice meant the planting of many crop species on the same piece of land to guard against crop failure in times of adverse climatic conditions. Mafongoya and Ajayi (2017) add that such indigenous knowledge systems are social capital for the poor (especially the vulnerable rural women) whose main asset is to ensure survival, produce food, and secure livelihoods. This also enabled women to contribute towards the national agenda on food security.

Returning to Old Ways of Farming: Conservation Farming

The women had also returned to the old ways of farming in which compost manure was used in the place of chemical fertilizers which was in tandem with conservative methods of farming. In this regard, planting holes were made which enabled them to collect a lot of rainwater. The planting holes were then filled with compost made from available resources (dry grass, leaves, and animal manure) (Radio Farmer, Breeze FM, 2016). The planting basins capture most of the rainwater which is made more available to the crops. The dry crop residues in the field also shield the soil from heavy raindrops on one hand, and reduces the evaporation of moisture in the soil because they shade it from direct sunlight on the other.

Kilonzo (2022: 85) also observed that to conserve soil in some parts of the continent, subsistence farmers, who are mainly women, are likely to use several approaches such as zero tilling, natural mulches to moderate soil temperatures and extremes, suppress diseases and harmful pests, and conserve soil moisture. By so doing, the women were playing their role to achieve the national goals on promoting conservation agriculture.

Nurturing Trees and Other Forms of Vegetation

Women in the valley were also planting trees and teaching others not to cut trees. The trees were not cut because apart from giving fruits, they also provided shade and were medicinal. For instance, the mango tree was valued for its numerous benefits. It was also seen as food for wild animals. This is clearly illustrated in the example of the palm tree whose trunks, leaves, and fibres serve to build houses, lay floors and weave baskets, sieves and hammocks that enable the people to store their produce at home, prepare food, and sleep at night (Virtanen 2011).

Besides the mango trees and the popular African fan palm trees in the Luangwa Valley, bamboos reeds were also considered valuable as they were used for weaving baskets and other items as recalled by the author's mother. This stresses the indigenous idea that trees, forests, and mountain forests are imbued with spirits. Accordingly, taboos surrounding the cutting and destruction of certain trees, shrubs, and forests abound. The care for the trees in the spirit of ubuntu is a symbol of solidarity between humans and nature, including the invisible world. Most importantly, it becomes an immoral act to cut trees anyhow such that firewood and building material is fetched from places other than sacred forests and trees. Pope (2005) also reports that fuel wood removal in the Luangwa Valley is done selectively, favouring the Mopane tree. Similarly, Barnes (2000) established that within the community, the majority of women recognize the need for sustainable principles of forest management in the collection of fuelwood:

> For example, some women stated—'it is better to cut branches so they sprout, and not burn trees', and 'we use dead and dry wood, and do not burn trees [in-situ] so Mopane trees sprout again'. Women also have an incentive to manage forests due to the many different uses of wood—'trees are used for houses, firewood, medicines, stop soil erosion, stop evaporation, feed wildlife and provide shade', and wood was considered the second most important natural resource, after water. (Barnes 2000: 5)

Santisteban (2020: 13) notes that indigenous women have for centuries been protecting nature and the territories not because they feel a subject-object bond but because they are a part of them (nature and territories). As such, nurturing vegetation went hand in hand with protecting the fields from bush fires: "…Our first task is to protect our fields from bushfires. Bush fires are a great problem because they burn all the crop residues in the field," (Mary Mumba Musangu, Chikowa village, Chief Jumbe, Mambwe on Radio Farmer programme, Breeze FM, n.d June 2016). This is not only a form of respect to nature and ancestors but also an action to be in solidarity with nature as related to the ideals of ubuntu. These acts in turn benefit the ecosystem.

Livestock Rearing and Harvesting Wild Fruits

Some women in the Luangwa Valley had embraced goat rearing which provided manure for their farming activities and meat for the households.

> We experience prolonged droughts, and it has been very difficult to grow maize, which was our main source of income. "So I decided that I join my fellow women who formed a cooperative," she said. She said during the last rainy season, the area experienced heavy rains only on five occasions, and it was dry for most of the period, which made it difficult for some crops to grow. "We also often slaughter some for home consumption," said Mrs Mackenzie. (Kunda, Daily Mail, 19 October 2022, np)

Goat rearing was not only a way to sustain the traditional protein for the people but also a source of revenue. It was also an adaptation measure in the context of climate change and a symbol of their heritage as observed by Pope (2005: 19) that traditionally the nutrition of the peoples of the Luangwa Valley combined a measure of rain-fed cultivation with protein supplied by fish, domestic livestock, and hunted antelope. As population densities increased, coupled with expansion of mono cropped cotton as cash crop and continued use of unsuitable hybrid maize varieties, food insecurity became a challenge. Thus the women's adoption of goat rearing could be seen as a way of reclaiming the livestock they had reared in the past and a strategy to promote food security:

> One woman joined a women's cooperative group that is rearing goats with the help of some non-governmental organisations (NGOs). Ms Zulu said she decided to engage in goat rearing to fight poverty and mitigate the impact of droughts. She said goat rearing is helping her to feed the family and provide other necessities. (Kunda, Daily Mail, 19 October 2022, np)

Other women started selling wild fruits to earn a living and support their families. This went hand in hand with drying seasonal vegetables and fruits for use in other seasons using simple techniques of solar drying and smoking. Through this, their skills of wild food identification, harvesting, and preservation were sustained and retained. This relates to diversification of the sources of livelihood amidst climate change. Whilst all these efforts are going on, women also tapped into ubuntu to mitigate climate change and promote food security by passing on the knowledge to young generations. Sithole and Lekorwe (2019: 116) affirm that for many centuries, women have been charged with ensuring that their families are secure in terms of food security and that they have achieved this by using the indigenous knowledge systems that has been orally handed down through the generations.

Kilonzo et al. (2009: 44) capture this well through the observation that there are a variety of taboos that protect plants, animals, snakes, sacred sites, and birds. It is therefore the responsibility of mothers (and fathers) to educate their children on all these taboos (Kilonzo 2022: 86). The generational transfer and preservation of traditional climate knowledge has made women climate experts, innovators, and problem-solvers given that as for women to take care of their family and society, they must take care of nature (The World Bank 2022). All this affirms the significant role played by African women in climate change mitigation and food security as informed by their IKS and African spirituality (Aluko 2018; Lukhele-Olorunju and Gwandure 2018; Matholeni et al. 2020).

Conclusion

The chapter explored the ways in which women farmers in the Luangwa Valley of Eastern Zambia were contributing towards climate change mitigation and food security. It has been demonstrated that the Luangwa Valley was characterized by recurrent impacts of different on-going climatic changes and food shortages. Using the notion of ubuntu as a tool for addressing climate change and food security, the chapter concludes that women were using indigenous knowledge to address climate change and food insecurity through conservation farming which restored soil health, boosted production, and enhanced food security. They also devised ways of water harvesting, reclaiming the seeds of the former generations, nurturing trees and other forms of vegetation, livestock rearing and harvesting wild fruits and passing on the heritage to future generations.

Given their active roles in climate change and food security interventions, the chapter concludes that women were active agents because they were in solidarity with nature. As such, the understanding of ubuntu as anchored on solidarity and equality remained a pillar for their engagement in climate change and food security interventions. This was reflected in the value which women drew from pro-climate and food security activities, such as being enabled to increase the nutrition of their households and the community at large, to sell the surplus for additional income, and uphold their identity as indigenous women farmers.

References

Aluko, Y. 2018. "Women's use of indigenous knowledge for environmental security and sustainable development in Southwest Nigeria," *The International Indigenous Policy Journal* 9(3). https://doi.org/10.18584/iipj.2018.9.3.2

Barnes, H., 2000. Women, Wood and Wildlife-The Role and Involvement of Women in CBNRM in the Luangwa Valley, Zambia. *The International Famine Centre*, *3*, pp.645–677.

Breeze FM. 2016. Farming Radio programme, June, n.d.

Carr, N. 1996. Kakuli: A Story About Wild Animals, Their Struggle to Survive and the People Who Live Among Them. CBC Publishing

Chibvongodze, D.T., 2016. Ubuntu is not only about the human! An analysis of the role of African philosophy and ethics in environment management. *Journal of Human Ecology*, *53*(2), pp. 157–166.

Chitando, E. 2017. "Praying for Courage: African Religious Leaders and Climate Change," *The Ecumenical Review*, 68(3), 425–435.

Chitando, E. 2022. Introduction: African perspectives on religion and climate change. Chitando, E., Conradie, E.M. and Kilonzo, S.M.(eds.) *African perspectives on religion and climate change*. London: Routledge, 1–21

Eneji, C.V.O., Ntamu, G.U., Unwanade, C.C., Godwin, A.B., Bassey, J.E., Willaims, J.J. and Ignatius, J., 2012. Traditional African religion in natural resources conservation and management in Cross River State, Nigeria. *Environment and Natural Resources Research*, *2*(4), pp. 45–53.

FAO, 2014. Adapting to climate change through land and water management in Eastern Africa. *Results of pilot projects in Ethiopia, Kenya and Tanzania.*

FAO. 2006. Food security Organization. Paris: FAO.

FAO. 2011. "Women in Agriculture: Closing the Gender Gap," *State of Food and Agriculture Report 2010–11*, Rome: FAO

Government of the Republic of Zambia. 2016. The Zambia National Policy on Climate Change. Lusaka: Government Printers.

Gyampoh, B.A., Amisah, S., Idinoba, M. and Nkem, J. 2009. "Using Traditional Knowledge to Cope with Climate Change in Rural Ghana," in *Unasylva* 231/232, Vol. 60, 70–74.

Ibnouf, F. O. 2012. The value of women's indigenous knowledge in food processing and preservation for achieving household food security in rural Sudan. *Journal of Food Research*, *1*(1), 238–53.

IFAD. 2009. Annual Report. IFAD: Rome, accessed from https://www.ifad.org/documents/38714170/39309762/print.pdf/41ccca4d-8dbd-4308-bc5d88b306fee339#:~:text=Women%20grow%20more%20than%20half,of%20basic%20foodstuffs%20consumed%20locally.

Kanene, K. M. 2016. Indigenous Practices of Environmental Sustainability in the Tonga Community of Southern Zambia. *Jàmbá: Journal of Disaster Risk Studies*, 8(1),1–7.

Kilonzo, S., Kurgat, G. and Omare, S. 2009. "The role of taboos in the management of natural resources and peace-building: A Case study of the Kakamega forest in Western Kenya," *Africa Peace and Conflict Journal* 2(1), 3–54.

Kilonzo, S.M., 2022. Women, indigenous knowledge systems, and climate change in Kenya. In E. Chitando, E.M. Conradie and S. Kilonzo (eds.) *African Perspectives on Religion and Climate Change* (pp. 79–90). Routledge.

Kunda, R. 2022.Climate Change affects Luangwa women. *Daily Mail*. 19 October.

Lukhele-Olorunju, P., & Gwandure, C. 2018. Women and indigenous knowledge systems in rural subsistence farming: the case of climate change in Africa. *Africa Insight*, *47*(4), 57–71.

Mafongoya, P.L. and Ajayi, O.C. (Eds.), 2017. *Indigenous Knowledge Systems and Climate Change Management in Africa*. Wageningen, The Netherlands: CTA.

Masekoameng, M.R. and Molotja, M.C., 2019. The role of indigenous foods and indigenous knowledge systems for rural households' Food Security in Sekhukhune District, Limpopo Province, South Africa. *Journal of Consumer Sciences* 4: 34–48

Matholeni, Nobuntu P., Georgina K. Boateng and Molly Manyonganise. Eds. 2020. *Mother Earth, Mother Africa and African Indigenous Religions*. Cape Town: Sun Press.

Metz, T., 2011. Ubuntu as a moral theory and human rights in South Africa. *African Human Rights Law Journal*,11(2), 532–559.

Ministry of Agriculture and Co-Operatives. 2011. *Zambia National Agricultural Policy*. GRZ: Lusaka.

Mugumbate, J. and A. Chereni., 2020. "Now, the Theory of Ubuntu Has Its Space in Social Work," *Journal of Social Work* 10(1): v-vx.

Mwale, N. 2016. "African Traditional Religion in the Context of Climate Change: A Zambian Perspective." *Journal of Humanities*. 13:1–24.

Mwale, N. 2021. Women and Sustainable Agriculture in the Context of the Kasisi Agricultural Training Centre: A Catholic Initiative in Zambia. In S. Daniel, M. D. Kebaneilwe, & A. Savala (eds.).*Mother Earth, Mother Africa and Mission.* Stellenbosch: African Sun Media, 141–158.

Mwale, N. 2022. The Mother Earth Centre: A Narrative of the Comboni Missionary Sisters' Contributions to Environmental Sustainability in Western Zambia. In: Sophia Chirongoma and Ven. Scholar Wayua Kiilu (eds.). *Mother Earth, Mother Africa: World Religions and Environmental Imagination.* Stellenbosch: Africa Sun Media, 95–110.

Nyawo, S., 2022. The climate crisis: Mitigation and control through Emaswati indigenous knowledge. In E. Chitando, E.M. Conradie and S. Kilonzo (eds.) *African Perspectives on Religion and Climate Change* (pp. 34–49). Routledge.

Oniang'o, R. K. (2005). Women are still the key in agriculture and food security in Africa. *South African Journal of Clinical Nutrition, 18*(2), 150–154.

Pope, A., 2005. Luangwa Safari Association Tourism Study. *WHYDAH Consulting Ltd, Lusaka.*

Radio Farmer. 2016. Breeze FM Radio programme, June, n.d.

Santisteban, R.S. (ed.) 2020. *Indigenous Women and Climate Change.* Lima: IWGIA.

Shoko, E. 2022. Indigenous Conflict Management and Contemporary Water Resource Governance in Rural Zimbabwe. *Journal of Peacebuilding & Development, 17*(2), 225–238.

Sithole, A., & Lekorwe, M. 2019. Women's Use of Indigenous Knowledge Systems to Cope with Climate Change. *Advances in Social Sciences Research Journal,* 6*(6),* 111–119.

Taringa, N. T. 2006. "How Environmental is African Traditional Religion?" *Exchange*, 35(2), 191–214.

Tarusarira, J. 2017. "African Religion, Climate Change and Knowledge Systems." *The Ecumenical Review*, 69(3), 398–410.

Taylor, S. D. 2006. *Culture and Customs of Zambia.* London: Greenwood Press.

The World Bank 2022. Indigenous Women Leaders Persevere Amid a Changing Climate. Feature Story, 8 August.

Thorpe, S.A., 1991, African Traditional Religions: An introduction, University of South Africa, Pretoria.

United Nations Development Programme. 2019. Reducing poverty, improving food security for farmers in Zambia. Accessed from https://www.adaptation-undp.org/reducing-poverty-improving-food-security-farmers-zambia 20 September 2022.

Virtanen, P.K., 2011. Guarding, Feeding, and Transforming: Palm Trees in the Amazonian Past and Present. *The Archaeological Encounter. Ethnographic Perspectives*, pp.125–73.

World Food Summit. 1996. The world free from hunger. Rome.

Index[1]

[1] Note: Page numbers followed by 'n' refer to notes.

L. Maseno et al. (eds.), *Religion, Climate Change, and Food Security in Africa*, Sustainable Development Goals Series, https://doi.org/10.1007/978-3-031-50392-4

Kolibri 360 GmbH
Ferdinand-Jühlke-Straße 7
99095 Erfurt, Deutschland
produktsicherheit@kolibri360.de